Perspectives and Challenges in Doctoral Research—Selected Papers from the 9th Edition of the Scientific Conference of the Doctoral Schools from the "Dunărea de Jos" University of Galaţi

Perspectives and Challenges in Doctoral Research—Selected Papers from the 9th Edition of the Scientific Conference of the Doctoral Schools from the "Dunărea de Jos" University of Galaţi

Editors

Eugen Rusu
Gabriela Rapeanu

MDPI • Basel • Beijing • Wuhan • Barcelona • Belgrade • Manchester • Tokyo • Cluj • Tianjin

Editors

Eugen Rusu
"Dunarea de Jos" University
of Galati
Romania

Gabriela Rapeanu
"Dunarea de Jos" University
of Galati
Romania

Editorial Office
MDPI
St. Alban-Anlage 66
4052 Basel, Switzerland

This is a reprint of articles from the Special Issue published online in the open access journal *Inventions* (ISSN 2411-5134) (available at: https://www.mdpi.com/journal/inventions/special_issues/Doctoral_Research).

For citation purposes, cite each article independently as indicated on the article page online and as indicated below:

LastName, A.A.; LastName, B.B.; LastName, C.C. Article Title. *Journal Name* **Year**, *Volume Number*, Page Range.

ISBN 978-3-0365-5395-5 (Hbk)
ISBN 978-3-0365-5396-2 (PDF)

Cover image courtesy of Gabriela Rapeanu

Contents

About the Editors

Eugen Rusu

Eugen Rusu received a diploma in Naval Architecture (1982) and a Ph.D. in Mechanical Engineering (1997). Between 1999 and 2004, he worked as a post-doc fellow at the at Hydrographical Institute of the Portuguese Navy, where he was responsible for wave modelling and provided environmental support for some major situations such as: the accident of the M/V Prestige (2002) and the NATO exercises Unified Odyssey (2002) and 'Swordfish' (2003). Eugen Rusu worked as a consulting Scientist at the NATO Centre for Maritime Research and Experimentation, La Spezia, Italy (2005), having modelling coastal waves and surf zone processes as main tasks.

Starting in 2006, in parallel with his activity as professor at the University Dunarea de Jos of Galati, he has been working as a professor collaborator at CENTEC (Centre for Marine Technology and Ocean Engineering), University of Lisbon, Portugal. Since 2012, he has also acted as an expert for the European Commission. Eugen Rusu has published more than 150 studies in the fields of renewable energy and marine engineering, and he received awards including Doctor Honoris Causa (2015) at the Maritime University of Constanta, Romania. Outstanding Contribution in Reviewing, *Renewable Energy* (2015) and *Ocean Engineering* (2016) journals; and Top 1% World Reviewers in the field of engineering (2018 and 2019). He is also the President of the Council of the Doctoral Schools in Galati University and of the Romanian National Commission of Mechanical Engineering. Since 2018, he has been a corresponding member of the Romanian Academy, the highest level scientific and cultural forum in Romania.

Gabriela Rapeanu

Gabriela Rapeanu is Professor at the Faculty of Food Science and Engineering, Dunarea de jos University of Galati. She received a diploma in Food Technology (1997) and a Ph.D. in 2005. Her PhD topic was to investigate the thermal and/or high-pressure inactivation of PPO from white grapes. The research activity was partially funded by an EU Project (FP5 - EU-Marie Curie Training Site: *Enzyme kinetics during thermal and non-thermal food processing* QLK1-CT-2000-60014) during the period of 2003–2004 at Laboratory of Food Technology, Katholieke Universiteit Leuven, Belgium. During the last 25 years had an intense publishing activity in journals with high impact factors, as well as books and chapters in books published in prestigious international publishing houses. She has been involved in collaborative projects merging food science, functional foods, food safety, food quality and biotechnology.

She is also member of the National Research Council (CNCS) (Chair of the Life Sciences and Biotechnology Committee (2011–2013 and 2020–present) and the Director of the Doctoral School of Fundamental Sciences and Engineering, Dunarea de Jos University of Galati, Galați, Romania.

Preface to "Perspectives and Challenges in Doctoral Research—Selected Papers from the 9th Edition of the Scientific Conference of the Doctoral Schools from the "Dunărea de Jos" University of Galaţi"

This Special Issue of the *Invention* journal include papers from different research fields, reflecting the sectional talks presented at **the 9th edition of the Scientific Conference of the Doctoral Schools from the "Dunărea de Jos" University of Galaţi**, which took place in June 2021. The objective of the 2021 Conference was to bring together perspectives and challenges in doctoral research in a common forum and to provide a platform enabling doctoral researchers to meet and share state-of-the-art developments in their fields.

The Special Issue include 16 papers presented at **the 9th edition of the Scientific Conference of the Doctoral Schools from the "Dunărea de Jos" University of Galaţi**, in several fields such as: mechanical and industrial engineering, food science and biotechnology, electrical/electronic engineering, systems engineering and information technologies, chemistry, electrochemistry, and economic models and strategies.

Eugen Rusu and Gabriela Rapeanu
Editors

inventions

MDPI

Editorial

Editorial for Special Issue "Perspectives and Challenges in Doctoral Research—Selected Papers from the 9th Edition of the Scientific Conference of the Doctoral Schools from the "Dunărea de Jos"

Eugen Rusu [1] and Gabriela Rapeanu [2,*]

1 Department of Mechanical Engineering, Faculty of Engineering, "Dunărea de Jos" University of Galați, Domnească Street, 800008 Galați, Romania; erusu@ugal.ro
2 Faculty of Food Science and Engineering, "Dunărea de Jos" University of Galați, Domnească Street 111, 800201 Galați, Romania
* Correspondence: gabriela.rapeanu@ugal.ro

Citation: Rusu, E.; Rapeanu, G. Editorial for Special Issue "Perspectives and Challenges in Doctoral Research—Selected Papers from the 9th Edition of the Scientific Conference of the Doctoral Schools from the "Dunărea de Jos". *Inventions* **2022**, 7, 33. https://doi.org/10.3390/inventions7020033

Received: 7 March 2022
Accepted: 21 March 2022
Published: 28 March 2022

This Editorial is dedicated to the 9th edition of the Scientific Conference organized by the Doctoral Schools of "Dunărea de Jos" University of Galati (SCDS-UDJG), which was organized in June 2021 in Galati (Romania). This Special Issue consists of papers from different research fields, reflecting the sectional talks presented at the SCDS-UDJG. This Conference and Special Issue aim to bring together new perspectives and challenges and to share state-of-the-art developments in doctoral research.

In total, 16 papers presented at the SCDS-UDJG conference were included in the Special Issue, in fields such as mechanical and industrial engineering, food science and bio-resources engineering, electrical/electronic engineering, systems engineering and information technologies, chemistry, electrochemistry, and economic models and strategies.

Most papers published in the Special Issue were devoted to studies in the mechanical and industrial engineering field, specifically six articles. Articles [1–3] focus on wind power resources and technologies as the main subject of research. Diaconita et al. (2021) [1] analyzed the behavior of circular and conical towers under the action of the same environmental conditions, both located in the Black Sea. The authors pursued the forces and moments resulting from the interaction of the wind with the towers' wind turbine. Girleanu et al. (2021) [2] used in situ measurements and the reanalysis of wind data to assess the wind energy potential that characterizes the Romanian coastal environment. Nedelcu and Rusu (2022) [3] also treated the Black Sea as an area of interest, the northwestern coast specifically, to assess an overview of the wind climate, using data from 2015 until 2020 on wind direction, speed, air temperature, and air pressure, provided by the National Institute of Marine Geology and Geoecology.

In Marcu et al. (2021) [4], the authors measured the impact of administrative burdens identified by ship-owners on technical measures implemented by the European Commission to optimize the flow of goods on the Danube using the Quality Function Deployment method.

Articles [5,6] studied the process of profiling in mechanical engineering. Baroiu et al. (2021) [5] present an algorithm for modifying rollers in order to generate the helical surface of the pump rotor with progressive cavities, utilized in different industries for the circulation of high-viscosity fluids. Moroşanu et al. (2021) [6] present an overview of the profiling methods for various tools, using the kinematic and decomposing helical movement methods.

Last but not least, Mihai and Rusu (2021) [7] present an overview of the actual practice of heating ventilation and air conditioning systems used on large ships, in the COVID-19 pandemic context.

Solcanu et al. (2021) [8] discuss a critical aspect of maritime safety, namely radio communications. This study focuses on the effectiveness of analog systems with a low signal-to-noise ratio in high-noise conditions, characteristic of the marine environment, using analytical and numerical methods. Also in the domain of system, engineering, and information technologies was the work of Minzu et al. (2021) [9]. This paper addresses the subject of using a metaheuristic algorithm to solve specific optimal control problems. Thus, a general harmonization between the controller of the closed loop and the evolutionary algorithm is needed.

Three articles of this Special Issue are devoted to different biologically active compounds from plants. The extraction optimization of betalain pigments from beetroot is discussed in Lazăr et al. (2021) [10], while Condurache et al. (2021) [11] and Roman et al. (2021) [12] consider the problem of the microencapsulation of biologically active compounds from eggplant peels and sea buckthorn fruits. All three studies are based on the need to replace chemical synthesis additives with natural ones in food and, at the same time, to close the economic value chain from the food industry by the valorization of different valuable by-products.

Two articles published in the Special Issue focused on the pharmaceutical field. In the work of Gunache et al. (2021) [13], using cyclic voltammetry, the electrochemical behavior of two screen-printed sensors was evaluated, in order to qualitatively and quantitatively determine the atorvastatin in pharmaceutical products. The study is considered by the authors to be a first step towards the development of a device for the detection of atorvastatin in biological samples. The work of Dinu et al. (2021) [14] also addresses the issue of developing a biosensor using cyclic voltammetry, but this time in order to determine and quantify the essential amino acid L-tryptophan from pharmaceutical products.

In the field of economic models and strategies, Anton et al. (2021) [15] analyzed the impact of mass tourism on the local community in Constanta, Romania. The economic, social, and environmental conditions were taken into account. The research continues with the work of Anton et al. (2022) [16], which aims to introduce an innovative concept that combines research, governance, and citizenship, using "Living labs" technology and knowledge to build a bridge between decision makers and stakeholders, promoting a shared vision of growth and innovation at a community level.

The articles presented above reflect the present state of the art in the research fields of mechanical and industrial engineering, food science and bio-resources engineering, electrical/electronic engineering, systems engineering and information technologies, chemistry, electrochemistry, and economic models and strategies. The Guest Editors of this Special Issue believe it to be a starting point for further studies, with the potential to open new research directions.

Author Contributions: Conceptualization, E.R. and G.R., validation, E.R. and G.R., resources, E.R., writing—original draft preparation, G.R., supervision, E.R.; project administration, E.R. and G.R.; funding acquisition, E.R. All authors have read and agreed to the published version of the manuscript.

Funding: All APC for papers published in this special issue were supported by Dunarea de Jos University of Galati.

Acknowledgments: This scientific event was supported by a grant from the Romanian Ministry of National Education.

Conflicts of Interest: The authors declare no conflict of interest.

References

1. Diaconita, A.I.; Andrei, G.; Rusu, E. Estimation of the tower shape effect on the stress–strain behavior of wind turbines operating under offshore boundary conditions. *Inventions* **2022**, *7*, 11. [CrossRef]
2. Girleanu, A.; Onea, F.; Rusu, E. Assessment of the wind energy potential along the romanian coastal zone. *Inventions* **2021**, *6*, 41. [CrossRef]
3. Nedelcu, L.I.; Rusu, E. An analysis of the wind parameters in the western side of the Black Sea. *Inventions* **2022**, *7*, 21. [CrossRef]
4. Marcu Turcanu, A.L.; Moga, L.M.; Rusu, E.V.C. Analysis of some essential aspects related to the navigation conditions on the Danube River. *Inventions* **2021**, *6*, 97. [CrossRef]

5. Baroiu, N.; Moroşanu, G.A.; Teodor, V.G.; Oancea, N. Roller profiling for generating the screw of a pump with progressive cavities. *Inventions* **2021**, *6*, 34. [CrossRef]
6. Moroşanu, G.A.; Baroiu, N.; Teodor, V.G.; Păunoiu, V.; Oancea, N. Review on study methods for reciprocally enwrapping surfaces. *Inventions* **2022**, *7*, 10. [CrossRef]
7. Mihai, V.; Rusu, L. An overview of the ship ventilation systems and measures to avoid the spread of diseases. *Inventions* **2021**, *6*, 55. [CrossRef]
8. Solcanu, V.; Gaiceanu, M.; Rosu, G. Study of resistance to disturbances of the main types of communication systems on board military ships used during interception or search and rescue missions. *Inventions* **2021**, *6*, 72. [CrossRef]
9. Minzu, V.; Riahi, S.; Rusu, E. Implementation aspects regarding closed-loop control systems using evolutionary algorithms. *Inventions* **2021**, *6*, 53. [CrossRef]
10. Lazăr (Mistrianu), S.; Constantin, O.E.; Stănciuc, N.; Aprodu, I.; Croitoru, C.; Râpeanu, G. Optimization of betalain pigments extraction using beetroot by-products as a valuable source. *Inventions* **2021**, *6*, 50. [CrossRef]
11. Condurache (Lazăr), N.N.; Turturică, M.; Enachi, E.; Barbu, V.; Bahrim, G.E.; Stănciuc, N.; Croitoru, C.; Râpeanu, G. Impact of wall materials on physico-chemical properties and stability of eggplant peels anthocyanin hydrogels. *Inventions* **2021**, *6*, 47. [CrossRef]
12. Roman, D.; Condurache (Lazăr), N.N.; Aprodu, I.; Enachi, E.; Barbu, V.; Bahrim, G.E.; Stănciuc, N.; Râpeanu, G. Insights of sea buckthorn extract's encapsulation by coacervation technique. *Inventions* **2021**, *6*, 59. [CrossRef]
13. Gunache (Roşca), R.O.; Bounegru, A.V.; Apetrei, C. Determination of atorvastatin with voltammetric sensors based on nanomaterials. *Inventions* **2021**, *6*, 57. [CrossRef]
14. Dinu, A.; Apetrei, C. Development of a novel sensor based on polypyrrole doped with potassium hexacyanoferrate (ii) for detection of l-tryptophan in pharmaceutics. *Inventions* **2021**, *6*, 56. [CrossRef]
15. Anton, C.; Micu, A.E.; Rusu, E. Multi-criteria analysis of the mass tourism management model related to the impact on the local community in Constanta city (Romania). *Inventions* **2021**, *6*, 46. [CrossRef]
16. Anton, C.; Micu, A.E.; Rusu, E. Introducing the Living Lab approach in the coastal area of Constanta (Romania) by using design thinking. *Inventions* **2022**, *7*, 19. [CrossRef]

inventions

Article

An Analysis of the Wind Parameters in the Western Side of the Black Sea

Laura-Ionela Nedelcu [1,2,*] and Eugen Rusu [2]

1 Department of Navigation and Naval Transport, Faculty of Navigation and Naval Management, "Mircea cel Batran" Naval Academy, 1st Fulgerului Street, 900218 Constanta, Romania
2 Department of Mechanical Engineering, Faculty of Engineering, "Dunarea de Jos" University of Galati, 47th Domneasca Street, 800008 Galati, Romania; eugen.rusu@ugal.ro
* Correspondence: laura.nedelcu@anmb.ro

Abstract: In the present research, an overview of the wind climate on the northwestern coast of the Black Sea basin is assessed, using a total of 6 years of data (2015–2020) provided by the National Institute of Marine Geology and Geoecology (GeoEcoMar). It is well known that the enclosed/semi-enclosed basins are complex environments and to accurately represent the features of wind and wave are necessary high resolution spatial fields. For the Black Sea, which is an enclosed basin with complicated regional geography, the main weather parameters reported (wind direction, wind speed, air temperature, air pressure) give a more comprehensive picture of how energetic the area of interest is, and represent the features of the Black Sea's diversified marine environment. Finally, the results obtained in this paper cover a broad range of applications in marine studies, being useful for future research in the area of wind climate in the Black Sea.

Keywords: Romanian coastal zone; Black Sea basin; wind parameters; historical data

Citation: Nedelcu, L.-I.; Rusu, E. An Analysis of the Wind Parameters in the Western Side of the Black Sea. *Inventions* **2022**, *7*, 21. https://doi.org/10.3390/inventions7010021

Academic Editor: Mohammad Hassan Khooban

Received: 6 December 2021
Accepted: 31 January 2022
Published: 2 February 2022

1. Introduction

Renewable sources of energy are not a new topic but it remains a controversial one. The 20th century energy system based on fossil fuels created an urgent need to experiment with alternatives leading to a sustainable energy future. Renewables are still far from overtaking fossil fuels in the global energy system; still, their growth is fast and contributes to a nascent and uncertain energy transition with major effects in the field of environment, economy, governance, and society. Renewable energy sources are crucial for long-term efforts to mitigate climate change and are expected to play a vital role in improving the entire planet's energy security [1–3].

According to the European Commission, by 2050, Europe must increase the offshore wind capacities between 230 and 450 GW. So far, Romania is the leader of Southeast Europe in terms of onshore developments, with an installed capacity of 3 GW [4]. The development of onshore wind farms was achieved mainly in the Dobrogea plateau. In this area are concentrated almost 80% of the existing wind turbines from Romania. The largest onshore wind farm in Europe is the Fantanele–Cogealac, also established in Dobrogea, having an installed capacity of 600 MW [5].

On contrary, there is still little information on the energetic offshore potential of the Black Sea basin. In their studies [6] the authors estimated a total potential natural capacity of 94 GW, out of which 22 GW could be installed as fixed turbines and 72 GW from offshore wind farms.

Previous studies based on the Black Sea wave wind conditions indicate that the western part of the sea is more energetic than the eastern sector [7–12]. Moreover, Onea and Rusu [13] affirmed that in terms of the spatial distribution of extreme events, most probable storm conditions occur in the western part, whereas in the eastern sector it is more likely to happen in a relatively short time window. In their work [7] the authors assessed the largest average power of 7 kW/m in the South Western part of the Black Sea, whereas the average mean wave power in the eastern part is about 3 kW/m, with a maximum of 4 kW/m. In terms of wind speed, the western side records significant seasonal variations, while the eastern area is characterized by more stable conditions [14,15]. In their research [15] the authors have conducted a comparison between average wind speed from coastal stations (4.9 m/s) and satellites data (6 m/s). In [16], based on the wind wave conditions evaluated over 50 years (1958–2007), it is concluded that the wind and wave climate of the western Black Sea shelf manifests large spatial and seasonal variability. The authors stated that wind speed exceeded 40 m/s in the open sea and 25 m/s near shore. The analysis made in [17,18] indicates that the Romanian coastal environment is more energetic during the winter season, with an average wind speed at 80 m of about 9.7 m/s and a power density of 870 W/m^2. Myslenkov et al. [19] pointed out that during summer months low quality of wind and wave were observed due to their low and weak speeds. In their work [20], Rusu, Raileanu and Onea obtained from the Gloria platform values of the wind fields at 10 m height from 6 m/s to 7 m/s, covering a time period from 1987 to 2009. Diaconita, Rusu, and Andrei have concluded in their study [21] that the mean wind speed at the height of 10 m was 6.7 m/s. Based on the literature review mentioned above, there is an increasing interest in marine renewable project development based on the wind and wave resources located in the northwestern part of the Black Sea.

In this context, the purpose of the present study is to contribute to the characterization of the wind climate along the western part of the Black Sea Basin over a period of 6 years (2015–2020) and to estimate ideal areas for exploiting this resource. Consequently, this study will focus on the wind parameters evolution based on the measurements provided by the 2 autonomous marine monitoring stations, part of the regional security system EMSO-EUXINUS. The present research can be considered as a first step for future renewable energy projects since there is significant growing interest from the Romanian authorities in energy transition towards renewables.

2. Materials and Methods

2.1. Target Area

The Black Sea represents an inland sea situated between Europe, Anatolia, and Caucasus, covering an area of 422,000 km^2. The basin is considered a semi-enclosed sea, connected to the Mediterranean Sea through the Straits of Dardanelles and Bosporus and to the Sea of Azov through the Kerch Strait. The sea is defined by an irregular shape being divided between Romania, Ukraine, Russia, Georgia, Turkey, and Bulgaria. The average water depth is about 1240 m, while a maximum of 2210 m is observed offshore. In terms of bathymetry, wide shelf area can be noticed that occupies up to 25% of the total seafloor area and, on average, being restricted to sea depths of 100–200 m and a width of 200 km. The remaining area is associated with the continental slope and the abyssal plain, where depths below 2000 m can be found. The configuration of the bathymetric line, the presence of depressions and canyons, has an important influence on the distribution of water masses, direction and speed of currents. The bathymetric chart of the Black Sea can be observed in Figure 1 [22–24].

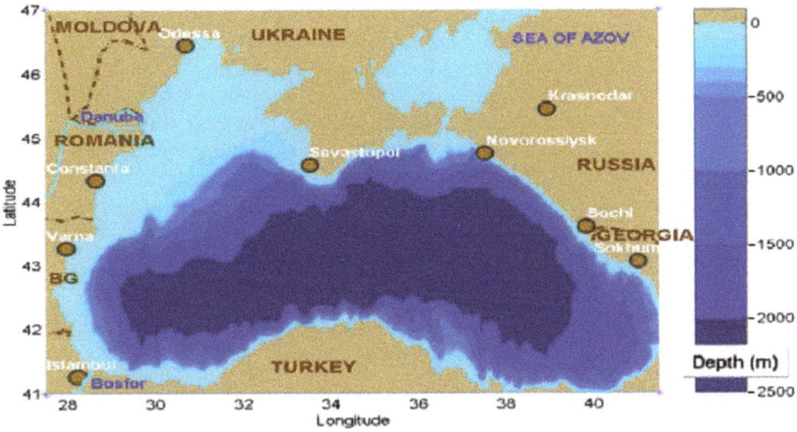

Figure 1. The bathymetric chart of the Black Sea.

The wind and wave climate are determined by the geographical position and the landscape configuration, which have temperate-continental characteristics. During winter months frequent storms and strong N-NE winds are accompanied by waves 2–4 m high, but in summer wave heights are generally less than 1 m. The tide is almost imperceptible, for it reaches scarcely 10 cm in height [23,24].

One of the greatest particularities of the Black Sea is the inability of the water layers to mix. The hydro-chemical configuration of the sea is primarily influenced by the fluvial inputs and the basin topography, which result in a strongly stratified vertical structure and a positive water balance. The first layer, located on the sea surface over 50 m depth, is generally cooler, less salty, and less dense than the deeper waters, as it is fed by large fluvial systems. The second area, known as a cold layer, is a sub-surface water mass located at 50 to 180 m depth [23,24].

The salinity of the Black Sea is determined by the balance between the fresh water and water exchange through the Bosporus Strait. At the surface water of the Black Sea, an average salinity of 16–18 PPT (two times less than the salinity of the World Ocean surface waters) was recorded. The salinity increases to 21 to 22.5 PPT (parts per thousand) at a depth below 120–200 m. This difference is motivated because the water layers do not mix, and high salinity water from Mediterranean origin penetrates into the sea through the Bosporus Strait [25].

The water temperatures vary depending on the season. In summer the surface water warms to 25 °C, (occasional extremes at the shores of up to 28 °C). In winter in the open sea, it gets to 6–8 °C. The northwestern coast has ice in the winter while the southeastern stays around 9 °C [26].

The Black Sea is considered a vital connection between Europe and Asia regarding maritime activities. Most of the research effort is focused on the Romanian coastal zone, while the eastern side characteristics are less discussed.

2.2. Wind Dataset

In the present work, we will analyze the wind dataset covering a time period from 2015 to 2020 as provided by the National Institute of Marine Geology and Geoecology—GeoEcoMar.

Romania is represented in the European Multidisciplinary Seafloor and Water Column Observatory-European Research Infrastructure Consortium (EMSO ERIC) through GeoEcoMar. The National Institute provides EMSO ERIC real-time data on the physico-chemical and biological parameters of the water column, at different depths, data on early warning in case of marine hazards, as well as the experience of the personnel involved in the proper functioning of the entire system. The data are provided from the EUXINUS network of marine and coastal observatories (3 offshore and one coastal observatory), part of the Black Sea Security System. EMSO-EUXINUS is a regional security system destined for the Western Black Sea with real-time alerting to marine geo-hazards (e.g., submarine landslides, active fault movements, submarine earthquakes, tsunami waves, etc.). It monitors the characteristics and dynamics of coastal water bodies, including several meteorological parameters. The amount of data is stored within the National Center for Monitoring and Alarming to Marine Natural Hazards, which operates within the Constanta branch of INCD (National Center for Research and Development) GeoEcoMar [27].

The monitoring and early warning system consisting of 3 autonomous marine monitoring stations (Figure 2) is composed of:

1. Surface buoy:
    - solar recharging system for accumulator batteries;
    - power supply management system of the station;
    - switchable satellite/radio communication system;
    - data acquirer with integrated management;
    - remote control system and two-way real-time radio/satellite communication;
    - weather sensor system;
    - active and passive radar system;
    - optical signaling system;
    - acoustic system for managing the communication and data transmission with the installed underwater equipment [27].
2. Multiparameter buoy fixed at 5 m depth:
    - Doppler sensor for current;
    - conductivity sensor;
    - temperature sensor;
    - pressure sensor;
    - oxygen sensor;
    - turbidity sensor;
    - chlorophyll sensor [27].
3. Multi-parameter buoy fixed to the seabed with bidirectional acoustic communication modem with surface buoy (approximately 2 years autonomy and remote triggering system):
    - Doppler sensor for current;
    - conductivity sensor;
    - temperature sensor;
    - pressure sensor;
    - battery system for powering the sensors with a duration of approx. 1 year.
4. Tsunameter assembled on the seabed, near the buoy:
    - anchoring system;
    - Tsunameter—positioned on the seabed with two-way acoustic communication modem with the surface buoy, with approximately 2 years autonomy and remote trigger system;
    - battery system for powering the sensors with approximately 2 years autonomy [27].

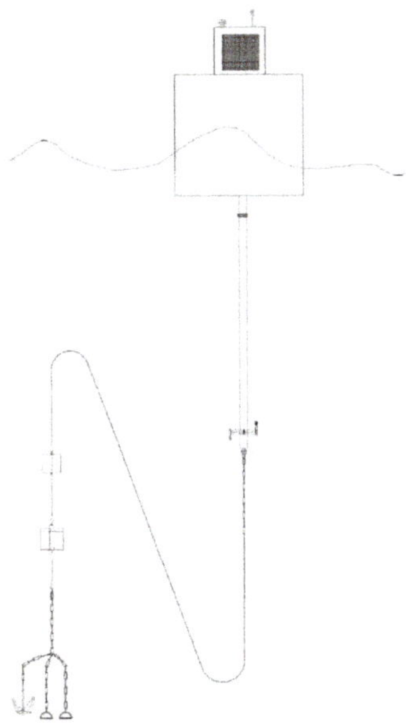

Figure 2. The scheme of one autonomous marine monitoring station.

The three complex buoys (EuxRO 1, EuxRO 2, and EuxRO 3) are located on the Romanian continental shelf, in the territorial waters of Romania, at a distance of approximately 160 km from the coast and were approved by the Ministry of National Defense, through the Romanian Naval Forces, represented by the Maritime Hydrographic Directorate (DHM). Their location can be seen in Figure 3 and also in Table 1.

Figure 3. Positions of the GeoEcoMar buoys in the Black Sea.

Table 1. Positions of the buoys in the Black Sea.

Station	Latitude	Longitude	Altitude
EuxRO 1	φ = 44°42′28.92″ N	λ = 030°46′46.80″ E	−100 [m]
EuxRO 2	φ = 44°19′37.40″ N	λ = 030°25′30.63″ E	−128 [m]
EuxRO 3	φ = 43°58′51.14″ N	λ = 029°56′12.15″ E	−115 [m]

The transmission of the data collected from the installed sensor systems is acoustically done, in real time. The data are transmitted to the surface buoys, then satellite, to the National Monitoring and Alarm Center for Marine Natural Hazards EUXINUS, in Constanța. The weather station used for measuring the wind parameters is AIRMAR WHEATHER STATION 200WX, being installed at an approximate height of 2.5 m above the water. The system is capable of recording the following data: air pressure, air temperature, water current amplitude, water current direction, water temperate, wind speed and wind direction, making 24 observations per day, one value of each parameter per hour. EuxRO 2 only records the water current direction, water current amplitude, and water temperature data. Consequently, this research focuses on the data coming from EuxRO 1 and EuxRO 3 for the period May 2015–December 2020. More details regarding the parameters and the characteristics of the databases considered in this work are presented in Table 2 [27]. It is worth mentioning that certain data are missing due to the maintenance periods of the offshore buoys, unfavorable weather factors or certain technical problems.

Table 2. Characteristics of the datasets considered for evaluation.

	EuxRO 1		EuxRO 3	
Parameters Recorded/Year	**Wind Direction**	**Wind Speed**	**Wind Direction**	**Wind Speed**
2015	07.05.2015–08.12.2015		09.05.2015–31.12.2015	
2016	09.06.2016–30.10.2016		01.01.2016–26.04.2016 09.06.2016–03.11.2016	
2017	17.03.2017–31.12.2017		19.03.2017–31.12.2017	
2018	01.01.2018–31.12.2018		01.01.2018–31.12.2018	
2019	01.01.2019–31.12.2019		01.01.2019–31.12.2019	
2020	01.01.2020–24.11.2020		01.01.2020–31.12.2020	

3. Results

The study is based on the wind parameters (speed and direction) analysis while considering the total data for the 6-year period (2015–2020). Figure 4 provides a first observation of the wind conditions in the Romanian coastal zone. From the analysis of the mean wind speed computed from the 2 autonomous marine monitoring stations, we may notice that there are no significant differences in the reported period. From the analysis performed for each year it resulted that the 2018 year is the most energetic year (7.1 m/s), whereas the lowest energetic is 2015 (6.3 m/s). From the analysis made in the entire period recorded, a mean value of the wind speed of 6.7 m/s was obtained, at a height of 2.5 m above the water.

Figure 4. The evolution of the wind speed considering the GeoEcoMar data for the time interval from May 2015 to December 2020.

3.1. Annually Evaluation of the Wind Data

A more detailed analysis for every year can be seen in Figure 5. As it can be observed from Figure 5a, the wind speed maximum average value is located in December (EuxRo 3–8.0 m/s), which is closely followed by November with an average of 7.8 m/s. On the opposite, the lowest value is found in July (a mean value of 4.9 m/s). From Figure 5b, a maximum value of the wind speed of 9.7 m/s in January 2016 can be observed, closely followed by 8.7 m/s in February, whereas a minimum of 4.8 m/s is seen in July 2016. For Figure 5c a major value is reported in December, during which an average of 9.6 m/s may be observed, while a minimum value of 4.9 m/s was found in June. In Figure 5d it is presented the 2018 year, the most energetic year, where we can witness a maximum average value of 9.8 m/s in January and a minimum of 3.4 m/s in July. From the values analyzed in 2019 (Figure 5e), it can be noticed that January (9.3 m/s) and February (9.1 m/s) present in general much higher values compared to July and June, where the values are located in the range 4.1–4.5 m/s. The 2020 data (Figure 5f) seems to be in good agreement with the data from above, a maximum average value has been noticed in winter (February—8.9 m/s) and a minimum in summer (June—4.7 m/s). Based on the analysis made for each month in the 6-year interval considered, it resulted that the greatest average values were encountered in January 2018—9.8 m/s (Figure 5d), followed by January 2016—9.7 m/s (Figure 5b).

3.2. Maximum Wind Values Recorded per Year

From the amount of analyzed data, the maximum values that were recorded should be highlighted. In 2015 the maximum wind speed value was registered by the EuxRO 1 on the 1st of December at 20.00. The wind had a speed of 17.1 m/s and a direction of 340°. In the next year, on the 18 January 2016 at 20.00 a value of 18 m/s and a direction of 336° were indicated by the EuxRO 3. Further, in the next years, all of the maximum values were stored by the EuxRO 3. In 2017, a value of 17.3 m/s and a direction of 272° were present on the 30 October. In 2018 and 2019 values over 20 m/s were observed. On 18 January 2018, wind speed of 22.5 m/s and direction of 327° were seen at 19.00, while on the 8 January 2019 at 05.00 the wind speed reached 25.5 m/s and direction 344°. The last year that was analyzed shows a maximum wind speed value of 17.2 m/s and a direction of 299° on 30 April at 24.00.

3.3. Seasonal Average Values

The seasonal average values of the wind speed were computed by following partitions: winter—DJF (December-January-February), spring—MAM (March-April-May), summer—JJA (June-July-August), and autumn—SON (September-October-November). As can be seen from Figure 6, winter is the most dynamic season. In this season, wind speed values were registered in the range of 8.6 m/s (December)–9.3 m/s (January). Further, the next season that presents dynamic variations is autumn. During these months average values of 6.1 m/s in September, 7.0 m/s in October and 8.0 m/s in November were recorded. In spring lower values of the wind speed were noticed: March—7.4 m/s, April—6.3 m/s, May—5.5 m/s. Nevertheless, the lowest energetic season is summer. The lowest wind speed average value was recorded in July (4.6 m/s), which is closely followed by June (4.9 m/s) and August (5.5 m/s).

3.4. Wind Direction

Comparing the average wind direction from the selected buoys, it can be observed that all the values are in good agreement. The mean value wind direction obtained for the entire study period is 167.1°. This result is based on the observation made from the 2 buoys, being more detailed in Table 3. The annual values are in the same range from 152.8° in the 2018 year to 176.9° measured in 2017. In terms of monthly average values, the results are also grouped by seasons with approximately equal directions. During winter, the wind direction is 191.5°, followed by autumn with a mean value of 151.9°. During spring, the average wind direction obtained is 171.2°, whereas in summer the direction is 157.7°.

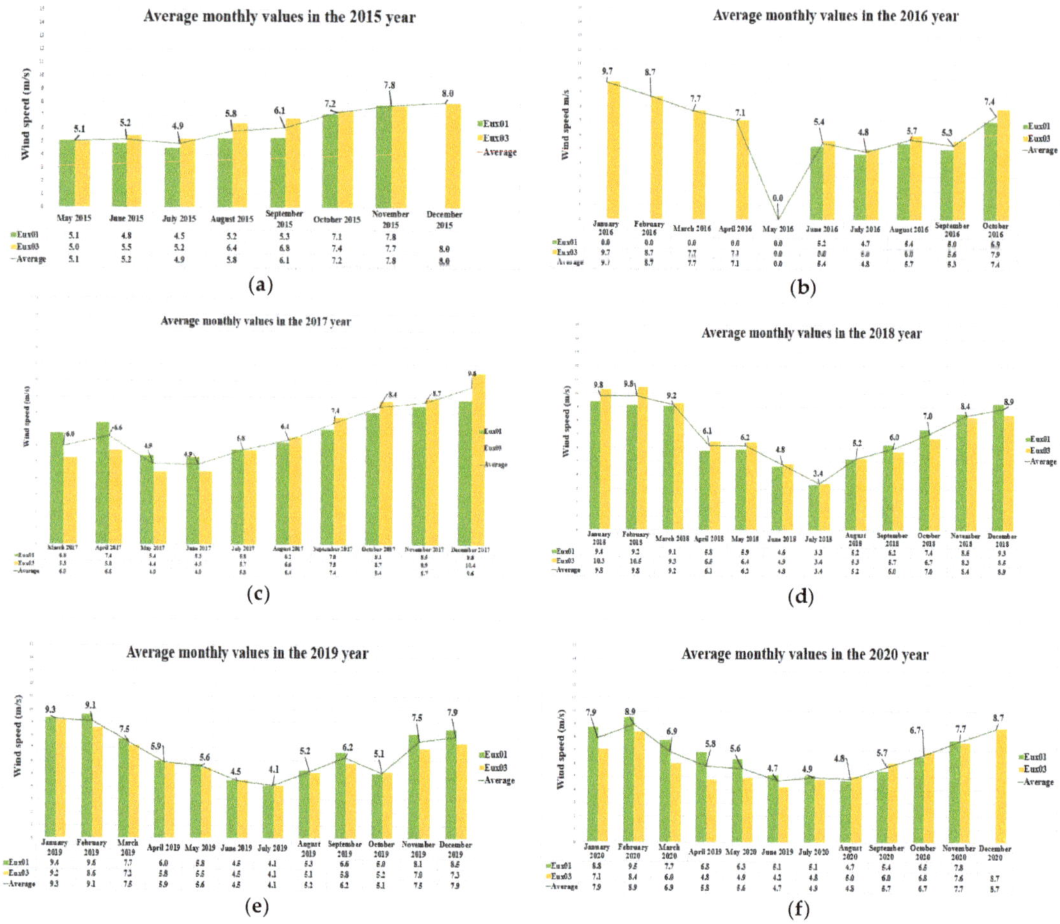

Figure 5. Annual evaluation of the wind data; (**a**) 2015; (**b**) 2016; (**c**) 2017; (**d**) 2018; (**e**) 2019; (**f**) 2020.

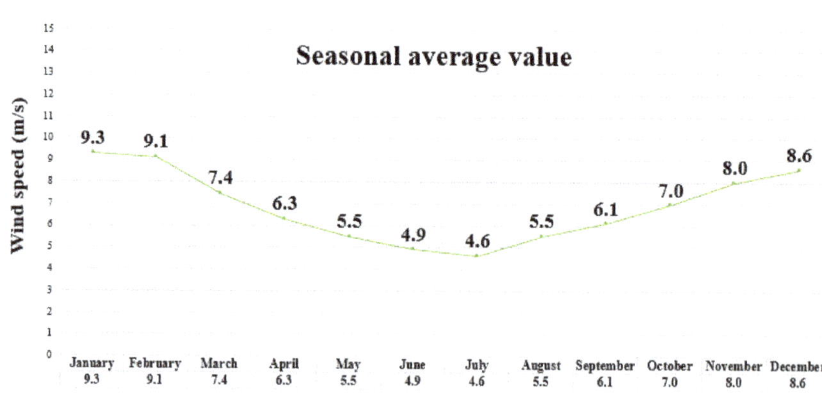

Figure 6. Seasonal average values of the wind speed.

Table 3. Statistics for wind direction (°).

	2015	2016	2017	2018	2019	2020
January	-	234.8	-	155.9	200.5	213.1
February	-	164.6	-	154.4	178.9	190.4
March	-	167.6	189.2	173.1	188.7	160.5
April	-	178.1	183.1	141.2	145.7	166.5
May	185.1	-	167	136.4	186.7	200.1
June	188.6	174.1	199	141.7	150.5	181.7
July	177	172.2	161.2	214.9	185.2	162
August	101.8	125.7	133.3	92.6	149.6	128.8
September	111.5	167.2	158.5	170.4	134	140
October	144.1	150.1	192.9	140.7	153.4	183.1
November	217.1	-	152.5	94.1	133.6	142.9
December	219.9	-	232.6	218	201.6	134.6
Average	168.1	170.5	176.9	152.8	167.3	167

4. Discussion

It is highly important to acknowledge that wind resources are directly related to wind speed not to mention that their productivity will be influenced by the climate. In order to assess the dynamics of the wind energy resources in the Black Sea, the 6 years of GeoEcoMar data (2015–2020) were considered for evaluation. To identify relevant trends, analysis of mean and maximum wind speed, as well as their seasonal variations were conducted. Moreover, an evaluation of the strongest winds recorded is presented.

In order to highlight the distribution of the strongest winds recorded in the Black Sea region, the maximum wind speeds were considered for discussion in order to identify some energetic peaks. On the 18 January 2018, a maximum wind speed of 22.5 m/s was seen at 19.00. Signs of a cyclone were seen on the 9 January when the wind started to blow with 8 m/s from N-NE. On the 11 January, the wind speed was around 14 m/s with a direction of 67°–126°. The wind speed had an average of 13 m/s in this period although the pressure continued to decrease. On 18th at 19.00 the maximum peak of wind speed was 22.5 m/s, direction of 327° and pressure of 983.7 mbar. On the next day, the average wind speed was 12 m/s from the western sector. On the 20th, 21th, 22th, 23th above-normal wind speed values were recorded. On the 24 January, the cyclone was no longer present in the western sector of the Black Sea. To sum up, the cyclone started on the 9 January (8 m/s), had a maximum peak on the 18th (22.5 m/s) and finished on 24th (12.5 m/s).

Following the same principle, the highest value of wind speed recorded from the entire study should be detailed. In this instance, the period is much shorter and with not so many well-defined severe weather parameters. On 8 January 2019 at 05.00, a maximum value of 25.5 m/s was recorded. Before this day, on the 5th, the wind speed reached a value of 10.2 m/s. The next day, the wind speed average was 10.375 m/s from the north-western sector. On the 7th, the wind had a similar direction and a speed average of 13.308 m/s. The pressure was in standard parameters. On the following day, the wind speed started to rise substantially, recording an average of 17.9 m/s. On the last day, the wind speed decreased to usual values resulting in completion of the extreme event reported.

Analyzing the data obtained for the mean wind parameters in the present study, it can be seen that the overall wind speeds for each year are close, indicating that no significant variations will be present in the near or distant future, proving that the Romanian coastal zone can be a reliable solution for the offshore wind farms development.

5. Conclusions

This research paper aimed to analyze the dynamics of the wind parameters along the west coast of the Black Sea, based on the weather data provided over a 6-year period (2015–2020) by GeoEcoMar Institute.

The present study outlines relevant trends in wind dynamics, resulting from the analysis of mean and maximum values of the wind, but also its seasonal and monthly variations. Moreover, two periods of severe weather were discussed.

Wind climate was analyzed covering the period 2015–2020. Two offshore autonomous marine stations data were studied in terms of their wind parameters by computing monthly and mean wind speed and direction. A mean wind speed value of 6.7 m/s was obtained from the analysis. Taking into account the seasonal and monthly variation of the wind, the climatic picture of the Black Sea is divided into two opposite seasons, winter and summer, the principal cause being the temperature increase or decrease during these periods. However, the other two seasons give a better transition from one extreme to another and also acquire some of their characteristics.

The data analyzed in this report support that the offshore regions are suitable for developing wind farms. Taking into consideration the data recorded from the buoys located at 160 km from the coast, and the fact that deep-water areas are more satisfactory for wind farms, it is highlighted that floating turbines are a feasible solution for the northwestern part of the Black Sea's offshore development. Moreover, given the fact that wind speeds increase along with the distance to the shore, and a large part of Romania's Exclusive Economic Zone consists of a deep-water area, prove once again that the area discussed represents a viable solution for the development of offshore wind farms [6]. The results that were achieved in this study contribute to the wind climate characterization in the discussed area and represent one of the first steps in future studies related to wind variability considering various approaches. Finally, it can be concluded that the Black Sea is a dynamic environment with rich sources for the development of renewable energy projects.

Author Contributions: L.-I.N. collected and processed the wind data, carried out the statistical analysis and interpreted the results. E.R. guided the entire research, wrote parts of the manuscript and handled the review process. The final manuscript has been approved by all authors. All authors have read and agreed to the published version of the manuscript.

Funding: This work was carried out in the framework of the research project DREAM (Dynamics of the REsources and technological Advance in harvesting Marine renewable energy), supported by the Romanian Executive Agency for Higher Education, Research, Development and Innovation Funding—UEFISCDI, grant number PN-III-P4-ID-PCE-2020-0008.

Institutional Review Board Statement: Not applicable.

Informed Consent Statement: Not applicable.

Data Availability Statement: The data that were used in the present study are openly available. The wind data were obtained from the National Institute of Marine Geology and Geoecology—GeoEcoMar.

Conflicts of Interest: The authors declare no conflict of interest.

References

1. Aklin, M.; Urpelainen, J. *Renewables: The Politics of a Global Energy Transition*; The MIT Press: Cambridge, MA, USA, 2018.
2. Ellabban, O.; Abu-Rub, H.; Blaabjerg, F. Renewable energy resources: Current status, future prospects and their enabling technology. *Renew. Sustain. Energy Rev.* **2014**, *39*, 748–764. [CrossRef]
3. Bull, S.R. Renewable energy today and tomorrow. *Proc. IEEE* **2001**, *89*, 1216–1226. [CrossRef]
4. Available online: https://www.interregeurope.eu/agrores/news/news-article/10859/opportunities-of-black-sea-offshore-wind/ (accessed on 15 January 2022).
5. Dragomir, G.; Serban, A.; Nastase, G.; Brezeanu, A.I. Wind Energy in Romania: A Review from 2009 to 2016. *Renew. Sustain. Energy Rev.* **2016**, *64*, 129–143. [CrossRef]
6. Bălan, M.; Dudău, R.; Cătuți, M.; Covatariu, A. *Romania's Offshore Wind Energy Resources: Natural Potential, Regulatory Framework, and Development Prospects*; Energy Policy Group: Bucharest, Romania, 2020.
7. Aydogan, B.; Ayat, B.; Yüksel, Y. Black Sea wave energy atlas from 13 years hindcasted wave data. *Renew. Energy* **2013**, *57*, 436–447. [CrossRef]
8. Rusu, L.; Bernardino, M.; Guedes Soares, C. Wind and wave modelling in the Black Sea. *J. Oper. Oceanogr.* **2014**, *7*, 5–20. [CrossRef]
9. Fulya, I.; Yalcin, Y. Inter-comparison of long-term wave power potential in the Black Sea based on the SWAN wave model forced with two different wind fields. *Dyn. Atmos. Ocean.* **2021**, *93*, 101192.

10. Rusu, L.; Ganea, D.; Mereuta, E. A joint evaluation of wave and wind energy resources in the Black Sea based on 20-year hindcast information. *Energy Explor. Exploit.* **2018**, *36*, 335–351. [CrossRef]
11. Girleanu, A.; Onea, F.; Rusu, E. Assessment of the Wind Energy Potential along the Romanian Coastal Zone. *Inventions* **2021**, *6*, 41. [CrossRef]
12. Davy, R.; Gnatiuk, N.; Pettersson, L.; Bobylev, L. Climate change impacts on wind energy potential in the European domain with a focus on the Black Sea. *Renew. Sustain. Energy Rev.* **2018**, *81*, 1652–1659. [CrossRef]
13. Onea, F.; Rusu, L. A Long-Term Assessment of the Black Sea Wave Climate. *Sustainability* **2017**, *9*, 1875. [CrossRef]
14. Staneva, J.V.; Stanev, E. Oceanic response to atmospheric forcing derived from different climatic data sets, Intercomparison studies for the Black Sea. *Oceanol. Acta* **1998**, *21*, 393–417. [CrossRef]
15. Atodiresei, D.; Toma, A.; Cotorcea, A.; Samoilescu, G. Opportunities for harnessing wind potential on board ships in the northwest area of the black sea. *Rev. Roum. Sci. Tech. Ser. Electrotech. Energetique* **2017**, *62*, 265–268.
16. Valchev, N.; Davidan, I.; Belberov, Z.; Palazov, A.; Valcheva, N. Estimation of wind wave climate of the Western Black Sea during the last 50 years. In Proceedings of the Ninth International Conference on Marine Sciences and Technologies Black Sea, Varna, Bulgary, 23–25 October 2008; pp. 232–239.
17. Onea, F.; Rusu, E. An Evaluation of the Wind Energy in the North-West of the Black Sea. *Int. J. Green Energy.* **2014**, *11*, 465–487. [CrossRef]
18. Ganea, D.; Mereuta, E.; Rusu, L. Estimation of the near future wind power potential in the Black Sea. *Energies* **2018**, *11*, 3198. [CrossRef]
19. Myslenkov, S.; Zelenko, A.; Resnyanskii, Y.; Arkhipkin, V.; Silvestrova, K. Quality of the Wind Wave Forecast in the Black Sea Including Storm Wave Analysis. *Sustainability* **2021**, *13*, 13099. [CrossRef]
20. Rusu, L.; Raileanu, A.; Onea, F. A comparative analysis of the wind and wave climate in the Black Sea along the shipping routes. *Water* **2018**, *10*, 924. [CrossRef]
21. Diaconita, A.I.; Rusu, L.; Andrei, G. A Local Perspective on Wind Energy Potential in Six Reference Sites on the Western Coast of the Black Sea Considering Five Different Types of Wind Turbines. *Inventions* **2021**, *6*, 44. [CrossRef]
22. Divinsky, B.V.; Kosyan, R.D. Spatiotemporal variability of the Black Sea wave climate in the last 37 years. *Cont. Shelf Res.* **2017**, *136*, 1–19. [CrossRef]
23. Rusu, E.; Onea, F.; Toderascu, R. Dynamics of the Environmental Matrix in the Black Sea as Reflected by Recent Measurements and Simulations with Numerical Models. In *Black Sea: Dynamics, Ecology and Conservation*; Nova Science Publishers, Inc.: Hauppauge, NY, USA, 2011.
24. Nedelcu, A.T.; Atodiresei, D.; Buzbuchi, N. Oceanographic research and recordings made on the North-West Coast of the Black Sea. *Int. J. Conserv. Sci.* **2020**, *11*, 615–624.
25. Amarioarei, A.; Tusa, I.; Sidoroff, M.; Pauna, M. Classification of Romanian salt water lakes by statistical methods. *J. Environ. Prot. Ecol.* **2018**, *19*, 1–14.
26. Available online: https://www.fao.org/3/n7372e/N7372E02.htm (accessed on 25 November 2021).
27. Available online: https://www.euxinus.eu/ (accessed on 26 November 2021).

inventions

Article

Introducing the Living Lab Approach in the Coastal Area of Constanta (Romania) by Using Design Thinking

Catalin Anton *, Angela Eliza Micu and Eugen Rusu

Doctoral School of Socio-Human Sciences, Dunarea de Jos University of Galati, 47 Domneasca Street, 800008 Galati, Romania; angela-eliza.micu@365.univ-ovidius.ro (A.E.M.); eugen.rusu@ugal.ro (E.R.)
* Correspondence: catanton@gmail.com

Abstract: Living Labs are an innovative concept that combines research, governance, and citizens, using technology and knowledge. Using design thinking techniques as a method of approach, this innovative idea builds a bridge between decision makers and stakeholders, promoting a shared vision of growth and innovation at the community level. The coastal zone is an extremely dynamic area in terms of human and natural activities. This is a particularly sensitive area to climate change, necessitating ongoing adaptation and mitigating action. This paper aims to introduce the application of the Living lab concept in the management of the coastal area of Constanta (Romania). The concept of the Living Lab means involving citizens along with public bodies and research structures. This new coastal management model will use the design thinking approach and will consider the pressures that exist between the activities that occur in the examined coastal zone. In the study, "Multi-Criteria Analysis of the Mass Tourism Management Model Related to the Impact on the Local Community in Constanța (Romania)", published in *MDPI Inventions* on 28 June 2021, a coastal management model was built that took into consideration only the data given by the government. In this paper, the authors aim to expand their research by including data from independent sources, using the concept of a Living Lab.

Keywords: living labs; design thinking; management model; coastal zone; integrated coastal zone management (ICZM); Constanta; Black Sea; sustainability

Citation: Anton, C.; Micu, A.E.; Rusu, E. Introducing the Living Lab Approach in the Coastal Area of Constanta (Romania) by Using Design Thinking. *Inventions* **2022**, *7*, 19. https://doi.org/10.3390/inventions7010019

Academic Editor: Volker Hessel

Received: 31 October 2021
Accepted: 26 January 2022
Published: 29 January 2022

Publisher's Note: MDPI stays neutral with regard to jurisdictional claims in published maps and institutional affiliations.

1. Introduction

Unlike the typical structure of global society, the coastline is a vibrant socio-ecological environment. The purpose of integrated coastal zone management (ICZM) is to design a sustainable development model that focuses on, surrounds, and incorporates a number of essential elements (economic, social and environmental). The Romanian Black Sea coast is currently under threat, mainly due to human pressures that have accumulated over time. The agricultural and industrial sectors of the economy have often developed in a chaotic manner. Coastal erosion has now been discovered to be under severe stress, with the possibility of sea infiltration into land areas. Excessive resource use will result in the devastation of the natural terrestrial and aquatic ecosystems. In the future, while addressing the negative effects of climate change, it will be necessary to seek adaptive or alternative solutions [1–3].

Integrated coastal zone management is a complex strategic management mechanism that requires the participation of all stakeholders and the general public at the national, regional, and local levels. The "strategic integration" of coastal development plans will be established by creating the organizational and legal framework, tools, and procedures required to ensure the best combination (integration) of coastal development plans.

Integrated coastal zone management (ICZM) is a social, ecological structure with extremely complex characteristics due to differences in society, culture, finance, nature, and governance. The ICZM's mission is to promote integrated coastal zone management by investigating the relationship between coastal zone management and governance variables [4,5].

Traditional coastal management is generally achieved through a linear approach based on an up–down pyramid decision-making structure. The hierarchical structure is reduced to governmental political entities, while the other existing actors in the coastal zone (small and medium enterprizes, academia, research institutes, citizens, etc.) are called stakeholders, and the way of involving the latter is smaller or larger depending on the decision of the authorities. The linear approach has a predominantly hierarchical and less consultative decision-making role. Living labs have emerged in Europe in recent years to become closer with citizens and to facilitate collaboration between different stakeholder groups. These mechanisms collect creative ideas and work skills in order to involve private users, together with public bodies, to improve the environment within communities. Approaching the concept of a Living Lab in coastal management can be beneficial because this structure aims to involve citizens along with public bodies and research structures. Through the design thinking approach technique, coastal Living Labs will identify, define, develop, test, or validate new development policies, projects, and community strategies. The implementation framework of a Living Lab in the coastal area is a centric one, starting from the observations of the citizens and of the scientific environment, which define, conceptualize, and test the existing vulnerabilities in that area. The role of decision makers is to take these observations and research results and turn them into public policies. To create this functional framework, the authors consider it essential to involve citizens, through the data they provide, both in research and development. A Living Lab implemented in a coastal area can only benefit the community. As it is currently defined, integrated coastal management entails a collaboration between the scientific component and the public/human component. The public policies of the area are created by combining these two components, and the management component of ICZM is carried out (Figure 1) [6–10].

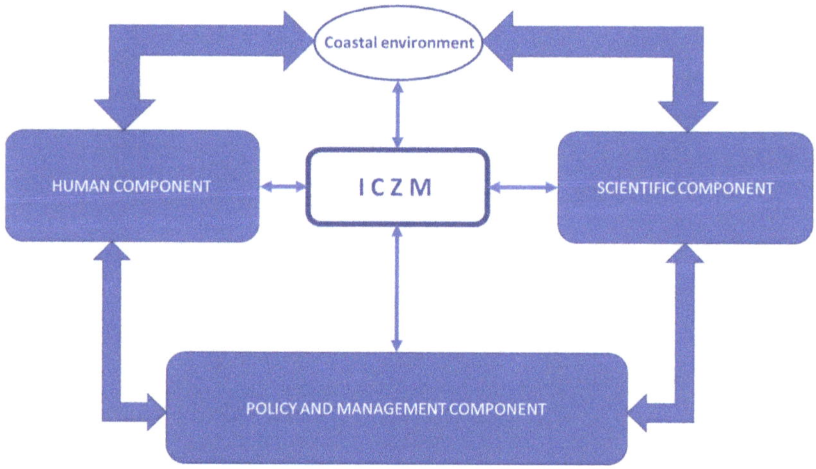

Figure 1. Diagram of the integrated coastal zone management process. Source: author after Rutger de Graaf-van Dinther (editor), Climate-Resilient Urban Areas' Governance, Design, and Development in Coastal Delta Cities.

Romania has taken small steps toward implementing integrated coastal management. Even though specific legislation has been in place since 2002, the organizational structure and decision-making process are limited to the adoption of consultative points of view that use authorities in the decision-making process [11–13]. As a result, establishing a Living Lab in the coastal area can help in actively involving stakeholders in decision making [14].

Given the foregoing, we may infer that the stage at which this paper is written represents a linear approach to integrated coastal management. The gap is the innovative Living Lab method in Europe, which may fill a void for the Constanta coastal area. The methodology we use is the design thinking method, and the innovation consists in enhancing data collection through citizen science and combining academic research resources as well as businesses development resources.

2. Materials and Methods

The creation of a "Living Lab" in coastal areas must consider a citizen-centered approach to research and development, taking into account technological innovations [10,12]. This type of approach will facilitate the interaction between stakeholders, bringing together all actors in the coastal area, whether we are talking about academia, the civic sector, state actors, or the business sector. A "Living Lab" is a new and dynamic body that brings technical research into real-world collaborative environments. The new concept has several principles by which it is defined [9,15–17], namely:

- Continuity;
- Opening;
- Realism;
- Empowering citizens;
- Spontaneity.

To explain these principles, it must be said that a Living Lab is based on creativity and innovation, but also on a good multidisciplinary collaboration. The innovative process must be credible and as open as possible, and at the same time it needs time to be structured. The multi-criteria approach of the newly created Living Labs is essential to provide a multi-perspective vision. The central element of the new structures is the citizen and his involvement in the innovation processes. Users are the ones who benefit from innovations and guide them to satisfy their own desires and needs. The more inventive the user communities are, the more efficient Living Labs become at achieving their intended purpose. Traditional coastal management systems have a linear approach, unlike the Living Labs structure, which has a centered approach [17]. The analysis of these systems is of mixed type, having qualitative elements but also emphasizing the quantitative side [18,19]. The aim of the authors is to show the usefulness of a mixed paradigm in coastal management research. We aim to maximize our strengths and weaken our strengths through this form of research and not to replace any of the particularly useful and important forms of classical research, including both quantitative and qualitative. The paradigm of mixed-method research, as in this paper, minimizes the schism between qualitative and quantitative approaches to research.

If we are to better understand this concept of the mixed research method, we can say that this new type combines methods and techniques, concepts, and qualitative or quantitative approaches in a single study.

The process involves the combination of three primary factors, namely, induction, deduction, and abduction. This means, in research practice, that models must be identified and theories tested, validated, and conceptualized. Therefore, traditional approaches to integrated coastal management are generally linear and often cyclical (Figure 2). Living Labs, on the other hand, takes a design-based approach with a focus on loop development (Figure 3) [20–24].

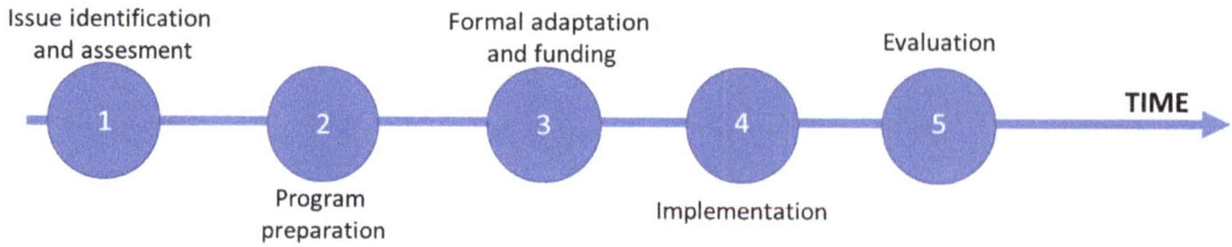

Figure 2. The conventional approach to integrated coastal management systems.

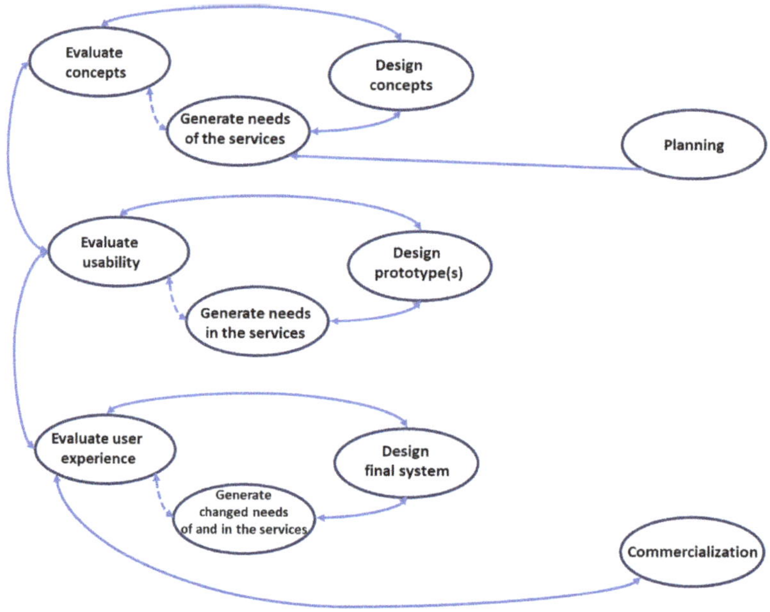

Figure 3. The in-system loop approach of Living Labs.

Perhaps the most obvious distinction between integrated linear coastal zone management and the Living Lab is based on conceptual realism. If the classical linear form has a theoretical and formal vision, a Living Lab is much more pragmatic, contextualizing real-world facts and attitudes. Even from the users' point of view, the two management systems are fundamentally different [16]. If, in the case of linear vision, users are formal and well-structured, with clear and precise attributes, in the case of Living Labs, these users are creative, informal, but deeply rooted in current reality. Therefore, in the case of Living Labs, users are urged to create in the most authentic contexts, under the observation of researchers and developers involved in the process [8,10,25]. The paradigm of mixed research in the case of Living Labs shows that users are considered as partners in the coastal management process, while in the case of traditional systems, they are considered as actors, the approach being organizational [26]. This element is a defining one in differentiating the two systems, showing that Living Labs has, as its central element, the citizen and his empowerment to the detriment of a formal organization [27–29]. This means that the new approach of Living Labs emphasizes the spontaneity and inventiveness of the citizens involved. Another important point is the relationship with the academic environment, which, in the case of the new approach, is a close one, with citizens using the research results, while in traditional systems, this type of connection is not always made [30]. That is why the new structures could use this to investigate both theoretical and practical problems.

Specifically, research is conducted in the Living Labs environment, while in traditional classical systems, the academic environment conducts research at the express request of decision makers, if any [31].

As previously stated, the approach to introducing a Living Lab in the coastal zone is a design thinking technique. This is also because design thinking involves a human-centered vision of innovation, which seeks to establish creative ideas and effective business models that are focused on people's needs [11,12]. The basic idea behind design thinking is that you apply designer approaches and methods to the development of innovations (this is what the word "design" means), while conducting a systematic, fact-based analysis of the economic feasibility and viability of these innovations—just like a researcher (this is what the "thinking" part of the term represents).

In the case of Living Labs, the research identifies citizens' problems or desires and considers them from the perspective of their target users. Designers use this knowledge to generate the first user-oriented ideas, visualize their creative solutions at an early stage, and then create prototypes. They quickly solicit community feedback and modify their concept as a result. Designers address the best solution for their target users' step by step. The project's approach and individual methods are supplemented by a mindset that intentionally analyzes the product's feasibility and economic viability during development [13].

To address social challenges such as climate change, population growth, food security, health, mobility, and energy supply, new ideas are required. These concepts are central to economic growth. Some ideas become global standards, while others fill niches in local and regional markets. Design thinking assists coastal managers in creative problem solving, regardless of the size of the problem. The method can be applied to a wide range of questions. These may include new products, services, business models, or social and organizational concepts [9,11].

Several principles must be followed in design thinking to ensure success [10]:

- Developing empathy: Investigating the emotions, thoughts, intentions, and actions of users;
- Illustration of ideas: Imagining ideas and illustrating them with a prototype for potential users to try;
- Learning from failure: Creating a culture that recognizes the value of mistakes, so that they are tolerated and learned from;
- Ensuring diversity: Age, gender, education, culture, and personality type all contribute to diversity;
- Creating creative and team-oriented workspaces: Spaces for individual and group work, along with spaces for the entire group, must be designed to be adaptable and inspiring;
- Process flexibility: The design mindset encourages a methodical approach. This procedure includes steps relating to problem analysis, such as formulating a task, developing possible initial solutions, testing them, and learning from feedback.

Figure 4 depicts the stages of the design thinking process.

If the design thinking process within a Living Lab can lead to the identification and evaluation of a coastal activity, these activities must be attributed to economic, social, or environmental factors. Because of the complexity of coastal areas, some activities are competitive while others are complementary. This type of activity, as well as the impact it generates, can be accurately identified using the design thinking process [12].

The basic identification of activities in terms of their social and environmental economic factors, in our opinion, is incomplete to carry out an effective analysis that will serve as the foundation for a sustainable coastal management model. By recognizing the interaction between indicators, one can demonstrate the dependence of some activities on others, as well as the areas where the interaction of these activities creates conflict that can result in coastal zone vulnerability [1,2].

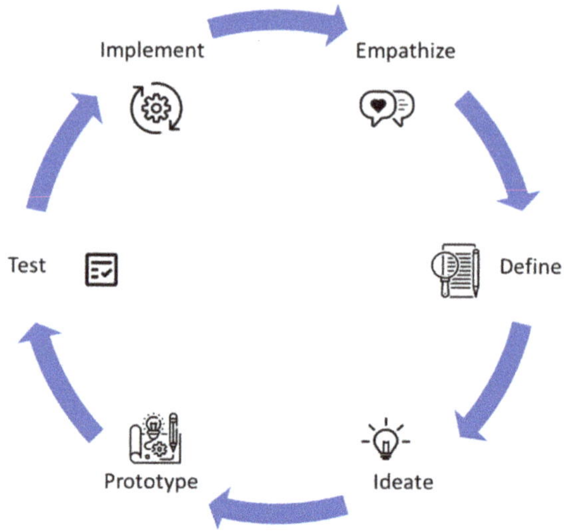

Figure 4. The stages of the design thinking process. Source: image processed after Christian Müller-Roterberg—Design Thinking.

We used the concept of a supportability factor to measure the interaction between these activities. This tool shows the value at which one of the activities can exist without causing the other to disintegrate. The supportability factor for activities that have affected each other reaches zero, depending on the greater or lesser impact between the activities. Furthermore, if one activity had a positive effect on the other, the supportability factor is also positive, on a scale of 0 to 1 [2–5].

The goal of a supportability analysis is to determine the context of the socio-economic and environmental conditions against which investment measures will be implemented, as well as the medium- or long-term evolution of an analyzed situation. These variables will effectively determine whether the proposed improvements are supportable for the community, particularly for vulnerable social strata [2].

The calculation of the rate of supportability is made according to the following formula [1,2]:

$$\mathrm{Rs} = \frac{\mathrm{i\ (ep)}}{\mathrm{i\ (sp)}}$$

where Rs indicates the rate of supportability, i (ep) represents the indicator that exerts pressure, and i (sp) is the indicator that supports pressure.

In the examples below, we will also use the calculation of the general supportability index, according to the following formula:

$$I_{sg} = \frac{\sum R_s}{n \times I_s} \tag{1}$$

where I_{sg} represents the general supportability index, R_s indicates the rate of supportability of the analyzed indicators, n represents the total number of indicators, and I_s represents the analyzed indicators. Similarly, the supportability index can be calculated on each factor [2].

The hypothesis is represented by the actual state of the pressures existing on the coastal areas. As we stated before, we will use citizen science in order to identify, test, and validate the hypothesis. The indicators of supportability were calculated based on a survey of 30 respondents who were specifically interested in coastal activities. This survey focused on 16 activities, including 7 economic, 5 environmental, and 4 social activities. This type of random sampling divided the subjects into two layers, and in each layer, a sample of 15 respondents was selected proportionally. The division followed the sizing of the sample, considering the population structure of the analyzed area. When dividing into groups, several criteria were considered, such as age, education, social status, religion, etc. Each subject has been assigned a number. The interviews were structured so that each subject could identify the interactions between coastal activities and determine the pressures exerted.

The stratification of the sampling ensured, in the analyzed case, a significant representation of the population, which determines a decrease of the sampling error. The present groups were adequately represented, and the division into subgroups led to a more pronounced heterogeneity. However, it must be said that, in this case, we started from a priori knowledge of the situation, data, and the multicriteria analyses of the coastal management model made by its classical approach. At the same time, in this case of stratified sampling, the procedure was more complex than in simple sampling, due to the need to divide the subjects and divide the subgroups.

3. Results

The implementation of a coastal management model in Constanta by introducing a Living Labs concept is the main theme of this research. Initially, the management model was developed in the multi-criteria analysis of the "Mass Tourism Management Model Related to the Impact on the Local Community in Constanta City (Romania)", published in *MDPI Inventions* on 28 June 2021. In the published paper, the coastal management model was based on the analysis of official statistical data. In this paper, we intend to put the Living Labs concept into action by utilizing data and information gathered from citizens. The hypothesis from which we started the current research is represented by the actual state of the pressures existing on the coastal areas. As we stated before, we used citizen science in order to identify, test, and validate the hypothesis. Thus, in Figure 5, you can see a comparison of the supportability factors related to the indicators provided by the citizens of Constanta and those provided by official sources. Figure 6 depicts a graphical analysis of the calculated coefficients of the supportability factors of coastal zone activities.

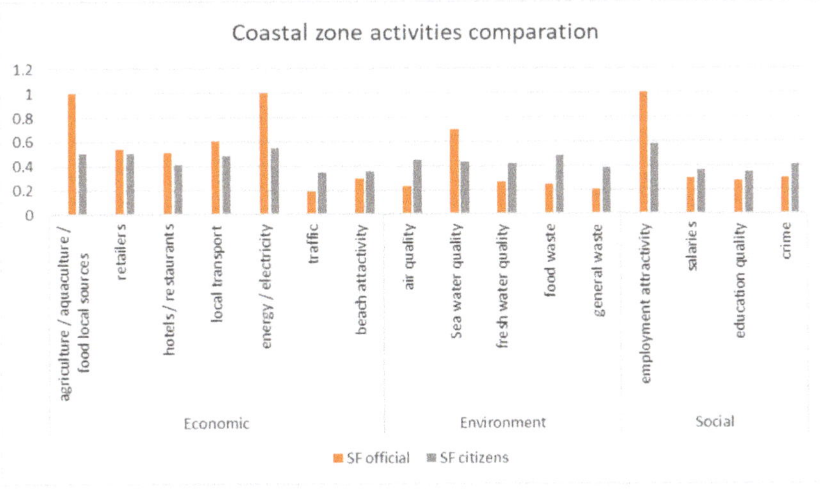

Figure 5. A comparison of the economic, social, and environmental activities that take place in the Constanta coastal area (Romania).

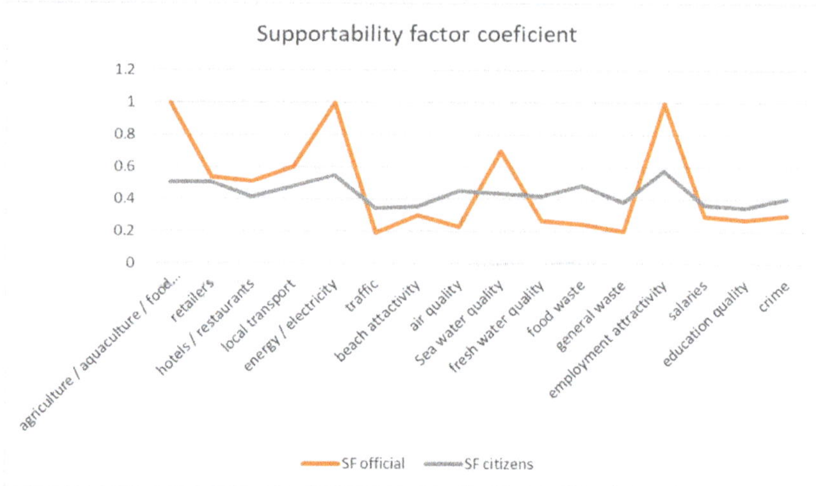

Figure 6. A graphical analysis of the activities in the Constanta coastal area (Romania).

Figures 7–9 show the activity analysis for each of the economic, social, and environmental factors.

Figure 10 depicts the share of the supportability factor calculated by citizens as well as sources provided by public institutions.

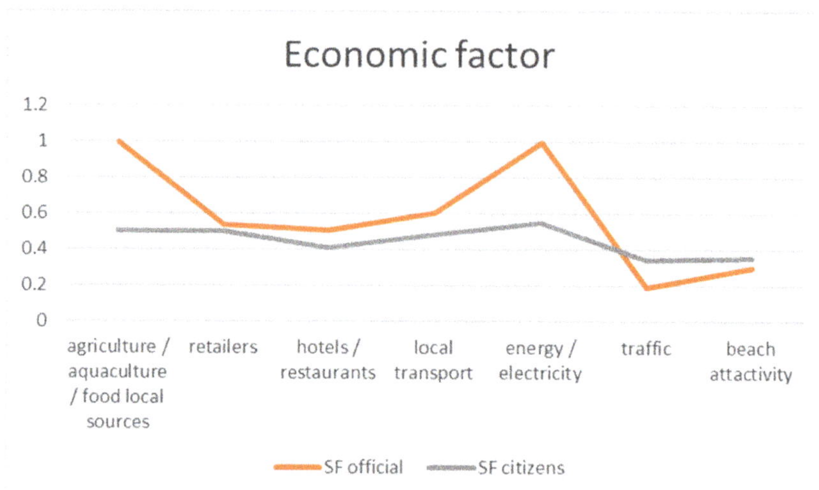

Figure 7. Analysis of economic activities by calculating the coefficient of the supportability factor.

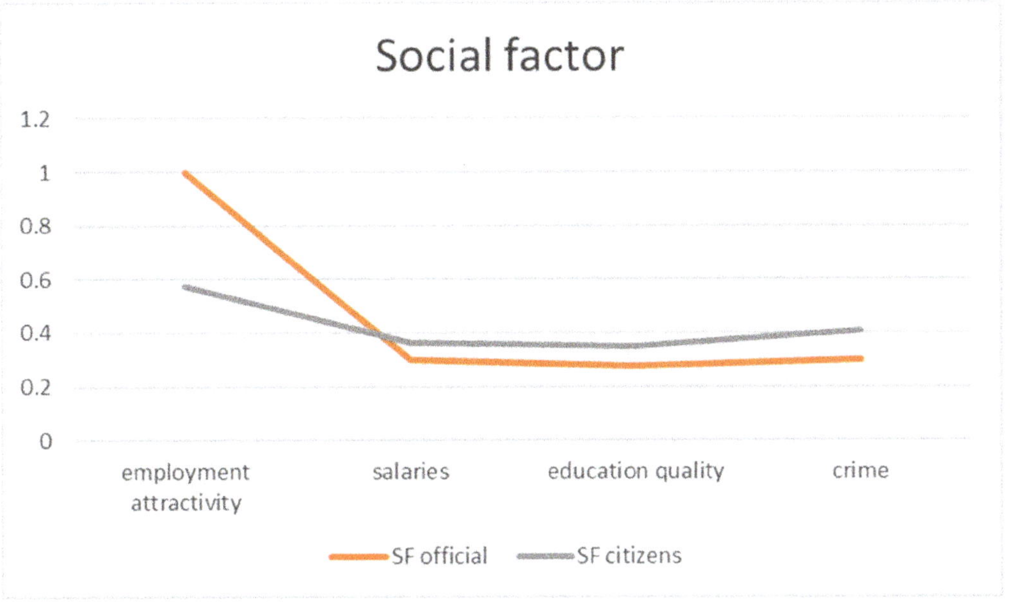

Figure 8. Analysis of social activities by calculating the coefficient of the supportability factor.

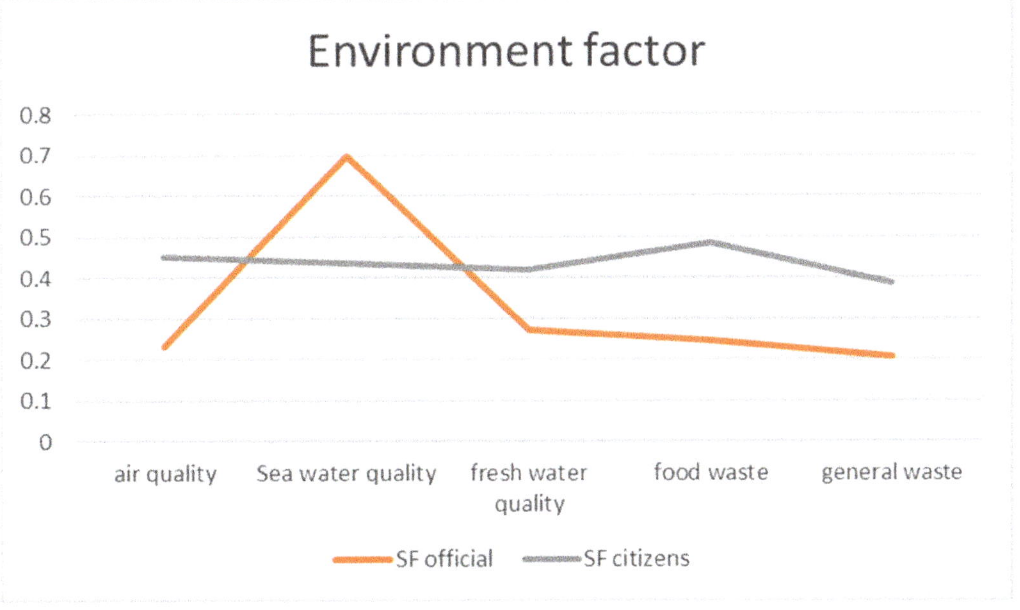

Figure 9. Analysis of environment activities by calculating the coefficient of the supportability factor.

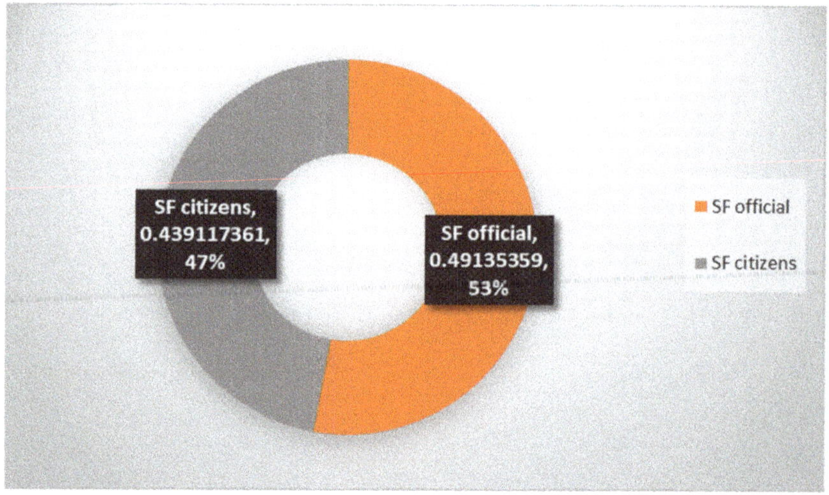

Figure 10. The share of the supportability factor resulting from the calculation of the two coefficients, from public sources and from citizen sources.

4. Discussion

This case study refers to the creation of a Living Lab concept in Constanta. The implementation framework of this new concept starts from the observations of the citizens and of the scientific environment, which define, conceptualize, and test the existing vulnerabilities in that area. To create this functional framework, the authors consider it essential to involve citizens, through the data they provide, both in research and development.

A Living Lab is a physical or virtual space where you can solve the economic, social, and environmental challenges for coastal areas by involving different stakeholders. The notion has received increased attention from academia, practitioners, and policymakers in recent years. Therefore, we conducted a systematic review of the literature on Living Labs to understand the complex approaches punctuated by the researchers. As mentioned before, the Living Laboratory is a multidisciplinary "organism" that encompasses multiple fields of research.

The contribution of the article to the scientific-technical knowledge emphasized in the multidisciplinary approach of the Living Laboratory that is discussed within the open innovation paradigms and of the users. Even the literature identifies Living Laboratory workers differently, who apply different methodologies, business models, or tools. This is normal, if we refer to the coastal areas and the fact that they differ by their particularities.

The analysis tried to demonstrate the importance of establishing a Living Lab for the management of the coastal model in Constanta (Romania). The calculation of the supportability coefficient of the analyzed activities reveals significant differences in many cases. Thus, when we examined indicators of employability (social) or bathing water quality, citizens' perceptions differed significantly from official data (environment). Moreover, in the case of economic activities, there were significant differences in opinion between authorities and citizens, particularly regarding the indicators of agriculture/aquaculture/local food and energy/electricity. It is impossible to say with certainty that one source is more credible than the other, especially because, in the field of official statistical data, there are frequently deficient methods of collection or insufficient sources to lend credibility to these types of data.

Even if the data differs in some cases, the analysis of the supportability factor shows that the trend is similar, in the sense that if the coefficient of the supportability factor is increased in the case of some activities or in the case of the analysis of the official data, this is also found in the case of the coefficient of the supportability factor of the data provided by citizens. This demonstrates that both authorities' and citizens' perceptions of certain community themes are similar, with the difference being the prioritization of activities within community themes.

In terms of the pressure on activities, according to the analysis, citizens believe that most economic indicators are under more pressure than the authorities believe. In terms of social indicators, both citizens and authorities believe that there is a lot of pressure on these indicators, with the exception of employability, where the authorities believe that jobs are sufficient. Citizens accept this, but believe that employability is not sufficiently diverse and of high quality. In terms of environmental indicators, official data are much more pessimistic than citizen trust, with the latter claiming that there is insufficient access for the cosmos.

A general analysis of the supportability factor reveals that citizens believe that the pressures exerted on coastal zone activities are significantly higher than what the data provided by the authorities show, but the difference between these two calculated values is not very large.

A hypothesis that can be deduced from the analysis made by the citizens, within the stratified sampling, led to results that can offer working tools and ways to improve the existing situations to the authorities but also to the researchers. Stratified sampling techniques are important for researchers who need to understand and design them appropriately. Our results indicate that, in the resulting coastal management model that is closer to the vision of the population, there are sometimes substantial differences from the approach of public institutions. Therefore, the model determined by Living Labs in coastal areas is a viable comparison model for government actions.

5. Conclusions

The analysis to introduce the application of the Living Labs concept in the management of the coastal area of Constanta (Romania) shows that these mechanisms can collect creative ideas and realistic approaches to some of the challenges that communities face. Citizens' participation in these structures is advantageous since it includes civil society in making development recommendations in an area. To prove this, we discovered that there were significant differences in opinion between authorities and citizens, particularly regarding the indicators of agriculture/aquaculture/local food and energy/electricity. More than that, a general evaluation of the supportability factor suggests that citizens think that the pressures exerted on coastal zone operations are substantially larger than what the statistics given by the government show, although the discrepancy between these two computed values is not particularly large. From these, we can easily conclude that the Living Labs concept is a useful approach in coastal management and aims to identify, define, develop, test, and validate new approaches to community plans and strategies.

The analysis of the integrated management model of the coastal area, based on economic, social, and environmental indicators, demonstrates the need for the establishment of a Living Lab in Constanta. This concept has the potential to lead to the harmonization of positions between citizens and authorities, as well as the inclusion of all stakeholders, who can contribute to the creation of a coherent framework in the coastal zone.

Decision makers must use innovative governance and encourage stakeholders and the general public to share the data they collect and hold in order for this to be possible. In turn, the authorities must make the databases transparent to citizens and make them easier to access.

Information and communication technology, as well as infrastructure, are required to facilitate new forms of collaboration and co-creation of new innovations among stakeholders. The digitalization of public services can aid in the development of citizen–authority relations.

A coastal Living Lab could be the community's response to the sensitivities and vulnerabilities caused by climate change. Following multi-criteria and multi-sectoral analyses, this new structure can provide dynamic and adequate solutions that are accepted by the community.

Author Contributions: Conceptualization, C.A. and A.E.M.; methodology, A.E.M.; validation, C.A., A.E.M. and E.R.; formal analysis, C.A.; investigation, A.E.M.; resources, A.E.M. and E.R.; data curation, C.A.; writing—original draft preparation, C.A.; review and editing, E.R.; visualization, C.A.; supervision, A.E.M. and E.R.; project administration, A.E.M.; funding acquisition, E.R. All authors have read and agreed to the published version of the manuscript.

Funding: This work was carried out in the framework of the research project DREAM (Dynamics of the REsources and technological Advance in harvesting Marine renewable energy), supported by the Romanian Executive Agency for Higher Education, Research, Development and Innovation Funding–UEFISCDI, grant number PN-III-P4-ID-PCE-2020-0008.

Conflicts of Interest: The authors declare no conflict of interest.

References

1. Anton, C.; Gasparotti, C.; Anton, I.A.; Rusu, E. Implementation of a Coastal Management model in Kinvara Bay at the North Atlantic Ocean. *J. Mar. Sci. Eng.* **2019**, *8*, 71. [CrossRef]
2. Anton, C.; Micu, A.E.; Rusu, E. Multi-Criteria Analysis of the Mass Tourism Management Model Related to the Impact on the Local Community in Constanta City (Romania). *Inventions* **2021**, *6*, 46. [CrossRef]
3. Yüksel, I. Developing a Multi-Criteria Decision Making Model for PESTEL Analysis. *Int. J. Bus. Manag.* **2012**, *7*, 52. [CrossRef]
4. Richardson, J.V., Jr. The library and information economy in Turkmenistan. *IFLA J.* **2006**, *32*, 131–139. [CrossRef]
5. Ståhlbröst, A.; Holst, M. *The Living Lab Methodology Handbook*; LuleGrafiska: Luleå, Sweden, 2012.
6. Rusu, E.; Soares, C.G. Wave energy assessments in the coastal environment of Portugal continental. In Proceedings of the 27th International Conference on Offshore Mechanics and Arctic Engineering—OMAE, Berlin, Germany, 9–13 June 2008; Code 76693. Volume 6, pp. 761–772.
7. Bergvall-Kareborn, B.H.M.S.A.; Hoist, M.; Stahlbrost, A. Concept Design with a Living Lab Approach. In Proceedings of the 42nd Hawaii International Conference on System Sciences, Waikoloa, HI, USA, 5–8 January 2009.
8. Fuglsang, L.; Hansen, A.V.; Mergel, I.; Røhnebæk, M.T. Living Labs for Public Sector Innovation: An Integrative Literature Review. *Adm. Sci.* **2021**, *11*, 58. [CrossRef]
9. Marine Institute. Characterisation Report Number V Clarinbridge Kinvarra Bay Shellfish Area County Galway. Available online: https://www.housing.gov.ie/sites/default/files/publications/files/filedownload22059en.pdf (accessed on 14 December 2019).
10. Anton, I.A.; Rusu, L.; Anton, C. Nearshore wave dynamics at Mangalia beach simulated by spectral models. *J. Mar. Sci. Eng.* **2019**, *7*, 206. [CrossRef]
11. De Graaf van Dinther, R.E. (Ed.) Climate Resilient Urban Areas Governance, design and development in coastal delta cities. In *Palgrave Studies in Climate Resilient Societies*; Springer International Publishing: Cham, Switserland, 2021; ISBN 978-3-030-57536-6 or 978-3-030-57537-3.
12. Diderich, C. *Design Thinking for Strategy, Innovating towards Competitive Advantage*; Springer: Richterswil, Switzerland, 2020.
13. Müller-Roterberg, C. *Design Thinking for Dummies*; John Wiley & Sons, Inc.: Hoboken, NJ, USA, 2020.
14. Rusu, E.; Conley, D.; Ferreira-Coelho, E. A hybrid framework for predicting waves and longshore currents. *J. Mar. Syst.* **2008**, *69*, 59–73. [CrossRef]
15. Ivan, A.; Gasparotti, C.; Rusu, E. Influence of the interactions between waves and currents on the navigation at the entrance of the Danube Delta. *J. Environ. Prot. Ecol.* **2012**, *13*, 1673–1682.
16. Anton, C.; Gasparotti, C.; Rusu, E. Identification of the economic pressure on environmental factors in the Romanian coastal zone-case study Eforie. In Proceedings of the 18th International Multidisciplinary Scientific GeoConference Surveying Geology and Mining Ecology Management, SGEM, Albena, Bulgaria, 24 August–2 September 2018; Volume 18, p. 461.
17. Anton, C.; Gasparotti, C.; Rusu, E.; Anton, I.A. Approach to the analysis and evaluation of the strategic intervention options in the Romanian coastal zone taking into account economic, social and environmental factors. In Proceedings of the International Multidisciplinary Scientific GeoConference Surveying Geology and Mining Ecology Management, SGEM, Albena, Bulgaria; 2018; Volume 18, p. 67. Available online: https://www.sgem.org/index.php/elibrary?view=publication&task=show&id=1571 (accessed on 30 October 2021).
18. Anton, C.; Gasparotti, C.; Rusu, E. A challenge for the inland navigation—A connection between the Baltic and the Black seas. In Proceedings of the ICTTE Conference, Belgrad, Serbia, 20–21 July 2018.
19. Cheng, E.W.L.; Li, H. Application of ANP in process models: An example of strategic partnering. *Build. Environ.* **2007**, *42*, 278–287. [CrossRef]
20. Dare, C. The UK tour-operating industry: A competitive analysis. *J. Vacat. Mark.* **2006**, *6*, 357–367.
21. Saaty, T.L.; Takizawa, M. Dependence and independence: From linear hierarchies to nonlinear networks. *Eur. J. Oper. Res.* **1986**, *26*, 229–237. [CrossRef]
22. Shilei, L.; Yong, W. Target-oriented obstacle analysis by PESTEL modeling of energy efficiency retrofit for existing residential buildings in China's northern heating region. *Energy Policy* **2009**, *37*, 2098–2101. [CrossRef]
23. Shyur, H.J. COTS evaluation using modified TOPSIS and ANP. *Appl. Math. Comput.* **2006**, *177*, 251–259. [CrossRef]

24. Tzeng, G.H.; Chiang, C.H.; Li, C.W. Evaluating intertwined effects in e-learning programs: A novel hybrid MCDM model based on factor analysis and DEMATEL. *Expert Syst. Appl.* **2007**, *32*, 1028–1044. [CrossRef]
25. Vitkiene, E. Questions that have to be answered by a well-managed strategy of coastal recreation and tourism enterprises development. *TILTAI* **2009**, *4*, 117–124.
26. Wu, W.W. Choosing knowledge management strategies by using a combined ANP and DEMATEL approach. *Expert Syst. Appl.* **2008**, *35*, 828–835. [CrossRef]
27. Elomda, B.M.; Hefny, H.A.; Hassan, H.A. An extension of fuzzy decision maps for multi-criteria decision-making. *Egypt. Inform. J.* **2013**, *14*, 147–155. [CrossRef]
28. Frynas, J.G.; Mellahi, K. *Global Strategic Management*; Oxford University Press: Oxford, MS, USA, 2015.
29. Gray, S.A.; Gray, S.; Cox, L.J.; Henly-Shepard, S. Mental modeler: A fuzzy-logic cognitive mapping modeling tool for adaptive environmental management. In Proceedings of the 46th Hawaii International Conference on System Sciences, Wailea, HI, USA, 7–10 January 2013.
30. Healey, N.M. The transition economic of central and eastern Europe: A political, economic, social and technological analysis. *Columbia J. World Bus.* **1994**, *29*, 62–70. [CrossRef]
31. Kosko, B. Fuzzy cognitive maps. *Int. J. Man-Mach. Stud.* **1986**, *24*, 65–75. [CrossRef]

inventions

Article

Estimation of the Tower Shape Effect on the Stress–Strain Behavior of Wind Turbines Operating under Offshore Boundary Conditions

Alexandra Ionelia Diaconita *, Gabriel Andrei and Eugen Rusu

Department of Mechanical Engineering, Faculty of Engineering, "Dunarea de Jos" University of Galati, 47 Domneasca Street, 800008 Galati, Romania; gabriel.andrei@ugal.ro (G.A.); erusu@ugal.ro (E.R.)
* Correspondence: alexandra.diaconita@ugal.ro

Abstract: The metal tower, or the pylon, is one of the most important elements in the construction of a wind turbine. It has the role of supporting the entire wind turbine, and it also allows access for inspection and conducting planned maintenance and repairs. Moreover, the tower ensures support for the structure and strengthens the whole assembly. It has a particularly important role, as it has to face very severe weather conditions. The present study aims to analyze the forces and moments resulting from the action of the wind on the tower of a wind turbine. Two important load cases are considered, namely, the load under operating conditions and the ultimate load under 50 year wind conditions. For this purpose, cylindrical and conical geometric shapes of the tower were chosen. These were analyzed under the action of both normal and extreme wind speeds. Then, the behavior of the two towers under the action of the wind speed for a location in the Black Sea was analyzed. Finally, in an attempt to make the structure more economical, the thickness of the shell was reduced.

Keywords: tower; wind turbine; load; offshore wind; marine renewable energy

Citation: Diaconita, A.I.; Andrei, G.; Rusu, E. Estimation of the Tower Shape Effect on the Stress–Strain Behavior of Wind Turbines Operating under Offshore Boundary Conditions. *Inventions* **2022**, *7*, 11. https://doi.org/10.3390/inventions7010011

Academic Editor: Emin Bayraktar

Received: 8 November 2021
Accepted: 3 January 2022
Published: 7 January 2022

1. Introduction

Global warming is a set of changes that affect the planet. Among them is an increase in temperature, especially in polar areas, where such an increase has exceeded 3 degrees Celsius since 1906, while for the rest of Earth, it increased by approximately 0.9 degrees Celsius [1]. Climate change leads to melting glaciers, the occurrence of extreme weather events, rising water levels, impacts on animal habitats, the extinction of several animal species, and many other effects. In addition to the negative effects on the environment, greenhouse gas emissions also harm human health [2], resulting in respiratory diseases caused by pollution and smog [3].

Reducing greenhouse gas emissions is a point of interest in the 21st century because of the negative effects caused by the exploitation of conventional resources to produce electricity. Carbon dioxide is responsible for approximately three-quarters of greenhouse gas emissions, which can remain in the atmosphere for thousands of years. These emissions have gradually increased, and in 2021, they reached a value of 33 Gt CO_2, despite the significant decrease that was noticed in 2020 (having a value of 31.5 Gt CO_2), which was considered the largest decrease recorded globally, even exceeding the decline in 2009 when the global financial crisis hit [4]. Due to the pandemic, not only have CO_2 emissions decreased, but energy demand has also fallen the most since World War II and to date, decreasing by approximately 4.5% [5].

One of the best ways to eliminate greenhouse gas emissions is to replace the sources that produce the highest amounts of CO_2. Globally, the sources producing the highest amounts of CO_2 emissions are the burning of materials such as coal, oil, gas, and wood to obtain energy. Therefore, attention has focused on renewable resources [6] that have lately gained increasing importance. According to the International Renewable Energy

Agency (IRENA) [7], the installed capacity from renewable sources increased by 10% in 2020, having the largest increase so far despite the COVID-19 pandemic. Currently, the total installed capacity reaches a value of 2799 GW, the highest percentage of which is allocated to the hydraulic renewable energy sector (approximately 44% of the total installed capacity) [8]. This spectacular growth makes the renewable energy sector the new normal in the 21st century, and it is expected to add more than 280 GW by the end of 2021. In addition, the energy generated in the first quarter of 2021 sustains the idea of eliminating conventional sources, and energy from fossil fuels and that from renewable sources recorded the same value of 34.7 TWh [9]. In terms of wind energy, this was doubled in 2020 compared to 2019 due to the technological advances in wind turbines. The offshore industry aims to exploit deeper waters where wind speeds are higher. Currently, there are wind farms located with the help of floating structures in waters with depths over 100 m [10]. Apart from the desire to advance as far as possible, the design of higher capacity turbines is also expected with the largest offshore turbine being Haliade-X at 14 MW. Obtaining high power is closely related to turbine components such as large rotors and high towers. Compared to onshore wind turbines, offshore wind turbines do not require very high towers, for instance, a height of 22 m between the lowest point of the rotor sweep and the average height of the water springs is quite enough [11]. On the other hand, the towers placed offshore are much more exposed to water corrosion and, for this reason, they must meet specific standards. Through the technological development of wind turbines, such as the growth of the rotor [12,13] and the tendency to place it in the deepest possible waters, the field of offshore renewable energy attracts increasing amounts of investments. Each location implies certain costs; therefore, it is necessary to consider a site-specific design so that the structure lasts approximately 25 years and, thus, the LCOE for each farm may vary [14]. The optimization of a wind turbine according to the weather conditions at a specific location takes into account two aspects. The first aspect is related to the behavior of the structure, and the second aspect refers to the identification of the optimal design variables [15]. In order for wind energy to be competitive with renewable energy sources, the trend is to optimize these already developed structures so that costs are minimized [16]. Among the methods for optimizing the tower of a wind turbine is the replacement of conventional materials, i.e., steel with carbon or glass fiber. These materials improve fatigue performance and have better corrosion resistance [17–19]. Another method of minimizing costs is to reduce the thickness of the shell and reduce the number of ring-stiffeners [20].

Since the towers for wind turbines of no more than 4 MW have diameters between 3 and 5 m, the present study aimed to analyze the behavior of two types of towers, one circular and one conical, under the action of the same environmental conditions. The forces and moments applied on the two towers were given only by the wind speed and the weight of the components that were part of the wind turbine, and those given by the waves were omitted. The structure was divided into four sections of different thicknesses. In addition, we considered the wind speed from a location in the west of the Black Sea, close to the Romanian nearshore. For this purpose, the wind data were taken from the EURO-CORDEX database and then processed using a MATLAB computational environment. From this data, the maximum and average wind speeds were taken into consideration to provide a more comprehensive image of the wind potential from the reference point considered. Then, a study was performed to analyze the behavior of the two structures under the action of the maximum wind. The possibility of reducing the thickness of the two towers by 25% and 50%, respectively, was also considered.

Similar studies were conducted by Lin Wang et al. [21] in a paper entitled "Structural Optimization of Wind Turbine Towers Based on Finite Element Analysis and Genetic Algorithm", and they conducted a structural analysis on an 87.6 m tower using as a reference the 5 MW turbine designed by NREL (National Renewable Energy Laboratory) [22], with the final goal of this work being to optimize the tower. Another study dealing with this subject is that of Zhu Rensheng et al. [23], who performed a finite element analysis on a turbine with a horizontal axis of 1.5 MW with the tower having a conical construction. A similar study was conducted by Chile Li and Limin Ren [24], who carried out a finite element analysis also for a 1.5 MW turbine. The two cases studied took into account both normal and storm conditions to establish whether the structures were safe. The analyses showed that the structures were sufficiently secure, and there was no strength failure. Yadneshwar S. Joshi and Pratibha Alandkar [25] conducted a study with the purpose of performing a finite element analysis of the tower but also of the connections between the sections of the tower in order to understand exactly how these connections behave under the action of environmental forces. The analysis was performed on a conical tower that was divided into 16 sections and which had a height that was considered to be 80. For this study, a Suzlon S88-2.1 MW turbine was considered.

2. Materials and Methods

2.1. Characteristics of the Wind Turbine and Tower

The turbine chosen for this study was a Vestas V112-3 MW which corresponds to IEC Class I. This type of wind turbine has a horizontal axis, with an upwind orientation. The turbine is equipped with a rotor with three blades that are controlled by a microprocessor pitch control system. The characteristics of the turbine are presented in Table 1 and Figure 1.

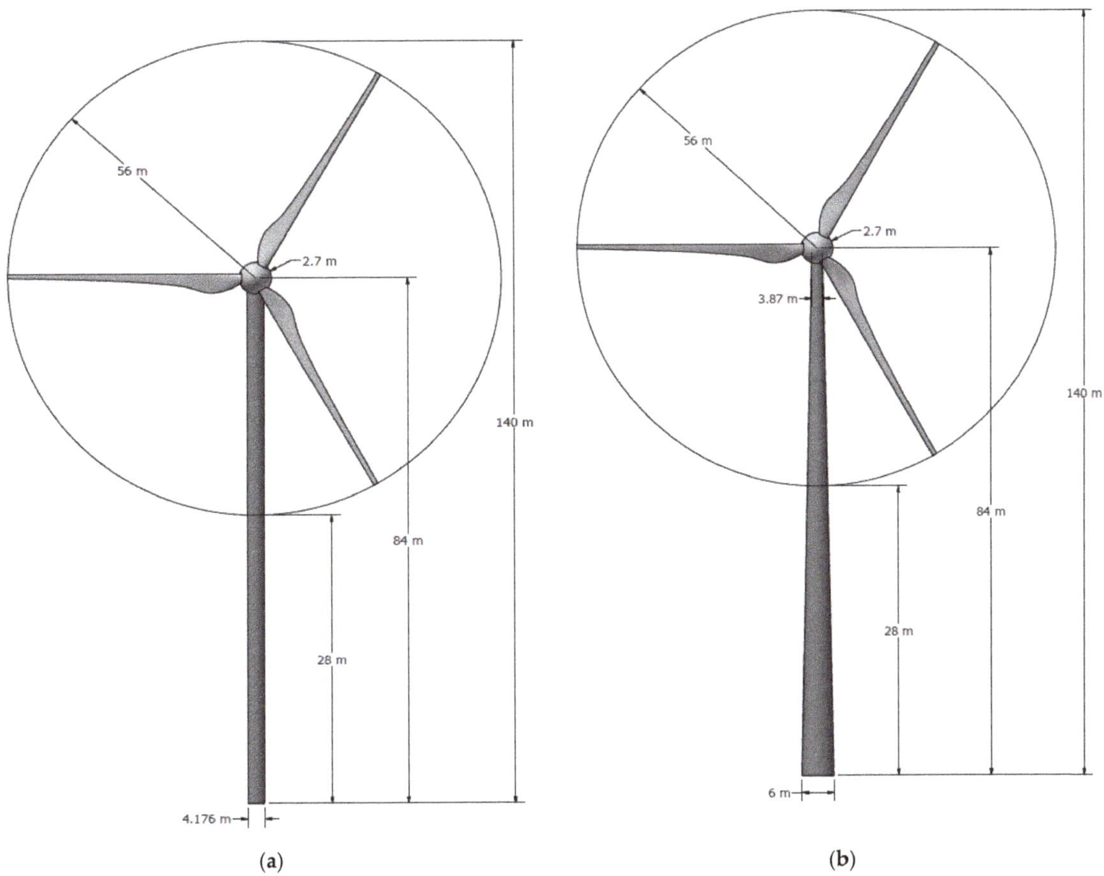

Figure 1. 2D model of the wind turbine with a (**a**) cylindrical tower; (**b**) conical tower.

Table 1. Wind turbine characteristics.

Properties	Values		
Rated capacity		3 MW	
Wind class		IEC Ia	
Rotor diameter		112 m	
Swept area		9852 m^2	
Mass (blades, rotor, nacelle)		155 t	
Hub height		84 m	
Material		Steel S355JR	
Type of tower	Cylindrical		Conical
Tower diameter	4.176 m		6 m at base 3.87 at top

The material characteristics of the Vestas V122-3 MW turbine tower are given in Table 2.

Table 2. Vestas V122-3 MW material characteristics.

Properties	Values
Young's modulus (E)	$200{\cdot}10^3$ MPa
Density (ρ)	7800 kg/m^3
Poisson's ratio (ν)	0.3

The power curve and the thrust coefficient are presented in Figure 2. The power curve shows how much power can be generated under the action of a certain wind speed. The power curve in Figure 2 was obtained using a quadratic equation that is related to the wind speed [26,27].

$$\mathrm{P_{wg}}(V) = \begin{cases} \mathrm{q(V)} & V_{\mathrm{cin}} < \mathrm{V} < V_{\mathrm{rat}} \\ \mathrm{P_r} & V_{\mathrm{rat}} < \mathrm{V} < V_{\mathrm{cou}} \\ 0 & \mathrm{V} \le V_{\mathrm{cin}} \text{ and } \mathrm{V} \ge V_{\mathrm{cou}} \end{cases} \quad (1)$$

$$\mathrm{q(V)} = \mathrm{P_r}\frac{V^2 - V_{\mathrm{cin}}^2}{V_{\mathrm{rat}}^2 - V_{\mathrm{cin}}^2}$$

where V (in m/s) is the wind speed at the height of 84 m, V_{cin} (in m/s) is the cut-in wind speed (having the value of 3 m/s), V_{cou} (in m/s) is the cut-out wind speed (having the value of 25 m/s), V_{rat} (in m/s) is the rated wind speed (having the value of 12 m/s), and $\mathrm{P_r}$ is the rated power in MW.

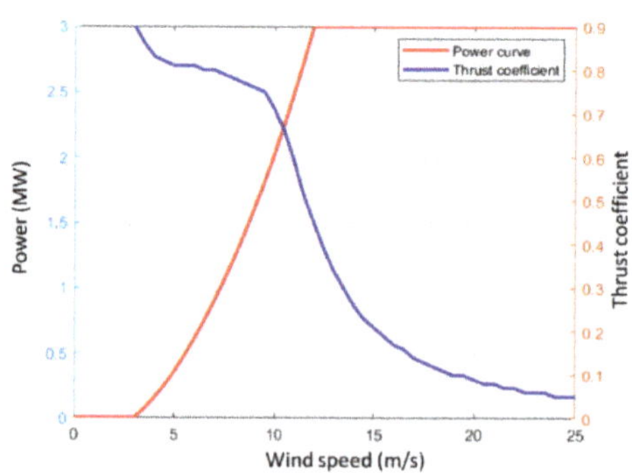

Figure 2. Power curve and thrust coefficient of the turbine Vestas V112-3 MW.

2.2. Basic Assumptions

- The tower was considered to be a cantilevered beam that was fixed to the structure. At the top, a mass was applied that was equal to the mass of the nacelle, blades, and rotor;
- The material of the tower was considered to be isotropic and homogeneous. One tower had a constant circular section along the height, and the other one had a decreasing section from the base to the top, and both of them featured small wall thicknesses;
- The secondary effects (such as axial and shear deformations and moment of inertia) were neglected;
- The aerodynamic distributed loads were caused only by drag forces.

2.3. Loads Calculation

The main loads acting on the tower were gravitational loads and aerodynamic loads given by the action of the wind, which can be drag and lift forces. The aerodynamic loads were grouped into two categories: those that act directly on the tower and those that act on the rotor and are further transferred to the upper part of the tower [28]. The forces and moments acting on the turbine are illustrated in Figure 3.

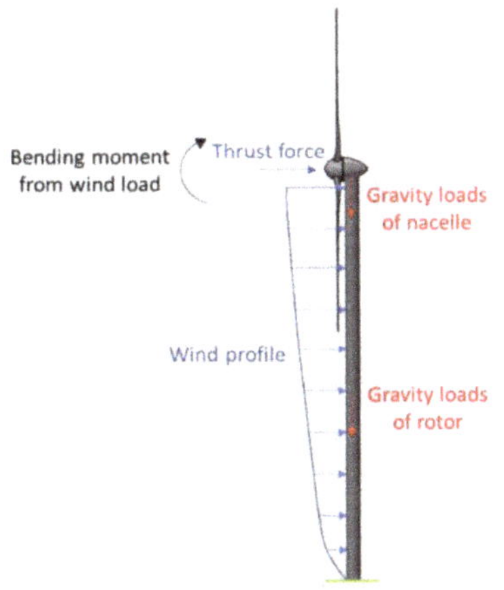

Figure 3. The forces and moments considered in this study.

2.3.1. Gravitational Loads

The gravitational loads are given by the weight of the supported components. These components are the nacelle, rotor, and blades. Their approximate weight was 155 tons. This load contributed to the compression loads on the tower's structure. The representation was made with the help of a mass point at the top of the tower.

2.3.2. Aerodynamic Loads

The wind loads acting on the tower arise due to the airflow and its interaction with the wind turbine tower. The distributed wind load along the tower height per unit length, z, can be calculated using the following formula (Figure 4) [29,30]:

$$F_d = \frac{1}{2}\rho V(z)^2 C_d D(z) \quad (2)$$

where ρ is the air density (having the value of 1.225 kg/m^3), V(z) is the wind velocity at height z (in m/s), C_d is the drag coefficient (this parameter has a value of 0.7).

As the wind speed varies along the tower height, under normal wind conditions, the wind will be calculated using the following formula [31]:

$$V(z) = V_{hub}\left(\frac{z}{z_{hub}}\right)^{\alpha} \tag{3}$$

where V_{hub} is the reference wind speed in m/s at height z_{hub} (the hub height), and α is the power law exponent (this parameter is also known as the Hellman exponent) [32,33]. The value of this coefficient was estimated to be 1/7 (approximately 0.143) [34] and evaluates the wind profile under normal weather conditions.

For the extreme wind model, the wind was calculated using the formula below, which was according to the IEC 61400-1 standard [35]:

$$V_{e50} = 1.4 \cdot V_{ref} \tag{4}$$

Offshore wind turbines are classified depending on wind speed and turbulence parameters, according to the IEC standard. In the above formula, V_{e50} is the expected extreme wind speed with a recurrence time interval of 50 years (in m/s). The chosen turbine belongs to class IEC Ia, and according to the standard, the reference speed V_{ref} is 50 m/s.

Using the wind profile power law, the equation above becomes:

$$V_{(e50)_z} = 1.4 \cdot V_{ref}\left(\frac{z}{z_{hub}}\right)^{\alpha} \tag{5}$$

The Hellman exponent in this case had a value of 0.11 [36], because the value of 0.143 can cause computational errors when it comes to the field of offshore wind energy; therefore, the value of 0.11 was much more suitable for open water.

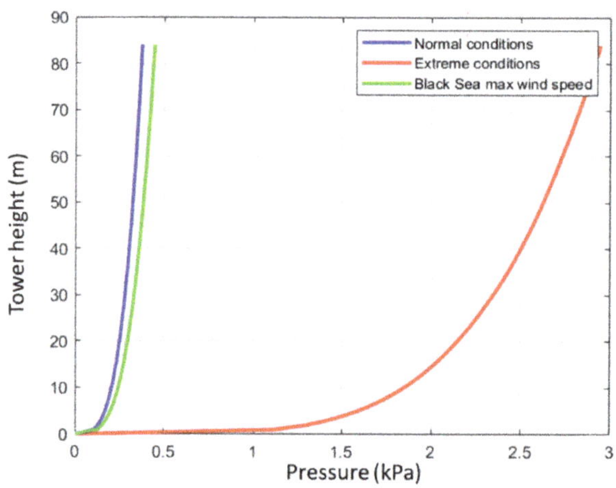

Figure 4. The pressure applied along the two towers.

Apart from the loads that were given by the wind on the tower, its upper part was also affected by a series of forces and moments given by the action of the wind on other components of the turbine assembly [37]. The IEC61400-1 [35] design standard defines twenty-two load cases for turbine design, covering all operating conditions, from start-up to extreme conditions. The typical cases used were those in extreme wind conditions and those in normal operating conditions. The aerodynamic force given by the action of the wind on the rotor can be calculated using the formula [24,38]:

$$F_x = \frac{1}{2} C_T \rho V^2 A \quad (6)$$

where C_T is the thrust coefficient, ρ is the air density (having the value of 1.225 kg/m^3), V is the wind speed in m/s, and A is the swept area in m^2.

2.4. Analysis of the Behavior of the Two Towers in a Location in the Black Sea and Possibilities to Reduce the Thickness of the Shell

In the study in [39], conducted on the potential of the Black Sea for renewable wind energy, it was established that of the six points analyzed, the highest wind speeds that were interpolated at a height of 80 m were for the point that is represented in Figure 5.

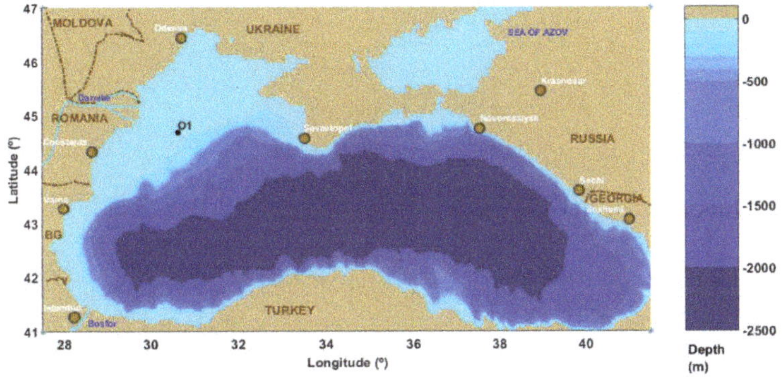

Figure 5. The location of the chosen point.

The algorithm used in Reference [39] was also considered in this case, but the interpolation was made at a height of 84 m. All calculations were conducted using MATLAB software. The data used to obtain wind speed at 84 m were obtained from the EURO-CORDEX database, and the period analyzed was over 30 years, from 1 January 2021 to 31 December 2050. The location chosen has an elevated percentage of high wind speed occurrence, which correspond to wind class 7 (this class is recommended for the offshore industry), which is approximately 34%, and 30% of speeds were lower but still suitable for extracting renewable energy. Wind speeds in the Black Sea are not strong, and for this reason there are currently no wind farms. The highest wind speeds are to be found on the west coast of the Black Sea near Romania. We further considered the maximum wind speeds in these areas and analyzed the behavior of the two towers, simulating the forces and moments given by them. The location and wind characteristics for the chosen point are presented in Table 3.

Table 3. Data regarding the chosen location.

Data	Values
Location	Black Sea
Longitude (°)	30.62
Latitude (°)	44.68
Depth (m)	75.29
Maximum wind speed (m/s)	27.17
Mean wind speed (m/s)	8.00
Annual energy production (MWh)	11,712
Capacity factor (%)	45

Since the chosen reference point had a maximum wind speed value of 27.17 m/s, we tried to reduce the tower wall thickness, once by 25% and once by 50%, to see what would happen.

3. Results and Discussion

The most common method of designing wind turbine towers is by manufacturing sections between 20 and 30 m coupled with flanges at both ends and then bolting them to the location. In this paper, only the sections of the tower were considered without simulating the connections. The towers were divided into four sections, three of which had a length of 20 m and one with a length of 24 m. Each section had a certain thickness allocated, starting from 50 mm at the bottom and reaching 19 mm at the top. The two towers were modeled as shell structures in ANSYS. The main characteristics are detailed in Tables 1 and 2. The thickness of the two towers are indicated in Table 4. The 3D model is presented in Figure 6.

Table 4. Tower wall thickness.

Segment	Thickness
Segment 1	50 mm
Segment 2	30 mm
Segment 3	19 mm
Segment 4	19 mm

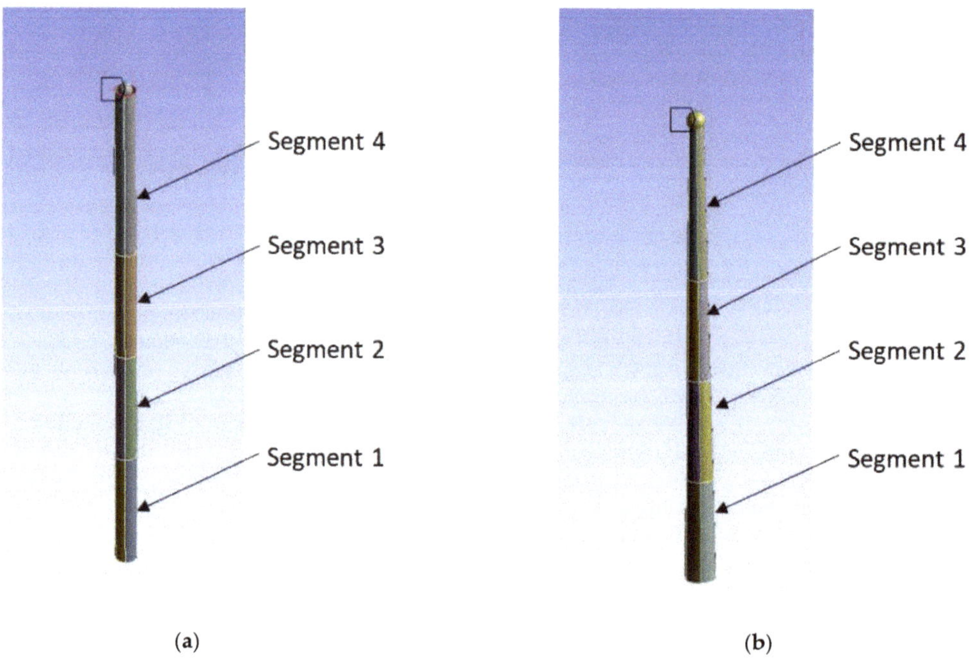

Figure 6. Tower geometry: (**a**) cylindrical; (**b**) conical.

The mesh was generated through quadrilateral elements, as they best reproduce the stress distribution, and the results given by the finite element analysis are dependent on the chosen discretization solution. The size of applied quadrilateral elements was 0.5 m.

To determine the size of the mesh elements, a mesh sensitivity study was performed. For this, the two towers were fixed at the bottom and allowed to vibrate freely without adding forces and moments. The results obtained for the six modal frequencies are presented in Table 5.

Table 5. The modal frequencies for the four sizes of the mesh elements.

Type of Tower	Modal Frequencies (Hz)	Mesh Size			
		2 m	1 m	0.5 m	0.25 m
Cylindrical	1	0.35984	0.36079	0.36082	0.36084
	2	0.35985	0.36079	0.36082	0.36084
	3	2.97390	2.91420	2.90540	2.90390
	4	2.97410	2.91750	2.90670	2.90450
	5	3.13870	2.97780	2.97790	2.97790
	6	3.15560	2.97780	2.97790	2.97790
Conical	1	0.50980	0.51150	0.51194	0.51206
	2	0.51066	0.51174	0.51199	0.51207
	3	2.89480	2.79470	2.79080	2.79140
	4	2.95560	2.79860	2.79230	2.79200
	5	3.45660	3.36670	3.36330	3.36410
	6	3.51160	3.37100	3.36490	3.36480

For the meshes of 0.5 (Figure 7) and 0.25 m, respectively, for both the cylindrical and conical towers, the differences were imperceptible. For the cylindrical structure, the difference between the natural frequency for mode 1 for the 0.5 and 0.25 m mesh was approximately 0.006%. For the conical structure, the difference between the natural frequency for mode 3 for the mesh of 0.5 and 0.25 m was approximately 0.023%.

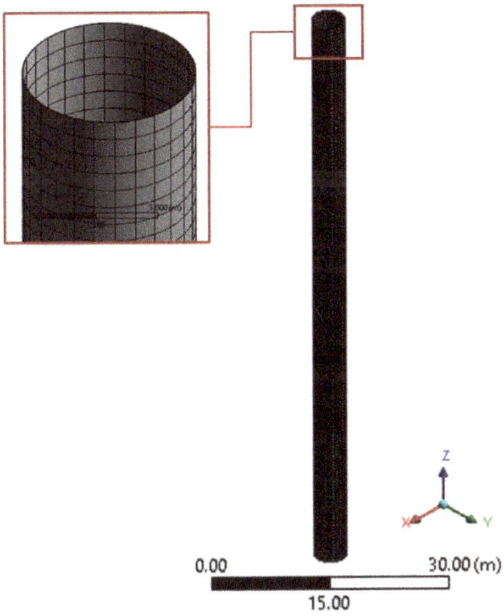

(a)

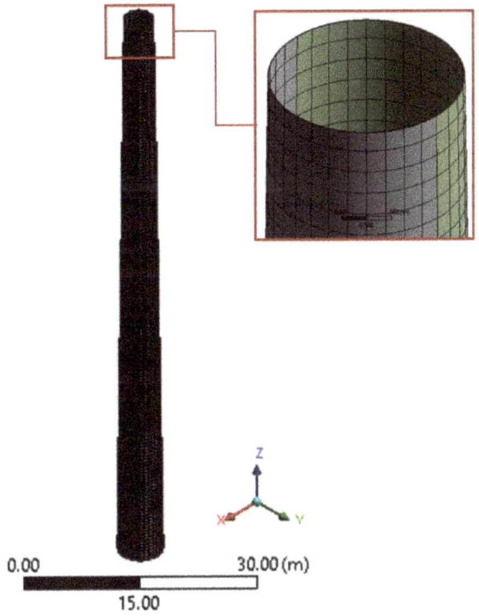

(b)

Figure 7. The mesh for the two towers: (a) cylindrical tower with 4368 elements; (b) conical tower with 5344 elements.

To avoid resonance caused by the vibrations, the frequency of the tower must be different from the harmonic vibrations associated with the rotor. The rotor had a value of harmonic vibrations between $1P = f_{rot} \cdot S_f$ and $3P = \frac{3 \cdot f_{rot}}{S_f}$, which corresponded to the frequency of the rotor. 1P is the primary excitation frequency of the rotor caused by the rotor spinning at a given velocity which induces mass imbalances. The second excitation frequency was caused by the blade passing the tower; this wind turbine had three blades, so it is written as 3P. This interdependence can be expressed by the inequality below. This range is a soft–stiff one and represents the optimum range for the best possible design [40]:

$$f_{rot} \cdot S_f \leq f_{tower} \leq \frac{3 \cdot f_{rot}}{S_f} \tag{7}$$

where S_f is the safety factor for frequency (has the value of 1.05 according to the GL standard [41]), f_{tower} is the first natural frequency of the tower, and f_{rot} is the frequency associated with the rotor. Vestas V112-3 MW has a rotor rotational speed of 12.8 rpm, which corresponded to the frequency of 0.213 Hz.

Thus, the values obtained for the first natural frequency of the tower must be between the values:

$$0.244 \leq f_{tower} \leq 0.640 \tag{8}$$

As shown in Figure 8, the first frequency mode for the cylindrical structure was 0.361 Hz, and for the conical structure, it was 0.512 Hz. Both values were within the range of 0.224–0.640 Hz.

Figure 9 illustrates the two deformed towers under the action of the forces and moments in normal wind conditions. A tower is considered safe when the maximum deformation (marked with d_{max}) does not exceed the allowable deformation (marked with $d_{allowable}$). It is considered that the structure is globally stable and is out of any danger when the following relationship is fulfilled [21]:

$$d_{max} \leq d_{allowable} \tag{9}$$

The allowable deformation can be calculated with the formula [42]:

$$d_{allowable} = 1.25 \cdot \frac{L}{100} \tag{10}$$

where L is the height of the tower and is measured in m.

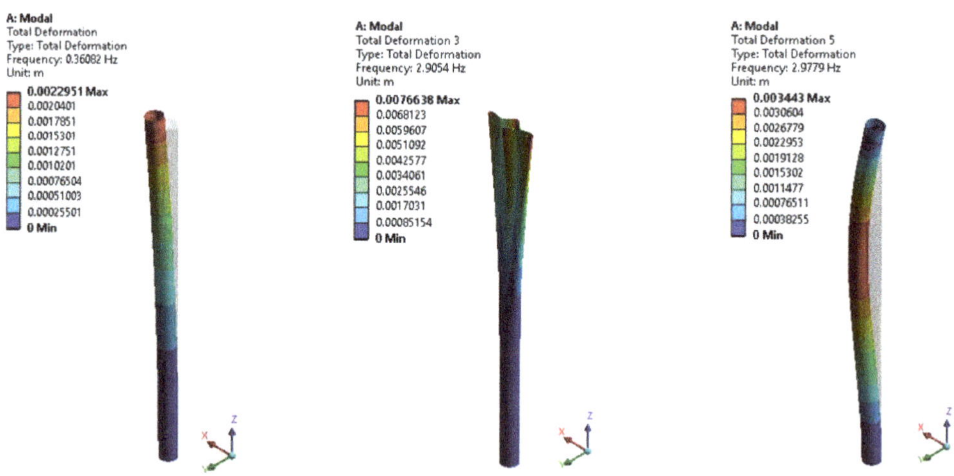

(a) (b) (c)

Figure 8. *Cont.*

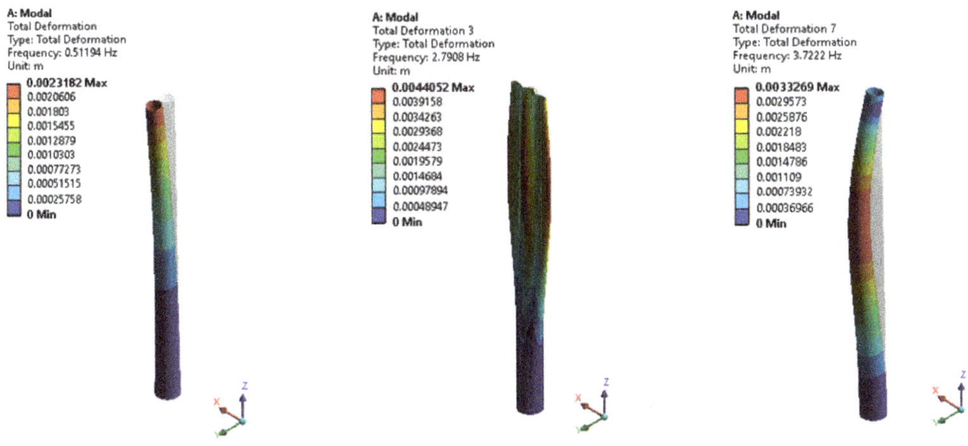

(d) (e) (f)

Figure 8. Vibration modes of the wind towers: (**a**) cylindrical tower—first mode; (**b**) cylindrical tower—third mode; (**c**) cylindrical tower—fifth mode; (**d**) conical tower—first mode; (**e**) conical tower—third mode; (**f**) conical tower—seventh mode.

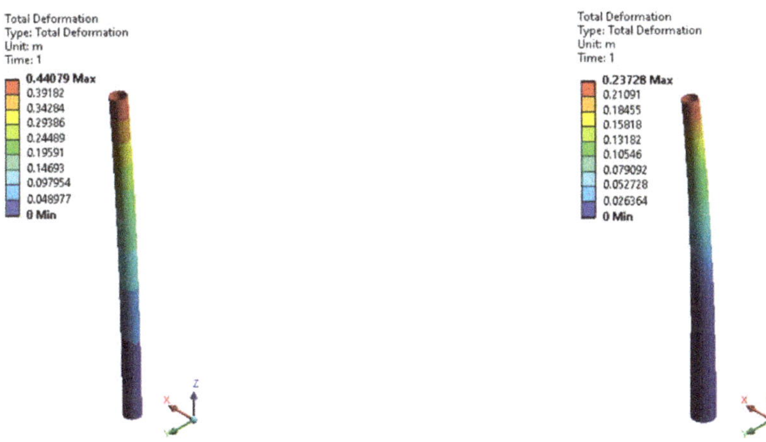

(a) (b)

Figure 9. Total deformation for the two structures in normal wind conditions: (**a**) cylindrical tower; (**b**) conical tower.

From the above equations, the value of 1.05 m for the admissible deformation was obtained. The deformations obtained for the two geometric shapes of the tower were below this value (Figure 9). The cylindrical structure had the maximum deformation at the top of the tower, and its value was approximately 0.44 m, while the conical structure was much smaller, i.e., approximately 0.24 m. Thus, the two structures avoided excessive motion. These values were much lower than the allowable value, which indicates that the structure was sufficiently rigid.

Figure 10 illustrates the deformation of the two towers under the action of forces and moments in extreme conditions. The maximum deformation for the cylindrical tower was 1.8907 m, which exceeds the value of the allowable deformation. This result indicates that the structure was not sufficiently rigid for these conditions. The maximum deformation of the conical tower was 1.0229 m. This value was 2.7% lower than the allowable value and indicates that the tower was sufficiently stiff and would not experience large deformations.

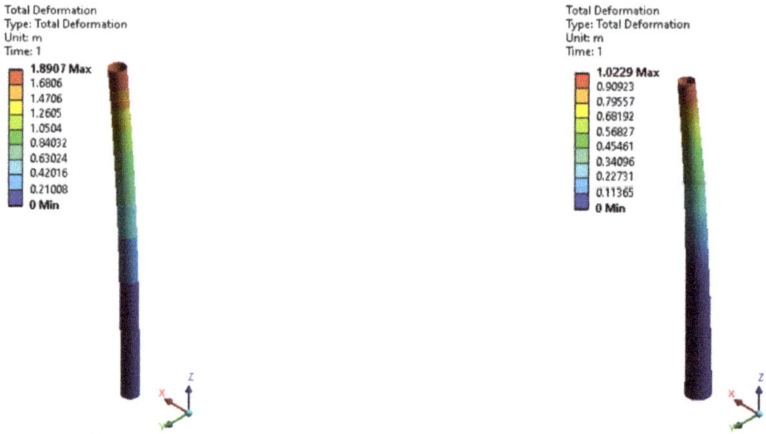

Figure 10. Total deformation for the two structures in extreme wind conditions: (**a**) cylindrical tower; (**b**) conical tower.

The von Mises stress, σ, generated by the loads cannot exceed the allowable stress. This can be expressed in the following inequality forms:

$$\sigma \leq \sigma_{\text{allowable}} \tag{11}$$

The allowable stress is given by:

$$\sigma_{\text{allowable}} = \frac{\sigma_y}{\gamma_m} \tag{12}$$

where σ_y is the yield strength, and γ_m is the material safety factor.

The material yield strength for S355 Steel is 335 MPa for thicknesses of $40 \leq t \leq 63$ mm, and for thicknesses of $16 \leq t \leq 40$ (mm), the yield strength is 345 MPa. In the case of the chosen towers, we considered the yield strength at the upper thickness to ensure that they met all the desired conditions.

The value of the safety factor was chosen as 1.1 [35]. This value is minimal compared to those used in practice but is a reference value that takes into account the instability of the material. For this reason, the values obtained from the calculation may be higher than the actual ones.

Using Equations (11) and (12), we obtained the value of the allowable stress of 305 MPa. The values of the von Mises stresses in the case of the cylindrical and conical structures under normal operating conditions were 100 and 65 MPa, respectively (Figure 11). These values are below the allowable stress value.

Figure 12 illustrates the distribution of von Mises stress in extreme wind conditions. We can see that the cylindrical tower had a von Mises stress value of 423 MPa; this exceeds the allowable value such that the cylindrical structure was not safe. The value of the conical tower had a von Mises stress of 277 MPa. The conical structure was safe both in normal operating conditions and in extreme conditions.

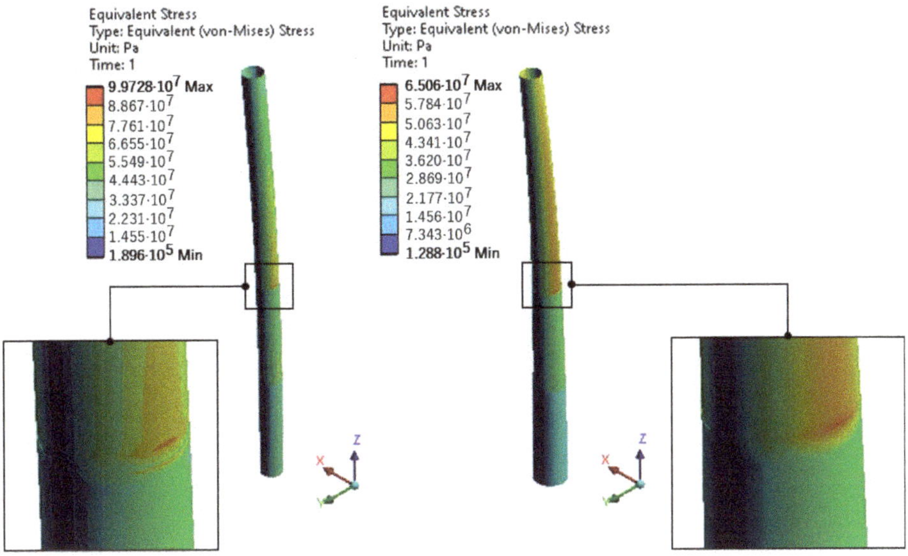

(**a**) (**b**)

Figure 11. Distribution of von Mises stresses on the two structures in normal wind conditions: (**a**) cylindrical tower; (**b**) conical tower.

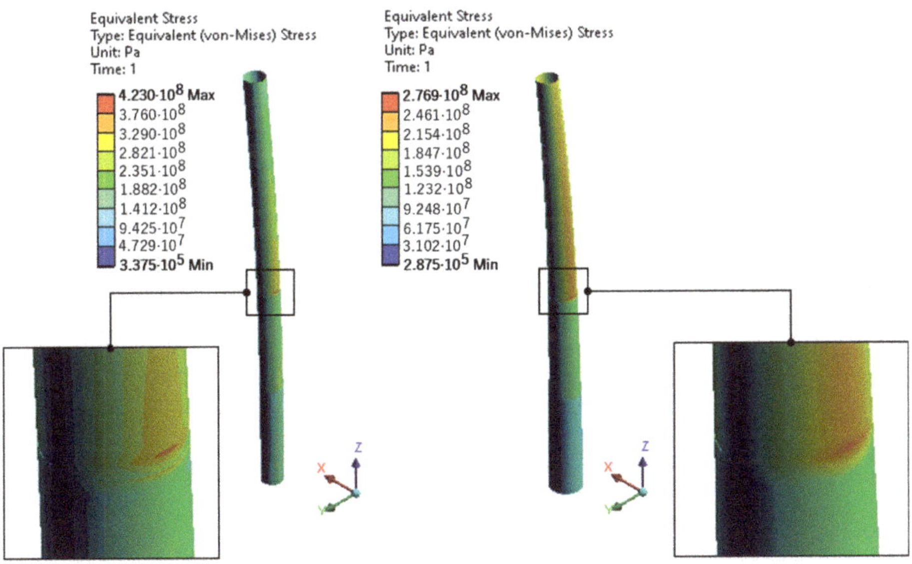

(**a**) (**b**)

Figure 12. Distribution of von Mises stresses on the two structures in extreme wind conditions: (**a**) cylindrical tower; (**b**) conical tower.

Figures 13 and 14 illustrate the total deformations and von Mises stresses. From them, it can be observed that the results are similar to those in Figures 9 and 11; this is because the wind speeds were also close.

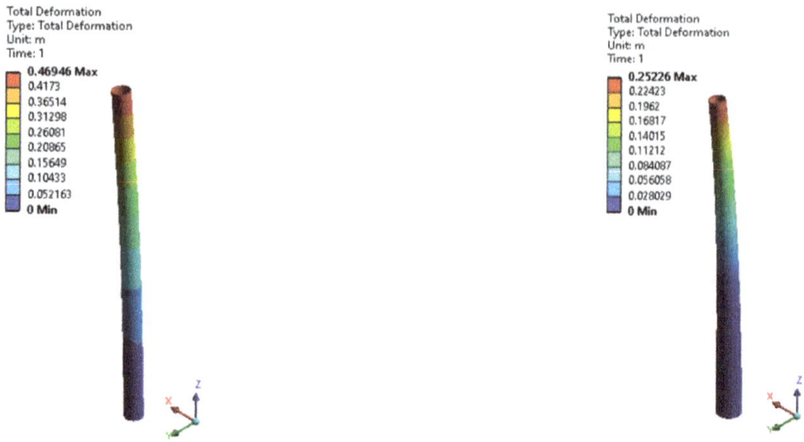

(a) (b)

Figure 13. Total deformation of the two structures under the conditions of the Black Sea: (**a**) cylindrical tower; (**b**) conical tower.

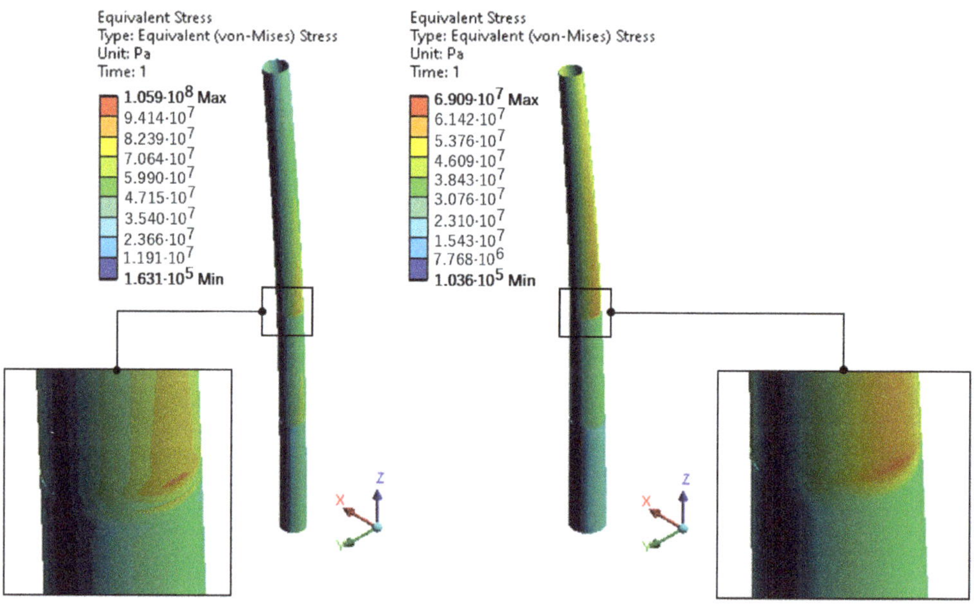

(a) (b)

Figure 14. Distribution of von Mises stresses on the two structures under the conditions of the Black Sea: (**a**) cylindrical tower; (**b**) conical tower.

As we reduced the thickness, the overall mass of the two towers reduced too, along with their frequencies. From Table 6, we can observe that there was a tendency for the natural frequencies to reduce as the thickness decreased too. For vibration mode 1, all of the values obtained relative to all of thicknesses were in the range mentioned in Equation (8), which means that the resonance effect was avoided regardless of the chosen thickness. In addition, the deformations tended to increase with the reduction in the tower's thickness, but no value exceeded the maximum allowable limit of 1.05 m. The highest value was obtained in the case of the cylindrical tower where the thickness was reduced by 50%. The von Mises stress also had an increasing tendency as well as deformations, but even these did not exceed the allowable limit established with the help of Equations (11) and (12).

Table 6. The results obtained for the three types of thicknesses.

Type of Tower	Parameters	Baseline	25% Thickness Reduction	50% Thickness Reduction
Cylindrical	Maximum total deformation	0.46946 m	0.62205 m	0.92772 m
	Maximum von Mises Stress	105.88 MPa	138.39 MPa	200.88 MPa
	Minimum von Mises Stress	0.16311 MPa	0.38075 MPa	0.56118 MPa
	1st Natural frequency	0.36082 Hz	0.32099 Hz	0.26951 Hz
Conical	Maximum total deformation	0.25226 m	0.32420 m	0.50011 m
	Maximum von Mises Stress	69.086 MPa	97.276 MPa	144.02 MPa
	Minimum von Mises Stress	0.10357 MPa	0.23201 MPa	0.20016 MPa
	1st Natural frequency	0.51194 Hz	0.45413 Hz	0.38008 Hz

4. Conclusions

This study aimed to evaluate two towers of a wind turbine: one with a cylindrical shape and the other one with a conical shape. The reference parameters were adopted from the Vestas V112-3 MW offshore wind turbine and are shown in Figure 1. The virtual models of the two towers were built using the ANSYS software.

Following the modal analysis, a 0.5 m mesh size was adopted, because the difference between the 0.25 m mesh and the 0.5 m mesh was almost imperceptible. The natural frequency for the cylindrical tower was 0.361 Hz, and for the conical tower it was 0.512, with both frequencies being between 0.224 and 0.640 Hz; thus, the resonance effect was avoided.

A static analysis was performed for two cases: when the turbine was operating in normal conditions, up to a cut-out speed of 25 m/s; when the turbine was no longer functional and the wind speed was 50 m/s. The forces applied on the tower, both for the normal case and the failure case, were given by the thrust force provided by the wind and acting on the blades, the bending moment of the blades, the distributed wind load on the tower, the gravitational force given the weight of the supported components (nacelle, blades, and rotor) and the weight of the tower.

The deformations of the two towers in normal operating conditions were below the allowable value, being smaller by more than 50%. In extreme wind conditions, the deformations for the conical structure were close to the admissible limit, and those for the cylindrical tower exceeded this admissible limit. The maximum deformations for both the conical and the cylindrical tower were located at the upper part of the tower, at a height of 84 m. If we consider a diameter for the cylindrical tower that is equivalent to the conical structure, the deformation obtained had a value of 1.18 m, a value that still exceeds the allowable deformation. Thus, the conical structure exhibited better resistance than the cylindrical one. The optimal diameter for the structure to deform below the allowable limit was 5.2 m (for this diameter, a maximum deformation of 1.02 m was obtained) a difference of 1 m from the diameter chosen for this work.

The von Mises stresses for the two towers under normal operating conditions were below the allowable value, the allowable stress being 305 MPa and those obtained for the cylinder and conical tower were 65 and 100 MPa, respectively; these values were four and three times lower than the allowable stress. In extreme wind conditions, the cylindrical structure exceeded the allowable value by approximately 118 MPa, while the conical tower was still below the allowable value. The highest stresses were identified in the coupling area of the tower sections, especially at the transition area from the wall thickness of 30–19 mm (at a height of 40 m). High stresses were also found at the junction between the 50 and 30 mm thick sections (at a height of 20 m). The reason why the stresses were found in these areas and not at the bottom of the tower, as would have been normal, is that the two towers were designed without flanges. If we consider the same diameter of 5.2 m, the von Mises stress would be below the allowable value, having a value of 287 MPa.

To validate the model in the paper, we compared the results obtained with the reference [43], which performed measurements on a turbine with a pillar that had a height of 80 m. This study consisted of placing sensors at different heights (on four levels, from level 0 to level 3) and measuring several types of signals, which were acceleration, strain, temperature, and inclination. The paper also presented a finite element model made with the help of the LUSAS software. This model was developed to determine whether it was viable and could be used in further analyses. The conclusion was that the results could be interpreted and computed using FE models. Following the study, a value for the natural frequencies for mode 1 was obtained being close to the value of 0.340 Hz. The values obtained from this paper were 0.361 Hz for the cylindrical tower and 0.512 Hz for the conical one. These differences were due to the thicknesses of the plates, which for the current study, were larger than those in the reference study; Table 6, however, showed a tendency of these frequencies to decrease with the decrease in thickness. Moreover, the differences are given by the diameters chosen for the towers. Taking into account all these aspects, we can say that the towers had similar frequencies, and the reference paper can serve as a method of validating the model developed in this paper. In addition, the value of the maximum stresses under the action of a wind speed of 25 m/s was approximately 70 MPa, and those for the towers in the study were approximately 100 and 65 MPa, respectively.

Following the study, a possible direction was identified towards the efficiency of this type of turbine, taking into account the possibility of placing it in the Black Sea, where the wind speed does not have such a high intensity, but the productivity of such a turbine can be approximately 45%. It was observed that by reducing the thickness of the sheets, the turbine would be strong enough for this location.

Author Contributions: A.I.D. carried out the statistical analysis and interpreted the results, wrote the final form of the manuscript, and drew the conclusions; E.R. conceived the paper's scope, methodology, and contributed to the final form of the manuscript in terms of editing; G.A. performed the literature review, editing, supervision, and contributed to the design of the article. The final manuscript was approved by all authors. All authors have read and agreed to the published version of the manuscript.

Funding: This work was carried out in the framework of the research project DREAM (Dynamics of the Resources and Technological Advance in Harvesting Marine Renewable Energy), supported by the Romanian Executive Agency for Higher Education, Research, Development, and Innovation Funding-UEFISCDI (grant number: PN-III-P4-ID-PCE-2020-0008).

Institutional Review Board Statement: Not applicable.

Conflicts of Interest: The authors declare no conflict of interest.

References

1. Post, E.; Alley, R.B.; Christensen, T.R.; Macias-Fauria, M.; Forbes, B.C.; Gooseff, M.N.; Iler, A.; Kerby, J.T.; Laidre, K.L.; Mann, M.E. The polar regions in a 2 C warmer world. *Sci. Adv.* **2019**, *5*, eaaw9883. [CrossRef] [PubMed]
2. Akbar, M.; Hussain, A.; Akbar, A.; Ullah, I. The dynamic association between healthcare spending, CO_2 emissions, and human development index in OECD countries: Evidence from panel VAR model. *Environ. Dev. Sustain.* **2021**, *23*, 10470–10489. [CrossRef]
3. Shakoor, A.; Ashraf, F.; Shakoor, S.; Mustafa, A.; Rehman, A.; Altaf, M.M. Biogeochemical transformation of greenhouse gas emissions from terrestrial to atmospheric environment and potential feedback to climate forcing. *Environ. Sci. Pollut. Res.* **2020**, *27*, 38513–38536. [CrossRef] [PubMed]
4. IEA. *Global Energy Review 2021*; IEA (International Energy Agency): Paris, France, 2021.
5. bp. *Statistical Review of World Energy 2021*; BP p.l.c.: London, UK, 2021.
6. Vivek, C.M.; Ramkumar, P.; Srividhya, P.K.; Sivasubramanian, M. Recent strategies and trends in implanting of renewable energy sources for sustainability–A review. *Mater. Today Proc.* **2021**, *46*, 8204–8208. [CrossRef]
7. Van de Graaf, T. Fragmentation in Global Energy Governance: Explaining the Creation of IRENA. *Glob. Environ. Politics* **2011**, *13*, 14–33. [CrossRef]
8. IRENA. *Renewable Capacity Statistics 2021*; International Renewable Energy Agency (IRENA): Abu Dhabi, United Arab Emirates, 2021.
9. National Statistics. *Energy Trends*; National Statistics: London, UK, 2021.

10. Higgins, P.; Foley, A. The evolution of offshore wind power in the United Kingdom. *Renew. Sustain. Energy Rev.* **2014**, *37*, 599–612. [CrossRef]
11. MCA. *MGN 372 (M+ F) Offshore Renewable Energy Installations (OREIs): Guidance to Mariners Operating in the Vicinity of UK OREIs*; Maritime and Coastguard Agency: Southampton, UK, 2008.
12. Navadeh, N.; Goroshko, I.; Zhuk, Y.; Moghadam, F.E.; Fallah, A.S. Finite element analysis of wind turbine blade vibrations. *Vibration* **2021**, *4*, 310–322. [CrossRef]
13. Bottasso, C.L.; Campagnolo, F.; Croce, A.; Dilli, S.; Gualdoni, F.; Nielsen, M.B. Structural optimization of wind turbine rotor blades by multilevel sectional/multibody/3D-FEM analysis. *Multibody Syst. Dyn.* **2014**, *32*, 87–116. [CrossRef]
14. Wang, L.; Liu, X.; Kolios, A. State of the art in the aeroelasticity of wind turbine blades: Aeroelastic modelling. *Renew. Sustain. Energy Rev.* **2016**, *64*, 195–210. [CrossRef]
15. Kaveh, A.; Sabeti, S. Optimal design of monopile offshore wind turbine structures using CBO, ECBO, and VPS algorithms. *Sci. Iran.* **2019**, *26*, 1232–1248. [CrossRef]
16. Muskulus, M.; Schafhirt, S. Design optimization of wind turbine support structures—A review. *J. Ocean Wind Energy* **2014**, *1*, 12–22.
17. O'Leary, K.; Pakrashi, V.; Kelliher, D. Optimization of composite material tower for offshore wind turbine structures. *Renew. Energy* **2019**, *140*, 928–942. [CrossRef]
18. Gutiérrez, E.; Primi, S.; Taucer, F.; Caperan, P.; Tirelli, D.; Mieres, J.; Calvo, I.; Rodriguez, J.; Vallano, F.; Galiotis, C. A wind turbine tower design based on the use of fibre-reinforced composites. In *MEGAWIND FP5 Project, ENK5-CT-2000-00328*; ELSA: Ispra, Italy, 2003; p. 163.
19. Young, A.C.; Goupee, A.J.; Dagher, H.J.; Viselli, A.M. Methodology for optimizing composite towers for use on floating wind turbines. *J. Renew. Sustain. Energy* **2017**, *9*, 33305. [CrossRef]
20. Uys, P.E.; Farkas, J.; Jarmai, K.; Van Tonder, F. Optimisation of a steel tower for a wind turbine structure. *Eng. Struct.* **2007**, *29*, 1337–1342. [CrossRef]
21. Wang, L.; Kolios, A.; Luengo, M.M.; Liu, X. Structural optimisation of wind turbine towers based on finite element analysis and genetic algorithm. *Wind Energy Sci. Discuss.* **2016**, *153*, 1–26.
22. Jonkman, J.; Butterfield, S.; Musial, W.; Scott, G. *Definition of a 5-MW Reference Wind Turbine for Offshore System Development*; National Renewable Energy Lab. (NREL): Golden, CO, USA, 2009.
23. Zhu, R.S.; Zheng, Z.Z.; Liu, Y.M.; Shen, J.; Xiao, Y.; Jiang, D.X.; Chen, J. Finite Element Analysis for MW Wind Turbine Tower. In *Applied Mechanics and Materials*; Trans Tech Publications Ltd.: Stafa-Zurich, Switzerland, 2012; Volume 130, pp. 124–127.
24. Li, X.L.; Ren, L.M. Finite Element Analysis of Wind Turbine Tower. In *Applied Mechanics and Materials*; Trans Tech Publications Ltd.: Stafa-Zurich, Switzerland, 2013; Volume 351, pp. 825–828.
25. Joshi, Y.S.; Alandkar, M.P.M. *Finite Element Analysis of Connections for Wind Turbine Towers*; ACADEMIA: San Francisco, CA, USA, 2017.
26. Diaf, S.; Notton, G.; Belhamel, M.; Haddadi, M.; Louche, A. Design and techno-economical optimization for hybrid PV/wind system under various meteorological conditions. *Appl. Energy* **2008**, *85*, 968–987. [CrossRef]
27. Diaconita, A.I.; Rusu, L.; Andrei, G. A Local Perspective on Wind Energy Potential in Six Reference Sites on the Western Coast of the Black Sea Considering Five Different Types of Wind Turbines. *Inventions* **2021**, *6*, 44. [CrossRef]
28. Shittu, A.A.; Mehmanparast, A.; Wang, L.; Salonitis, K.; Kolios, A. Comparative study of structural reliability assessment methods for offshore wind turbine jacket support structures. *Appl. Sci.* **2020**, *10*, 860. [CrossRef]
29. Chantharasenawong, C.; Jongpradist, P.; Laoharatchapruek, S. Preliminary design of 1.5-MW modular wind turbine tower. In Proceedings of the 2nd TSME International Conference on Mechanical Engineering, Krabi, Thailand, 19–21 October 2011.
30. Dagli, B.Y.; Tuskan, Y.; Gökkuş, Ü. Evaluation of offshore wind turbine tower dynamics with numerical analysis. *Adv. Civ. Eng.* **2018**, *2018*, 1–11. [CrossRef]
31. Liu, Y.; Chen, D.; Yi, Q.; Li, S. Wind Profiles and Wave Spectra for Potential Wind Farms in South China Sea. Part I: Wind Speed Profile Model. *Energies* **2017**, *10*, 125. [CrossRef]
32. Manwell, J.F.; McGowan, J.G.; Rogers, A.L. *Wind Energy Explained: Theory, Design and Application*; Wiley: Chichester, UK, 2009.
33. Hellmann, G. Über die Bewegung der Luft in den untersten Schichten der Atmosphäre. *Meteoral. Z.* **1916**, *34*, 273.
34. Albani, A.; Ibrahim, M.Z. Wind Energy Potential and Power Law Indexes Assessment for Selected Near-Coastal Sites in Malaysia. *Energies* **2017**, *10*, 307. [CrossRef]
35. International Electrotechnical Commission. *Wind Turbines—Part 1: Design Requirements*; Tech. Doc. IEC 61400-1; International Electrotechnical Commission: Geneva, Switzerland, 2005.
36. American Bureau of Shipping (ABS). *Guide for Building and Classing Offshore Wind Turbine Installation*; ABS: Houston, TX, USA, 2014.
37. Hsu, Y.; Wu, W.-F.; Chang, Y.-C. Reliability analysis of wind turbine towers. *Procedia Eng.* **2014**, *79*, 218–224. [CrossRef]
38. Liu, H.; Yang, S.; Tian, W.; Zhao, M.; Yuan, X.; Xu, B. Vibration reduction strategy for offshore wind turbines. *Appl. Sci.* **2020**, *10*, 6091. [CrossRef]
39. Diaconita, A.; Andrei, G.; Rusu, L. New insights into the wind energy potential of the west Black Sea area based on the North Sea wind farms model. *Energy Rep.* **2021**, *7*, 112–118. [CrossRef]
40. Gentils, T.; Wang, L.; Kolios, A. Integrated structural optimisation of offshore wind turbine support structures based on finite element analysis and genetic algorithm. *Appl. Energy* **2017**, *199*, 187–204. [CrossRef]

41. Lloyd, G.; Hamburg, G. *Guideline for the Certification of Wind Turbines*; Germanischer Lloyd Industrial Services GmbH: Hamburg, Germany, 2010.
42. Nicholson, J.C. Design of Wind Turbine Tower and Foundation Systems: Optimization Approach. Ph.D. Thesis, The University of Iowa, Iowa City, IA, USA, 2011.
43. Rebelo, C.; Veljkovic, M.; da Silva, L.S.; Simões, R.; Henriques, J. Structural monitoring of a wind turbine steel tower—Part I: System description and calibration. *Wind Struct.* **2012**, *15*, 285. [CrossRef]

inventions

MDPI

Article

Analysis of Some Essential Aspects Related to the Navigation Conditions on the Danube River

Andra Luciana Marcu Turcanu [1,*], Liliana Mihaela Moga [1] and Eugen Victor Cristian Rusu [2]

[1] Department of Economics, Faculty of Economics and Business Administration, University "Dunarea de Jos" of Galati, 800008 Galaţi, Romania; liliana.moga@ugal.ro

[2] Department of Mechanical Engineering, Faculty of Engineering, University "Dunarea de Jos" of Galati, 800008 Galaţi, Romania; eugen.rusu@ugal.ro

* Correspondence: andra.marcu@ugal.ro

Abstract: The European Union has emphasized the creation of an appropriate framework to optimize the internal market and inland transport waterways, and remove barriers to their wider use. Administrative barriers in the logistics of goods on the Danube waterway and its navigable tributaries constitute a significant obstacle to the efficient and sustainable use of the Danube as the region's central transport hub. The approach proposed in this paper was designed to identify and analyze the relationship between the main variables leading to problematic inland waterway traffic, in this case, on the Danube, and the measures taken by the European Commission to improve it. In terms of the applied research method, "Quality Function Deployment" (QFD), we assign global (overall) and local priority degrees. The proposed framework for adapting QFD as a tool for improving quality and, therefore, performance, aims to identify and prioritize directions for this improvement. The House of Quality (HOQ) is the tool that links areas for improvement to technical requirements. The disclosure of these connections helps identify and prioritize the technical features that will generate the most significant improvements.

Keywords: navigation conditions; Danube River; management tool; quality function deployment; house of quality

Citation: Marcu Turcanu, A.L.; Moga, L.M.; Rusu, E.V.C. Analysis of Some Essential Aspects Related to the Navigation Conditions on the Danube River. *Inventions* **2021**, *6*, 97. https://doi.org/10.3390/inventions6040097

Academic Editor: Francisco Manzano Agugliaro

Received: 10 September 2021
Accepted: 10 December 2021
Published: 17 December 2021

1. Introduction

A single European transport area should facilitate the movement of citizens and freight transport [1], reduce costs, and increase the sustainability of European transport [2]. Regarding maritime transport, the existence of a "Blue Belt" in European seas will simplify formalities for ships traveling between European Union (EU) ports. It is necessary to establish an appropriate framework for European inland waterway transport. Market access to ports needs to be improved further.

Water transport accounts for 90% of world trade, and cargo ships are part of the backbone of the global economy. Internationally, it has been found that transported goods have an estimated annual value of USD 4.5 trillion, and this value is expected to increase by about 4.7% in coming years, according to the Water Quality Assurance Union (2020) [3].

In addition, specialists have indicated that this means of transport is not exploited to its true potential [4]. Inland waterways transport should be developed because it plays a significant role in environmental protection and developing small enterprises [5]. However, this is a challenge for shipowners, and progress can be made only with the help of the European Commission.

International trade undertaken on the water is essential to provide direct links. Inland waterways are essential for the European States [6] because they connect these States, and for import–export activities to and from Asia. It is a significant priority for the European Commission to develop a unified navigation system to connect and maximize trans-European navigation potential.

The significant advantage of river navigation is the capability to transport a large quantity of heavy and oversized goods, simultaneously with price advantages [7]. Considerations of operationality also concern the estimated travel time and the related time costs. When calculating the estimated time, it is essential to consider crew costs, fuel, docking, transshipment, waste discharging, waiting times for customs formalities, and waiting times for unloading–loading. Economic conditions and territorial aspects can stimulate water transport [8], and these two factors must be in concordance with the needs of shipowners. The primary conditions for efficient water transport can be improved, and the bottlenecks addressed, by the relevant authorities.

The abundance of administration burdens reduces the economic and environmental potential of water transport, thus harming the economic growth and social welfare of the entire region. The international legal regulation of rivers represents one major problem [9]. Mitigating significant administrative barriers in a transnational coordinated manner is the primary step towards better governance of this mode of transport, and reduces the logistical costs of large industries that depend on inland waterway transport. The European Commission has sought to address administrative barriers and derivatives, and reduction of these factors is a significant priority in the agenda of its current transport policy. Consultation between the Directorate General for Mobility and Transport of the European Commission, which is responsible for the EU's mobility and transport policy (DG MOVE), and the Department of the European Commission, which is responsible for the EU policy on the development of European regions and cities (DG Regio), led to the development of a program for the implementation of the EU Strategy for the Danube Region (and its priority area 1a). This implementation—the Transnational Danube Program—is a platform on which stakeholders can efficiently identify essential, concrete, and specific solutions to the many existing barriers.

The European institutions responsible for the Danube Region have developed many strategies and projects in order to improve shipping quality. We consider that it is essential to prioritize the measures implemented by the European Commission considering the bottlenecks identified by the shipowners, as presented in Section 3 of this paper.

Quality Function Deployment (QFD) was selected for the analysis in this study. This method was used to measure the impact of administrative burdens identified by shipowners on technical measures implemented by the European Commission to optimize the flow of goods on the Danube. Implementing the quality function is an approach to quality assurance in the product development process, but is not necessarily applied at every stage [10].

Quality Function Deployment is an instrument that is used in many fields, including maritime transport. Many studies have been conducted to help decision makers to improve their activities. Zeiner Sener and Ece Ozturk (2015) proposed the use of QFD to select an appropriate ship considering the company's needs [11]. Ship routing is another vital element related to water transport. Many experts have used QFD for selecting suitable routes for the successful transfer of goods and passengers [12–14].

The results of the applied methodology indicate the essential point of intervention. In this study, we developed a set of recommendations to improve the navigation conditions on the Danube River.

2. Materials and Methods

Developed in Japan in 1960, Quality Function Deployment (QFD) was introduced into the United States in the early 1980s, and was adopted due to its popularity and history of success in the automotive industry [15]. Copying the model from manufacturers such as Toyota and Mitsubishi, the "Big Three" United States car manufacturers used QFD to better meet customer requirements in their industry. Once adopted, this method significantly shortened the design cycles and reduced the total number of employees required in the design process. Subsequent to its success in Japan, this method has been extensively developed [16,17].

The new standard of the International Organization for Standardization (ISO) 16355 -2015 [10] defines and demonstrates the dynamic nature of a customer-based approach. Since its establishment in 1966, the quality function has expanded and deepened the methods and tools to respond to changing business conditions of QFD users, management, customers, and products. Those who used QFD older models will find that these improvements QFD easier and faster to use. Methods and tools presented in the standards represent decades of improvements QFD list is neither exhaustive nor exclusive.

ISO 16355 describes the process of implementing the quality function, its purpose, users, and tools. It is not a standard management system, and it does not provide requirements or guidance for organizations to systematically develop and manage their policies, processes, and procedures to achieve specific objectives. Users of this part of ISO 16355 will include all organizational functions necessary to ensure customer satisfaction, including business planning, marketing, sales, research and development (R&D), engineering, information technology (IT) [18], production, procurement, quality, production, services, packaging and logistics, support, testing, regulation and other phases in hardware, software, services, and system organizations.

The primary tool developed in this method is the House of Quality. The "House of Quality" is obtained through the development of matrices that allow the identification of areas for improvement, their classification, and their weighting in a final matrix [16,19]. The quality house (HoQ) is the essential tool by this method [20]. The house consists of a matrix that links the input and output data [16]. This stage is the most important. Over time, this method has been applied and adapted in various products, tourism, medical services, electricity, and transportation.

The traditional HoQ matrix is composed of seven major parts, including shipowners' requirements (CR), priority requirements, technical measures (TM), correlation matrix, relationship matrix (between TM and CR), the sum of products of priority degrees and degrees of relationship, and, finally, the degree of priority of TMs [21]. With the help of this graphic representation, the needs of the shipowners concerning the technical measures will be visible.

In practice, it is both difficult, and unnecessary to include all HOQ elements and build different HOQ models involving different elements. The simplest but most widely used HOQ model contains only customer needs and their relative importance, technical measures and the relationships between the two elements, and assessments of the importance of technical measures. Some authors add correlation matrices to this simple model [22]. Fewer models include technical competitive assessment because this information is challenging to treat. As such, the objectives and probability factors for technical measures are rare in HOQ studies—even if they are included, they are challenging to incorporate into the calculation of importance ratings, which does not refer to the competitive technical assessment.

To avoid inconsistencies and facilitate applications, we propose in this paper a unified HOQ model in 5 steps (Figure 1 House of Quality) after the model of Chan and Wu [23] contains the frequently used HOQ elements. The needs of customers, in our case shipowners, which are listed in the left column and are connected to the technical requirements, in our case to the action pillars, listed at the top, completing the section called the Rij relationship matrix [24]. Here, the weighted matrix of customer requirements is used to determine the weighted characteristics of the product.

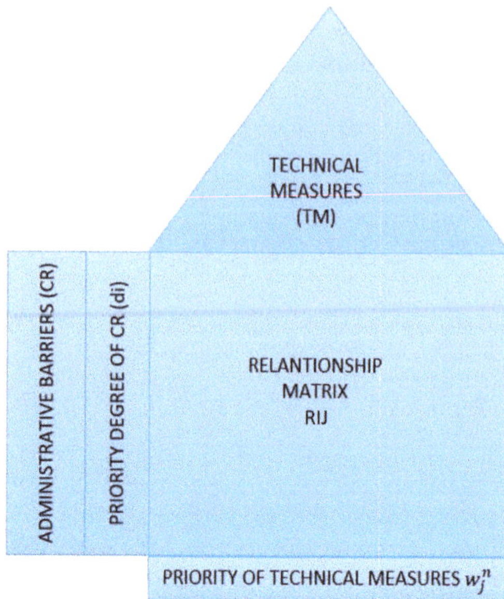

Figure 1. House of Quality.

The model applied by us have five steps presented in Figure 1 and detailed below:

(a) The left side is represented by the voice of customers (CR), in our specific case by the administrative barriers identified by the shipowners, detailed in Section 3.1.
(b) It is necessary to calculate a priority degree of shipowner's needs (di), detailed in Section 3.2. By prioritizing them, we can find out that the most critical shipowners' needs are not the most important regarding the technical measures taken.
(c) The roof is represented by the Technical Measures (TM) and is detailed in Section 3.3.
(d) Rij: This matrix is a systematic means of identifying the relationship level between CR and TM. Usually, these relationships are measured by the following scale (1), which places more weight on the solid relationship [25,26].

Without relationship	Low relationship	Moderate relationship	Strong relationship	
0	1	5	9	(1)

Either R or the matrix of the relationship between CR and TM, the element Rij indicates the level of impact of TM on CR satisfaction.

The correlation matrix allows the identification of conflicting design requirements, i.e., the rows or columns that remain blank indicate that there are no relationships between them. This means that shipowners' requirements do not affect any performance indicator selected. That step is detailed in Section 3.4. The numerical process for assigning the priority degree of TMs is as follows:

- Let m be the requirements of the shipowners indicated by CRi, (i = 1, 2, ..., m) and n technical measures indicated by TMi (i = 1, 2, ..., n).
- Let di (i = 1, 2, ..., m) be the priority degree of i between the whole set of CR, while we (j = 1, 2, ..., n) which indicates the relative weight of the importance j TM, is determined by the relationship between CR and TM.

The sum of the product of the priority degree, di and Rij is calculated as follows:

$$w_j = \sum_{i=1}^{m} d_i R_{ij} \quad \mathrm{J} = 1, 2, \ldots, n \tag{2}$$

So, barriers identified by shipowners will be marked with CR. The priority level of a requirement is derived through the sum-product measures the relative weight of the European Commission in terms of performance and degree of relationship corresponding technical measures provided. The pillars of the European Commission's action will be marked with TM (technical measures). A normalized value of the result will show the relative weight (the priority) technical measure to satisfy the need of the shipowner.

This method offers us an appropriate approach to prioritize the technical measures (TM). By developing this research method in this paper, we have the entire decision-making problem process and the possibility of prioritizing the best solution and even offering a set of recommendations by using a one-to-one relationship [27]. One of the most common issues in the decision-making process is the connectivity with the customers' desires (CR). After we browse all the steps mentioned in Figure 1, the technical measures will be prioritized by their impact on the system. We will have an objective point of view that will give the action direction.

3. Results

3.1. Administrative Barriers Identified

Ensuring clear visibility throughout the supply chain is vital to ensure safe and appropriate timely delivery from end to end. Without it, organizations can become victims of countless inefficiencies, leading to delays, transport damage, or loss.

The purpose of the project Improving Administrative Procedures and Processes for IWT Danube (DANTE), founded by the Transnational Danube, was to identify and eliminate administrative barriers that prevent the development of inland waterway transport (IWT) its full potential. Through this project was developed a global online reporting of administrative barriers. According to IWT Administration Danube Strategy and Action Plan [28], developed within the mentioned project, the main variables that determine the delay of freight on the Danube are presented in Table 1. Furthermore, the identified barriers received an identification number for better recognition in Appendix A.

Table 1. Presentation of the main administrative barriers and competent authorities.

Responsible Authority	Barrier Identified	Identification Number in Appendix A
Border police, taxes and customs	Misconduct of staff;	1
	Difficult formalities at the RO-BG border;	2
	Lack of software applications in the customs sector;	3
	Delayed transit in the Serbian area	4
Navigation/traffic control authorities	Communication with the police on public channels;	5
	Radio signal and speed issues;	6
	Temporary canal closures;	7
	Lack of maintenance work-depth;	8
Port authorities/administrations	Large number of documents requested in RO;	9
	Non-professional staff (RO; BG; RS);	10
	Additional taxes in RO;	11
	Theft/Smuggling in RO/RS;	12
Unknown authority / multiple authorities involved	Forms with the same content in each country;	13
	Additional border controls with BG;	14
	2017, 7 additional documents;	15
	Uploading data to BULRIS—additional delay;	16
Waterway and canal management	Low water level reported in HU;	17
	Radar signals from HU buoys;	18
	High taxes through the Danube–Black Sea Canal;	19
	Lack of coordination between RO-RS dispatchers;	20

Source: Danube IWT Administration Strategy & Action Plan, 2019.

3.2. Prioritization of Administrative Barriers

First of all, we identified the main factors that determine the delays in freight transport. The shipowner's requirements were the most important. The next step was to prioritize them by the multicriteria analysis method. This method is often used in transport problems [29]. After prioritizing the variables, the matrix analysis can be done.

Analytic hierarchy priority (AHP) is made using the online software provided by Business Performance Management Singapore, available at https://bpmsg.com (accessed on 10 May 2021). Klaus D. Goepel [30] implemented a free online AHP system with notable features, allowing detailed analysis of decision-making issues. The author intended to provide a complete and free software tool for education and research the calculations and algorithms are well documented, and all input data and results can be exported in an open format for further processing or presentation [30].

The resulting weights, presented in Appendix A (Normalized matrix) are based on the principal eigenvector of the decision matrix. A number of 190 comparisons were made, with the CR consistency ratio equal to 8.1%, which demonstrates the coherence of the judgment, which means that the AHP process was correctly performed and can be continued. The resulting weights for the criteria based on pairwise comparisons are presented in Table 2. It can be seen that the main barriers leading to traffic delays on the Danube are: lack of software applications, complex formalities at the RO-BG border, as well as inappropriate staff behavior, problems related primarily to the border police, taxes, and customs.

Table 2. Priority ranking of the variables.

Category	Priority	Rank
Inappropriate staff behavior	10.3%	3
Difficult formalities at the RO-BG border	11.4%	2
Lack of software applications in the customs sector	13.7%	1
Delayed transit in the Serbian area	7.4%	4
Communication with the police on public channels	1.4%	20
Radio signal and speed problems	1.5%	19
Temporary channel closures	7.3%	5
Lack of maintenance work-depth	6.1%	6
Large number of documents requested in RO	4.1%	11
Non-professional staff (RO; BG; RS)	5.6%	8
Additional taxes in RO	4.4%	10
Theft/Smuggling in RO/RS	5.9%	7
Forms with the same content in each country	2.5%	14
Additional border controls with BG	3.0%	12
2017, 7 additional documents	2.3%	16
Loading data in BULRIS—additional delays	2.3%	15
Low water level reported in HU	5.0%	9
Radar signals from HU buoys	1.7%	17
High taxes through the Danube-Black Sea Canal	2.8%	13
Lack of coordination between RO-RS dispatchers	1.5%	18

3.3. Technical Measures for Streamlining Freight Transport on the Danube

The Danube region can become a safe area where conflicts, marginalization, and crime are adequately addressed. According to European Commission directives, all citizens of this region should enjoy better prospects for higher education, employment, and prosperity in the areas where they live. The Danube Strategy should make this region, which truly belongs to the 21st century, safe and confident in its strengths and one of the most attractive in Europe [31]. The pillars of action proposed by the authority mentioned above are presented in Table 3, and in this research, technical measures are called, as they have the functional role of responding to the immediate needs of shipowners.

Table 3. Technical measures taken by the European Commission to streamline freight traffic on the Danube.

Main Objective	Proposed Measure
Connecting the Danube Region	Improving mobility and intermodality
Building prosperity in the Danube Region	Development of society through research, education and information technology
	Supporting the competitiveness of enterprises, including cluster development
	Investing in people and skills
Environmental protection in the Danube Region	Restoring and maintaining water quality
	Environmental risk management
	Preserving biodiversity, landscapes, reducing pollution
Consolidation of the Danube Region	Strengthening institutional capacity and cooperation
	Strengthen cooperation to promote security and fight crime

Source: European Union Strategy for the Danube Region.

3.4. Relationship between Administrative Barriers and Technical Measure

The relationship matrix was built on the premise that there is a strong relationship between "Inappropriate staff behavior" and "Investing in people and skills," as well as between "Difficult formalities at the RO-BG border" and "Strengthening institutional capacity and of cooperation, "respectively" Lack of software applications in the customs sector "and" Development of society, through research, education and information technology. "

At the same time, we considered that there is a moderate relationship between "Delayed transit in the area of Serbia" and "Restoration and maintenance of water quality," as well as with "Strengthening cooperation to promote security and fight crime." "Reduced relationship" was considered between "Communication with the police on public channels" and "Investing in people and skills," as well as between "Problems of radio signal and speed" and "Restoring and maintaining water quality." The relationships established can be seen in Appendix B (Relationship between administrative barriers in the navigation sector on the Danube and the action pillars proposed by the European Commission).

4. Summary and Discussion

Based on the correlation matrix and the prioritization performed on the needs of shipowners, the relative importance of the technical measures concerning the identified needs could be determined, applying Equation (2) exemplified above. This set of competitive priority ratings is presented in the last line of Appendix B. The relationship between administrative barriers in the Danube navigation sector and the action pillars proposed by the European Commission, presented in Figure 2, from which we note that "Improving mobility and intermodality" is the highest priority, followed by "Developing society through research, education and information technology" and "Investing in people and skills.

Of the four significant intervention directions in the Danube region, we can see in Figure 2 that the main direction is towards "Building prosperity in the Danube Region," with a need intensity of 45%. According to the measures proposed by European Commission in "Action Plan- European Union Strategy for the Danube Region", this direction of intervention refers primarily to:

1. Smart growth strengthens all three themes of this goal [32]: innovation by improving the framework conditions to turn ideas into marketable products or services; education by improving the human capital of the region; and the digital society by improving internet access and availability of electronic content.
2. Sustainable growth, as innovation and new technologies will help combat climate change, increase energy efficiency and reduce transport pollution. In addition, the priority area of business competitiveness will improve the business environment, especially for SMEs [32].

3. Inclusive growth strengthens human capital through education and training, and that issue will be complemented by a better functioning of the macro-regional labor market. It also emphasizes the fight against poverty, especially against marginalized communities [32].

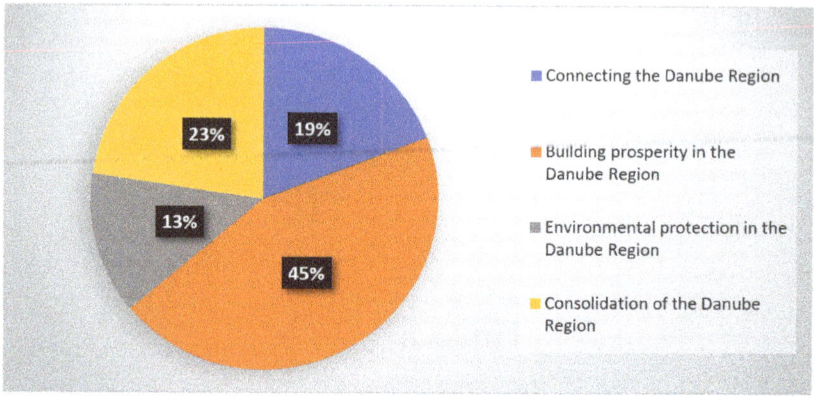

Figure 2. Major directions of intervention in the Danube region.

The second most crucial direction of intervention in the Danube Region is represented by "Consolidation of Danube Region." The main focus in that direction is to ensure the most substantial cooperation between institutions and increase security.

A harmonized administrative process will have a substantial contribution in optimizing the shipping time. Furthermore, in that direction, the European Authorities develop projects that fight against corruption. They consider that a well-governed and safer region is more attractive for people, and their standard of living will increase.

Through the research method applied, we have demonstrated that in shipowners' opinion, the priority directions should be those who have a central focus on the investment in human capacity and their standard of life. If the water transport is stimulated, the ports and the cities from riparian states will also be developed like commercial hubs. Part of the advantages will be:

- The framework conditions will be improved by innovation, the products and services will be developed in order to be merchandized;
- The education of the human capital will be improved by accessing new technologies and new markets;
- Sustainable growth will occur because the businesses will be more competitive and will improve their business environment.

Based on the applied method results, we think it is essential to show through diagrams the relationship based on the influence degree between the shipowner's needs and the authority position. For this, we used two excel tools, a spider chart (Figure 3) and a histogram chart (Figure 4). We will analyze the first three measures according to their relationship resulting from Appendix B, with the barriers identified by shipowners. For shipowner's needs, we maintain the same numeration as in Table 1.

From Figure 3, we can see the linear relationship between the 3 measures, so there are no significant differences between them, their degree of priority being similar. In contrast, from Figure 4, we can see that human resources have the most outstanding contribution in reducing existing barriers. The main direction of the action "Building Prosperity in the Danube Region" is given by investing in people's skills, improving access to information technology, and raising living standards. Several research projects have shown that organizational performance is greatly improved by paying special attention and giving human resources clear work directions.v

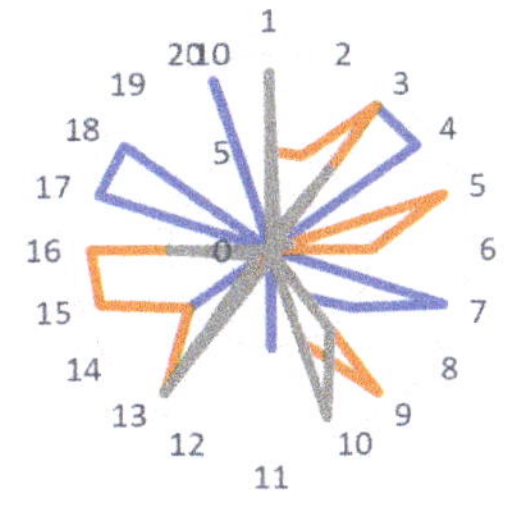

Figure 3. The relationship between technical measures taken by the European Commission.

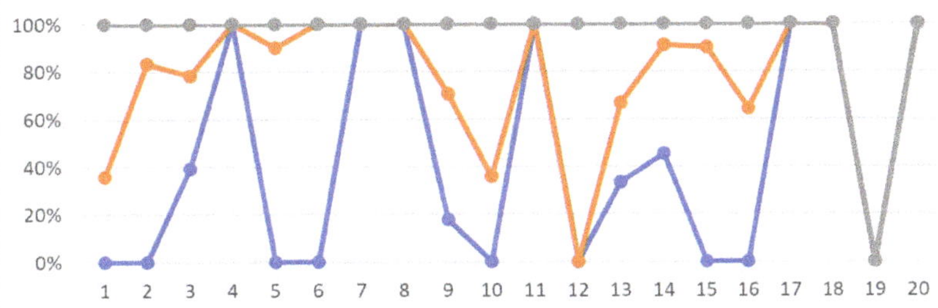

Figure 4. Contribution measures taken by the European Commission to remove barriers identified by shipowners.

QFD results lead to preventive action or corrective actions depending on whether efforts to improve were to address quality problems or improve a product or process. In case this method has helped to prioritize variables depending on the degree of relationship with technical measures. The tools used have shown that the main factor that all technical measures have in common and that could significantly optimize travel time is human resources. Certain but also uncertain factors are taken into account in the planning of river transport. Usually, uncertainty is given by weather conditions, but through this study, we have shown that human resources represent an equally important factor.

5. Conclusions

In this paper, we have prioritized the main measures to be taken in order to take advantage of the opportunities offered by the Danube River at the international level. These are in line with the directions set out in the "White Paper Roadmap to a Single European Transport Area—Towards a competitive and resource-efficient transport system."

The proposed approach uses the QFD methodology to identify and prioritize areas for improvement. HOQ links the areas for improvement to the technical barriers mentioned. The disclosure of these connections helps identify and prioritize the technical features that will generate the most significant improvement impacts.

Congestion, in the case of this type of transport, takes place in ports. If the operations were more efficient in the transit and transshipment points, the planned time and the time traveled would be the same or the diminished delays.

The main actors involved in the import–export process on inland waterways are shipowners. Their main objective is to be more efficient considering the travel time and the quality of the goods. That requirement comes to them from the customers.

Through the applied research method, we met their main requirements, with the technical measures provided by the European Commission, and thus a prioritization could be performed. The tools used have shown that the main factor that all technical measures have in common and that could significantly optimize travel time is human resources. Certain but also uncertain factors are taken into account in the planning of river transport. Usually, uncertainty is given by weather conditions, but through this study, we have shown that also an important factor is human resources. The goals established can be achieved through programs developed by well-trained and dedicated people. Human resources constitute an essential factor in improving the value chain process. It has been observed a necessity to strengthen collaboration through academic entities and shipping companies, and authorities. It could be necessary to develop specialized training pieces and ensure a technological transfer from academia to the workplace. The communication between those entities can be done bidirectional, and the curricula improvement should be made in partnership.

Another critical point observed in the study conducted in this paper is done by information technology. Digitalization of the process, considering the border, fiscal, loading, unloading formalities, will optimize the time spent by a ship in a port. That factor is essential to connect with the curricula developed to train people with advanced digital skills.

One of the most urgent demands of the shipowners is to harmonize the legislation in all Danube countries though the principle if we cross the same river is essential to have the same rules in all riparian countries. It is mandatory to have more transparent and efficient border control procedures, which could be achieved through digitalization. The new direction should be well-trained people in digitalized ports.

Author Contributions: Conceptualization, A.L.M.T. and L.M.M.; methodology, A.L.M.T.; validation, A.L.M.T., L.M.M. and E.V.C.R.; formal analysis, A.L.M.T.; investigation, A.L.M.T.; resources, A.L.M.T. and E.V.C.R.; data curation, A.L.M.T.; writing—original draft preparation, A.L.M.T.; review and editing, L.M.M. and E.V.C.R.; visualization, A.L.M.T.; supervision, L.M.M. and E.V.C.R.; funding acquisition, E.V.C.R. All authors have read and agreed to the published version of the manuscript.

Funding: This work was carried out in the framework of the research project DREAM (Dynamics of the Resources and technological Advance in harvesting Marine renewable energy), supported by the Romanian Executive Agency for Higher Education, Research, Development, and Innovation Funding—UEFISCDI, grant number PN-III-P4-ID-PCE-2020-0008.

Data Availability Statement: The data that support the findings of this study are available in the public domain.

Acknowledgments: This work was carried out in the framework of the Project ANTREPRENORDOC, in the framework of Human Resources Development Operational Programme 2014- 2020, financed from the European Social Fund under the contract number 36355/23.05.2019 HRD OP/380/6/13—SMIS code: 123847. The results of this work have been presented to the 9th edition of the Scientific Conference organized by the Doctoral Schools of "Dunărea de Jos" University of Galati (http://www.cssd-udjg.ugal.ro/ accessed on 5 September 2021), that was held on 10–11 June 2021 in Galati, Romania.

Conflicts of Interest: The authors declare no conflict of interest.

Appendix A

Table A1. Normalized Matrix.

	1	2	3	4	5	6	7	8	9	10	11	12	13	14	15	16	17	18	19	20
1	1	1	1	2	7	3	3	3	3	1	4	3	3	3	5	3	3	3	3	7
2	1	1	1	1	4	3	3	3	5	5	3	3	4	4	4	4	4	3	5	3
3	1	1	1	2	7	9	3	3	7	5	5	5	5	5	5	5	3	3	4	3
4	0.5	1	0.5	1	7	7	1	1	4	4	1	1	4	2	3	3	1	4	3	3
5	0.14	0.25	0.14	0.14	1	1	0.2	0.2	0.33	1	0.2	0.33	0.33	0.5	0.33	0.33	0.14	1	0.5	1
6	0.33	0.33	0.11	0.14	1	1	0.33	0.17	0.33	0.33	0.33	0.14	0.33	0.33	1	0.33	0.14	1	1	1
7	0.33	0.33	0.33	1	5	3	1	1	2	4	4	1	3	3	7	5	1	3	2	4
8	0.33	0.33	0.33	1	5	6	1	1	1	1	3	1	4	3	3	3	1	5	3	4
9	0.33	0.2	0.14	0.25	3	3	0.5	1	1	1	1	1	1	3	1	3	1	3	3	3
10	1	0.2	0.2	0.25	1	3	0.25	1	1	1	4	1	4	4	4	4	1	3	3	3
11	0.25	0.33	0.2	1	5	3	0.25	0.33	1	0.25	1	1	3	3	3	3	1	4	1	4
12	0.33	0.33	0.2	1	3	7	1	1	1	1	1	1	5	5	5	5	1	3	1	3
13	0.33	0.25	0.2	0.25	3	3	0.33	0.25	1	0.25	0.33	0.2	1	1	1	1	0.33	3	1	3
14	0.33	0.25	0.2	0.5	2	3	0.33	0.33	0.33	0.25	0.33	0.2	1	1	3	3	1	4	1	3
15	0.2	0.25	0.2	0.33	3	1	0.14	0.33	1	0.25	0.33	0.2	1	0.33	1	3	1	1	1	1
16	0.33	0.25	0.2	0.33	3	3	0.2	0.33	0.33	0.25	0.33	0.2	1	0.33	0.33	1	1	3	1	3
17	0.33	0.25	0.33	1	7	7	1	1	1	1	1	1	3	1	1	1	1	6	1	5
18	0.33	0.33	0.33	0.25	1	1	0.33	0.2	0.33	0.33	0.25	0.33	0.33	0.25	1	0.33	0.17	1	0.33	2
19	0.33	0.2	0.25	0.33	2	1	0.5	0.33	0.33	0.33	1	1	1	1	1	1	1	3	1	3
20	0.14	0.33	0.33	0.33	1	1	0.25	0.25	0.33	0.33	0.25	0.33	0.33	0.33	1	0.33	0.2	0.5	0.33	1

Appendix B

Table A2. Relationship between Administrative Barriers in the Navigation Sector on the Danube and the Action Pillars Proposed by the European Commission.

				Connecting the Danube Region	Building Prosperity in the Danube Region			Environmental Protection in the Danube Region			Consolidation of the Danube Region	
Priority Grade (AHP)				Improving Mobility and Inter-modality	Development of Society through Research, Education and Information Technology	Supporting the Competitiveness of Enterprises, Including Cluster Development	Investing in People and Skills	Restoring and Main-taining Water Quality	Environm-ental Risk Manage-ment	Preserving Biodiversity, Landscapes, Reducing Pollution	Strengthening Institutional Capacity and Cooperation	Strengthen Cooperation to Promote Security and Fight Crime
42.8	10.3	Border police, taxes and customs	Misconduct of staff;	0	5	5	9	0	0	0	1	1
	11.4		Difficult formalities at the RO-BG border;	0	5	5	1	0	0	0	9	5
	13.7		Lack of software applications in the customs sector;	9	9	5	5	0	0	0	1	0
	7.4		Delayed transit in the Serbian area	9	0	0	0	5	0	0	5	5
16.3	1.4	Navigation/traffic control authorities	Communication with the police on public channels;	0	9	5	1	0	0	0	5	0
	1.5		Radio signal and speed issues;	0	5	5	0	1	0	5	0	0
	7.3		Temporary canal closures;	9	0	0	0	5	1	1	0	0
	6.1		Lack of maintenance work-depth;	5	0	0	0	9	5	1	3	0

Table A2. *Cont.*

Priority Grade (AHP)				Connecting the Danube Region	Building Prosperity in the Danube Region			Environmental Protection in the Danube Region			Consolidation of the Danube Region	
				Improving Mobility and Inter-modality	Development of Society through Research, Education and Information Technology	Supporting the Competitiveness of Enterprises, Including Cluster Development	Investing in People and Skills	Restoring and Main-taining Water Quality	Environm-ental Risk Manage-ment	Preserving Biodiversity, Landscapes, Reducing Pollution	Strengthening Institutional Capacity and Cooperation	Strengthen Cooperation to Promote Security and Fight Crime
20	4.1	Port authorities/ administrations	Large number of documents requested in RO;	3	9	0	5	0	0	0	5	0
	5.6		Non-professional staff (RO; BG; RS);	0	5	0	9	0	0	0	3	0
	4.4		Additional taxes in RO;	5	0	9	0	0	0	0	3	0
	5.9		Theft/Smuggling in RO/RS;	0	0	0	1	0	0	0	5	9
10.1	2.5	Unknown authority/multiple authorities involved	Forms with the same content in each country;	9	9	5	9	0	0	0	3	0
	3		Additional border controls with BG;	5	5	1	1	0	0	0	9	5
	2.3		2017—a number of 7 additional documents;	0	9	5	1	0	0	0	3	0
	2.3		Uploading data to BULRIS —additional delay;	0	9	5	5	0	0	0	5	0

Table A2. *Cont.*

				Connecting the Danube Region	Building Prosperity in the Danube Region			Environmental Protection in the Danube Region			Consolidation of the Danube Region	
Priority Grade (AHP)				Improving Mobility and Inter-modality	Development of Society through Research, Education and Information Technology	Supporting the Competitiveness of Enterprises, Including Cluster Development	Investing in People and Skills	Restoring and Main-taining Water Quality	Environm-ental Risk Manage-ment	Preserving Biodiversity, Landscapes, Reducing Pollution	Strengthening Institutional Capacity and Cooperation	Strengthen Cooperation to Promote Security and Fight Crime
11	5	Waterway and canal management	Low water level reported in HU;	9	0	0	0	9	5	1	0	0
	1.7		Radar signals from HU buoys;	9	0	0	0	5	5	5	1	0
	2.8		High taxes through the Danube-Black Sea Canal;	0	0	9	0	0	0	0	3	0
	1.5		Lack of coordination between RO-RS dispatchers;	9	0	0	0	9	0	0	5	0
			Relationship CR-TM	4.317	3.957	2.7249	3.4339	1.969	0.713	0.344	3.394	1.724

References

1. Bendkowski, J.M.; Bendkowski, J.M. *Development of Freight Transport in the European Union—Selected Issues*; Silesian University of Technology Publishing House: Gliwice, Poland, 2020; pp. 23–45.
2. European Commission. *White Paper—Roadmap to a Single European Transport Area—Towards a Competitive and Resource Efficient Transport System*; Publications Office of the European Union: Brussels, Belgium, 2011.
3. Syndicate of Water Quality Insurance. Top Risks Facing the Cargo Shipping Industry. 2020. Available online: https://www.wqis.com/top-risks-facing-the-cargo-shipping-industry/ (accessed on 25 October 2021).
4. Miloslavskaya, S.; Panychev, A.; Myskina, A.; Kurenkov, P. Intermodal transportation using inland water transport in Russia and abroad. In *IOP Conference Series: Materials Science and Engineering*; IOP Publishing: Bristol, UK, 2019.
5. Laude, A. For the European Parliament, Sustainable Transport Goes through Inland Waterways (2021). September 2021. Available online: https://www.reneweuropegroup.eu/news/2021-09-14/pour-le-parlement-europeen-le-transport-durable-passe-par-les-voies-navigables-interieures (accessed on 25 October 2021).
6. Scholten, A.; Rothstein, B. *Navigation on the Danube—Limitations by Low Water Levels and Their Impacts*; Publications Office of the European Union: Luxembourg, 2017.
7. Backalic, T.; Maslaric, M. Navigation Conditions and the Risk Management in Inland Waterway Transport on the Middle Danube. *Transp. Probl.* **2012**, *7*, 13–24.
8. Dávid, A.; Madudová, E. The Danube river and its importance on the Danube countries in cargo transport. In Proceedings of the 13th International Scientific Conference on Sustainable, Modern and Safe Transport, Žilina, Slovakia, 29–31 May 2019; Transportation Research Procedia: Amsterdam, The Netherlands, 2019; pp. 1010–1016.
9. Vdovenko, T. Principles of Legal Regulation of Navigation on International Rivers in Europe. Problems and Solutions (on the Example of the Danube and Rhine Rivers). *J. Danub. Stud. Res.* **2020**, *2*, 206–216.
10. ISO. International Standard (ISO-16355-1). *Application of Statistical and Related Methods to New Technology and Product Development Process—General Principles and Perspectives of Quality Function Deployment (QFD)*; ISO Copyright Office: Geneva, Switzerland, 2015.
11. Sener, Z.; Ozturk, E. A QFD-Based Decision Model for Ship Selection in Maritime Transportation. *Int. J. Innov. Manag. Technol.* **2015**, *6*, 202–205. [CrossRef]
12. Celik, M.; Cebi, S.; Kahraman, C.; Er, I.D. An integrated fuzzy QFD model proposal on routing of shipping investment decisions in crude oil tanker market. *Expert Syst. Appl.* **2009**, *36*, 6227–6235. [CrossRef]
13. LeeLam, J.S. Designing a sustainable maritime supply chain: A hybrid QFD-ANP approach. *Transp. Res. Part E Logist. Transp. Rev.* **2015**, *78*, 70–81. [CrossRef]
14. Plazibat, V.; Krčum, M.; Skračić, T. Tools of Quality in Determining the Characteristics of Services in Maritime Passenger Transport. *Naše More* **2015**, *62*, 53–58. [CrossRef]
15. Chan, L.K.; Wu, M.L. Prioritizing the Technical Measures in Quality Function Deployment. *Qual. Eng.* **1998**, *10*, 467–479. [CrossRef]
16. Shabestari, S.S. Enhanced integrated sensitivity analysis in model-based OFD method. In Proceedings of the 21st International Conference on Engineering Design (ICED17), Vancouver, BC, Canada, 21–25 August 2017; pp. 317–326.
17. Zhang, F.E. Using integrated quality function deployment and theory of creative problem-solving approach for ergonomic product design. *Comput. Ind. Eng.* **2014**, *76*, 60–74. [CrossRef]
18. Mohamad, S.M. *Improvement of Take-Away Water Cup Design by Using Concurrent Engineering Approach*; University Malaysia Pahang: Gambang, Malaysia, 2012.
19. Chan, L.K.; Wu, M.L. A systematic approach to quality function deployment with a complete illustrative example. *Int. J. Manag. Sci.* **2015**, *33*, 119–139.
20. Duru, O.; Bulut, E.; Huang, S.; Yoshida, S. Shipping performance assessment and the role of key performance indicators (KPIs): "Quality Function Deployment" for transforming shipowner's expectation. In Proceedings of the Conference of International Association of Maritime Economists, Taipei, Taiwan, 15–18 July 2013; pp. 1–18.
21. Tsamboulas, D.; Yiotis, G.S.; Panou, K.D. The use of multicriteria methods for the assessment of transport infrastructure projects: A comparative analysis. *J. Transp. Eng. Am. Soc. Civ. Eng.* **1999**, *125*, 407–414.
22. Quality Function Deployment. Available online: https://www.productplan.com/glossary/quality-function-deployment/ (accessed on 2 September 2021).
23. Hauser, J.R.; Clausing, D. The House of Quality. *Harvard Business Review*. 1988. Available online: https://hbr.org/1988/05/the-house-of-quality (accessed on 14 December 2021).
24. Sharma, J. A cross-disciplinary approach to product development and design through quality function deployment, target costing and value engineering. *Int. J. Product. Qual. Manag.* **2012**, *9*, 309–331. [CrossRef]
25. Erdil, N.O.; Arani, O.M. Quality Function Deployment: More than a Design Tool. *Int. J. Qual. Serv. Sci.* **2018**, *11*, 142–166. [CrossRef]
26. Raharjo, H. On normalizing the relationship matrix in quality function deployment. *Int. J. Qual. Reliab. Manag.* **2013**, *30*, 647–661, ISSN: 0265-671X. [CrossRef]
27. Wu, H. Applying grey model to prioritise technical measures in quality function deployment. *Int. J. Adv. Manuf. Technol.* **2005**, *29*, 1278–1283. [CrossRef]

28. Danube IWT Administration Strategy & Action Plan. 2019. Available online: http://www.interreg-danube.eu/uploads/media/approved_project_output/0001/34/f39c91f7e4e7666bbba9f8e064bdbf97df1c1954.pdf (accessed on 29 July 2021).
29. Brunelli, M. *Introduction to the Analytic Hierarchy Process*; Springer: Helsinki, Finland, 2015.
30. Goepel, K.D. Implementation of an Online Software Tool for the Analytic Hierarchy Process (AHP-OS). In Proceedings of the International Symposium on the Analytic Hierarchy Process, Hong Kong, China, 13–15 July 2018.
31. Communication from the Commission to the European Parliament, the Council, the European Economic and Social Committee and the Committee of the Regions. *European Union Strategy for the Danube Region*; Communication from the Commission to the European Parliament, the Council, the European Economic and Social Committee and the Committee of the Regions: Bruxelles, Belgium, 2010; Available online: https://ec.europa.eu/regional_policy/sources/docoffic/official/communic/danube/com2010_715_danube_ro.pdf (accessed on 29 July 2021).
32. European Commission. *Action Plan—European Union Strategy for the Danube Region*; European Commission: Brussels, Belgium, 2010.

inventions MDPI

Article

Study of Resistance to Disturbances of the Main Types of Communication Systems on Board Military Ships Used during Interception or Search and Rescue Missions

Vasile Solcanu [1], Marian Gaiceanu [1] and Georgiana Rosu [2,*]

[1] Doctoral School of Fundamental Sciences and Engineering, 'Dunarea de Jos' University of Galati, 800008 Galați, Romania; vasilesolcanu@dedeman.ro (V.S.); Marian.Gaiceanu@ugal.ro (M.G.)
[2] Department of Military Electronic Systems and Equipment, 'Ferdinand I' Military Technical Academy, 050141 Bucharest, Romania
* Correspondence: georgianamarin01@gmail.com

Abstract: In addition to combat missions, military ships often participate in search-and-rescue missions or interception of ships with refugees or migrants. The communication systems and modes of work that may be used during these missions will be in accordance with the International Convention for the Safety of Life at Sea (SOLAS). This paper aims to demonstrate by theoretical methods (analytical and numerical) the noise stability of communication systems using analog modulation in high-noise conditions, characteristic of the marine environment. The stability of analog systems employing amplitude (AM), frequency (FM), and phase (PM) modulations is investigated. The analyzed systems are currently under use in distress maritime communications.

Keywords: modulation; electromagnetic disturbances; electromagnetic compatibility; military ship; communication systems

Citation: Solcanu, V.; Gaiceanu, M.; Rosu, G. Study of Resistance to Disturbances of the Main Types of Communication Systems on Board Military Ships Used during Interception or Search and Rescue Missions. *Inventions* **2021**, *6*, 72. https://doi.org/10.3390/inventions6040072

Academic Editors: Eugen Rusu and Gabriela Rapeanu

Received: 7 September 2021
Accepted: 22 October 2021
Published: 27 October 2021

1. Introduction

Radio communications technology was first used in maritime rescue operations in 1899 and since then, tens of thousands of lives have been saved [1]. Organizational deficiencies in the field of saving lives at sea were brought to light by the tragedy of the Titanic on April 15/16, 1912. The special media impact of this disaster led to the appearance of the first version of the SOLAS (International Convention for the Safety of Life at Sea) in 1914 [2]. The need for interstate cooperation for the adoption of common rules in the naval field led to the adoption, at the ONU conference in Geneva in 1948, of the convention for the establishment of the Inter-Governmental Maritime Consultative Organization—IMCO, a name that was changed to IMO in 1982. One of the first tasks of the IMO was to adopt a new version of SOLAS (in 1960), which thus became the most important international treaty dealing with maritime safety [3].

The current SOLAS convention contains, in 14 chapters, standards considered minimum and mandatory for the construction and operation of ships and equipment from the point of view of maritime safety. Chapter IV—Radiocommunications—incorporates the Global Maritime Distress and Safety System (GMDSS) which addresses all passenger ships and all cargo ships with a gross tonnage of over 300 tons sailing in international waters. The equipment, standards, procedures/protocols, and communication frequencies provided herein are designed to increase the chances of rescue in the event of a maritime accident. They were developed through cooperation between the IMO and the International Telecommunication Union (ITU) [4].

Chapter V—Safety of navigation—provides provisions of an operational nature that are applicable to all ships on all types of voyage and provides for the obligation of all ship-masters to provide assistance to those in difficulty. Additionally, in this chapter there are established the obligatory navigation safety services that must be provided by the governments of the signatory states as well as their obligation to ensure that all ships are efficiently and sufficiently equipped from the point of view of maritime safety [2,5].

A critical aspect of maritime safety is ensuring adequate radio communications in distress situations, in the conditions of an overcrowded electromagnetic spectrum and a noisy propagation environment. One proposed solution is the joint use of the communication and radar spectrum of various services [6]. Another solution signals the necessity of modernizing the current GMDSS system, by turning from analog to digital communications, and from voce to data transmissions [7]. This idea is supported by the development of several maritime safety systems, such as: AIS (Automatic Information System), DSC (Digital Selective Calling), or VDE (VHF Data Exchange) [7]. Other solutions rely on the use of artificial intelligence for increasing the quality of maritime communications [8]. However, in the era of digital communications, maritime distress transmissions still use analog communications, due to the paramount importance of ensuring an efficient communication framework in such situations—voice communications still remain critical in organizing SAR (search-and-rescue) operations.

Also, another critical aspect supports preserving the use of analog systems in distress communications, and this is related to the vulnerability to cyberattacks. One of the strongest cyberattacks was NotPetya and took place in 2017. It caused a total loss of USD 10 billion, of which USD 300 million was the losses of the shipping company Maersk, which also lost most of its data. However, in a maritime cyber security survey conducted in 2020 by Safety at Sea and BIMCO Maritime Cyber Security, only 42% of respondents said their organizations had taken measures to protect ships from cyber threats. Moreover, with the onset of the COVID 19 pandemic, the number of cyberattacks increased by 400% [9]. Thus, although the trend in maritime communications is to modernize the infrastructure to allow the interconnection of ships and shore facilities through broadband communications, "VHF voice digitalization, currently under studies, is one more expectation." [4].

Recent years have witnessed a dramatic increase in traffic and maritime activities, which has inevitably led to an increase in accidents and distress situations [8,10]. Another consequence of this phenomenon is the exponential increase of both functional and non-functional noise sources, implicitly of the increasing agglomeration of the frequency spectrum. Given the particular importance of the providing adequate transmission of distress calls, distress messages, and distress traffic, the question is whether the emission classes currently used are sufficiently resistant to disturbances or whether other technical solutions, including the transition to digital communications, must be adopted. This information is particularly useful because a possible change must be adopted at the same time by all states signing the SOLAS convention, for both civilian and military ships, which are among the first ones to intervene in distress situations. In addition, for these ships, the bureaucratic circuit of endowment/modification programs is particularly long and difficult. An optimal planning of the use of electromagnetic spectrum in military naval operations, requires the knowledge of emission classes and frequencies used in search-and-rescue communications to eliminate the risk of disrupting these communications both by common channel, adjacent channel, or secondary reception channel disturbances.

This paper aims to demonstrate by theoretical methods (analytical and numerical) the effectiveness of using analog systems in high-noise conditions—with a low signal-to-noise ratio (SNR), characteristic of the marine environment. The stability of analog systems employing amplitude (AM), frequency (FM), and phase (PM) modulations is investigated. The analyzed systems are currently under use in distress maritime communications. The paper is thus structured: Part 2 provides some technical details on how maritime communications are organized and regulated at international level; Part 3 is the main one and it performs a thorough analysis from the perspective of interference stability, on the behavior of analog modulation systems similar to those implemented in the communication systems on board ships and used in SAR operations. Part 4 verifies the theoretical results by simulating the analog systems in the Matlab-Simulink environment. Part 5 comprises the results discussion and Part 6 provides the paper conclusions.

2. Materials and Methods

2.1. Technical—Organizational Details

Technical details regarding working frequencies, bandwidths, gap between channels, etc. are established by the radio regulations issued by the ITU. These regulations are updated/revised every 3–4 years at World Radiocommunication Conferences (WRCs), and their impact extends over long periods of time, between 10 and 20 years after the conference [10]. The GMDSS concept from Chapter IV of the SOLAS convention stipulates the obligation to provide ships with different communications equipment depending on the 4 maritime areas in which they can sail. These areas are defined according to the radio coverage and illustrated in Figure 1a:

- Sea area A1 is defined as the area covered by VHF radio enabling Digital Selective Calling (DSC) and Radio Telephone (RT)—up to 20–30 nautical miles from shore;
- Sea area A2—covered by MF radio, also enabling DSC and RT—up to about 150 nautical miles from shore;
- Sea area A3 is covered by HF radio and Inmarsat geostationary satellite enabling DSC and satellite communications—global area between 70° N and 70° S;
- Sea area A4 is comprising polar regions above 70° N and below 70° S and is covered by radio communications in the HF radio range with DSC.

In [11–13] there are presented the frequencies, structure, and destination of the radio channels allocated for the maritime mobile service in the MF, HF and VHF maritime bands as well as the changes that came into force starting with 1 January 2017. Thus, distress calls, distress messages, and distress traffic have absolute priority over all other types of messages/communications. Emergency communications, safety communications, and other communications follow. Moreover, all ships are required to be able to transmit distress alerts by at least two independent methods.

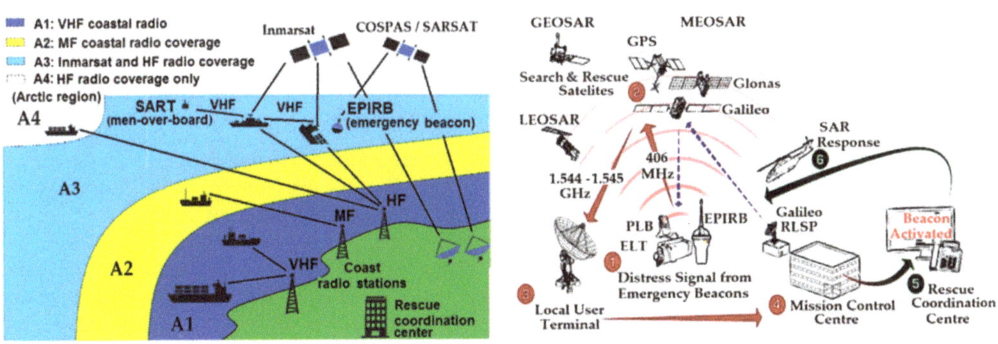

(a) (b)

Figure 1. (**a**) Area of the Global Maritime Distress and Safety System (GMDSS) [14]. (**b**) The concept of the Global Satellite System for Search and Rescue [15].

Vessels equipped with DSC systems will transmit distress alerts through this system on one of the following frequencies in the MF/HF ranges: 2187.5 kHz, 4207.5 kHz, 6312 kHz, 8414.5 kHz, 12,577 kHz, 16,804, 5 kHz using F1B emission class (frequency modulation with digital information, without a sub-carrier for automatic reception) with a bandwidth of 300 Hz or in VHF on the frequency 156.525 MHz using G2B emission class with a bandwidth of 16 KHz. The ETSI EN 300 338-1 V1.4.2 (2017-11) standard stipulates that the integrated equipment used for communications in the MF/HF range must also be able to operate in J2B mode ("Single Side Band SSB with digital information, with the use of a modulating sub-carrier, with the carrier suppressed to at least 40 dB below peak envelope power") [16].

According to Recommendation ITU-R M.489-2 (WRC-07), for ships not equipped with DSC systems, distress alerts will be transmitted on one of 4125 kHz; 6215 kHz; 8291 kHz; 12,290 kHz; 16,420 kHz using a J3E (single-sideband radiotelephony) emission class with a bandwidth of 3 kHz or 165.8 MHz using a G3E emission class with a bandwidth of 16 kHz [13].

Additionally, for the transmission of distress and safety messages through narrow-band direct-printing telegraphy (NBDP) are allocated the frequencies 4177.5 kHz, 6268 kHz, 8376.5 kHz, 12,520 kHz and 16,695 kHz, specifying that the transmission speed is not must exceed 100 Baud for FSK and 200 Baud for PSK. The bandwidth occupied for NBDP communications is 304 Hz.

DSC equipment must comply with the standards IEC 61097-3, IEC 61097-8, IEC 62238, EN 300 338, EN 301 025, EN 301 033 and maritime radiocommunication equipment and systems must comply with IEC 60,533 (Electrical and electronic installations in ships—Electromagnetic compatibility), IEC 60945, IEC 61097-7, IEC 61162, ETS 300 162, ETS 300 225 [17].

A special category of signals transmitted by radio waves are location signals. These signals are issued for the location of ships in distress, lifeboats, or survivors. In this regard there are: Emergency Position Indicating Radio Beacons (EPIRB), Emergency Locator Transmitters (ELT), Personal Locator Beacons (PLB), and Maritime Survivor Locator Devices (MSLD). Common frequencies assigned to these devices are:

- 121.5 MHz and 243 MHz with the use of emission class A3X (amplitude modulated with full carrier and both sidebands) with a modulation factor between 85 and 100%, not allowing overmodulation and N0N. The carrier must be identifiable separately from the 2 sidebands and the level of its carrier must be at least 30% above the sidebands. During the A3X broadcast, depending on the mode of variation of the audio signals that modulate the carrier, a distinction can be made between EPIRB, ELT, PLB or MSLD [18];
- 406.025 MHz using G1B emission class (phase modulation using $\pm$ 1.1 + 0.1 radians peak, referenced to an unmodulated carrier) for satellite emergency position indicating radio beacons [19];
- 156.525 MHz for EPIRBs operating on VHF Channel 70 (G2B).

Communications performed during search-and-rescue (SAR) operations are called "on scene communication" [20]. The frequencies used for these communications are 2182 kHz (J3E) and 158.6 MHz (G3E). In communications with SAR aircraft, in addition to the frequencies listed above, the following may be used: 3023 kHz, 4125 kHz, 5680 kHz, 121.5 MHz, 123.1 MHz and VHF channel 6 (156.30 MHz). For passenger ships it is mandatory to have radiocommunication equipment that can make calls on the aero-nautical air/sea VHF frequencies of 121.5 MHz and 123.1 MHz. To perform distress, emergency, and safety communications with seagoing vessels, aircraft will be able to communicate in J3E mode when using the carrier frequency 2182 kHz or 4125 kHz and in G3E mode when using the frequency 156.8 MHz and, optionally, frequency 156.3 MHz. Additionally, during SAR operations, the aircraft can use DSC equipment on the frequency 156.525 MHz and automatic identification system (AIS) equipment on the AIS frequencies 161.975 MHz and 162.025 MHz [17].

2.2. Theoretical Considerations

2.2.1. Formulation of the Problem

One method to analyze the interference resistance of communication systems is to sum up at the reception, over the signal modulated with modulation signal $x(t)$, a white Gaussian noise, uncorrelated, with zero mean value $z(t)$ [21–24]. In this sense, the improvement factor of the demodulator (IFD) is defined in Equation (1) as the ratio between the SNR (signal-to-noise ratio) at the demodulator output and the SNR (P_{Sinp}/P_{Zinp}) at the demodulator input:

$$\rho = (P_{Sout}/P_{Zout})/(P_{Sinp}/P_{Zinp}), \quad (1)$$

Here, P_{Sout} and P_{Zout} represent the signal power and the noise power, respectively, at the demodulator output, whereas P_{Sinp} and P_{Zinp} represent the signal power and the noise power, respectively, at the demodulator input. The quantities are noted on the schematic diagram illustrated in Figure 2.

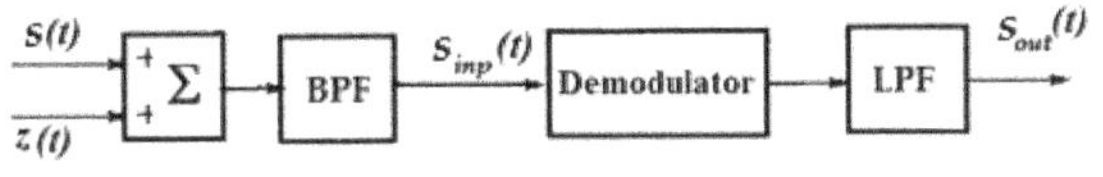

Figure 2. Equivalent scheme for demodulation analysis in the presence of noise.

The demodulator input noise power P_{Zinp} is considered equal to the output noise at the band-pass filter (BPF) with bandwidth B and the central frequency, fc, i.e., a narrow-band noise that can be written as Equation (2):

$$z(t) = n_1(t)\cos\omega_c t - n_2(t)\sin\omega_c t, \quad (2)$$

where ω_c is the central angular frequency, t is the time, and $n_1(t)$ and $n_2(t)$ are white, Gaussian, uncorrelated, low frequency noises with zero mean value, $(\overline{n_1(t)} = 0, \overline{n_2(t)} = 0)$. If we denote by N_0 the power spectral densities of $n_1(t)$ and $n_2(t)$, the power of the noise at the input to the demodulator P_{Zinp} can be calculated with Equation (3):

$$P_{Zinp} = \frac{1}{2\pi}\int_{\omega_c-\pi B}^{\omega_c+\pi B} N_0 d\omega = N_0 B, \quad (3)$$

The signal and noise power at the demodulator output P_{Sout} and P_{Zout}, respectively, are evaluated according to the ideal low-pass filter (LPF) with a transfer factor equal to the unit in the passband and the cut-off frequency equal to the maximum modulation frequency (f_{mM}). Additionally, in this analysis there is considered that the amplitude of the signal is higher in relation to the noise for most of the time (with a maximum probability).

2.2.2. Determining the IFD for Emission Classes A3X

If the modulation signal is represented as $x(t) = X_m f(t)$ in which X_m—is the amplitude of the modulation signal, $f(t)$—represents the normalized modulation signal with zero mean value $(\overline{f(t)} = 0)$, and m represents the degree or index of modulation, the amplitude-modulated signal $s_{A3X}(t)$ with double sideband can be described by Equation (4):

$$s_{A3X}(t) = A_p[1 + m \cdot f(t)]\cos\omega_p t, \quad (4)$$

where A_p is the carrier amplitude, and ω_p is the carrier angular frequency. The signal power P_{Sinp} at the demodulator input is given in Equation (5):

$$P_{Sinp} = \overline{s^2{}_{A3X}(t)} = \frac{1}{2}A_p{}^2\overline{[1 + mf(t)]^2} = \frac{1}{2}A_p{}^2\left[1 + m^2\overline{f(t)^2}\right], \quad (5)$$

Considering that the efficiency of the detector (demodulator) is η_d, the following signal in Equation (6) is obtained at its output:

$$s_{out}(t) = \eta_d\left[A_p m \cdot f(t) + n_1(t)\right], \quad (6)$$

At the output, the signal power, P_{Sout}, and the noise power, P_{Zout}, are expressed as in Equations (7) and (8):

$$P_{Sout} = \eta_d^2 A_p^2 m^2 \cdot \overline{f^2(t)}, \quad (7)$$

$$P_{Zout} = \eta_d^2 \cdot \overline{n_1^2(t)} = \eta_d^2 N_0 B, \quad (8)$$

It turns out that the IFD is computed with Equation (9)

$$\rho_{A3X} = \frac{P_{Sout}/P_{Zout}}{P_{Sinp}/P_{Zinp}} = \frac{A_p^2 m^2 \cdot \overline{f^2(t)}/N_0 B}{A_p{}^2\left[1 + \overline{m^2 f(t)^2}\right]/2NB_0} = \frac{2m^2 \cdot \overline{f^2(t)}}{1 + m^2 \cdot \overline{f^2(t)}}, \quad (9)$$

2.2.3. Determining the IFD for Emission Classes J3E and J2B

In the following paragraph, the case of a J3E/J2B signal that uses the upper sideband is analysed, considering the case of coherent detection (the carrier generated at the receiver is perfectly aligned in phase with the carrier generated at the transmitter). Therefore, the signals belonging to these emission classes can be expressed as in Equation (10):

$$s_{J3E}(t) = (1/2)x(t)\cos\omega_P t - (1/2)\,\hat{x}(t)\sin\omega_P t, \quad (10)$$

where $\hat{x}(t)$—represents the Hilbert transform of the modulation signal $x(t)$.

The signal power at the input of the demodulator is as follows in Equation (11):

$$P_{Sinp} = \overline{s_{J3E}^2(t)} = (1/4)\overline{x^2(t)}, \quad (11)$$

At the output of the product detector (with the same efficiency η_d as in the previous case), a signal proportional to the amplitude of the phase component is obtained in Equation (12):

$$s_{out}(t) = \eta_d[(1/2)x(t) + n_1(t)\cos\pi Bt - n_2(t)\sin\pi Bt], \quad (12)$$

such that the output signal power is expressed in Equation (13):

$$P_{Sout} = (1/4)\eta_d^2\overline{x^2(t)}, \quad (13)$$

At the demodulator input, P_{Zin} is expressed according to Equation (3), and P_{Zout} is given by Equation (8). Therefore, IFD can be deducted as mentioned below in Equation (14):

$$\rho_{J3E} = (P_{Sout}/P_{Zout})\,/\,(P_{Sinp}/P_{Zinp}) = \left[(1/4)\eta_d^2\overline{x^2(t)}/\eta_d^2 N_0 B\right]/\left[(1/4)\eta_d^2\overline{x^2(t)}/N_0 B\right] = 1, \quad (14)$$

2.2.4. Determining the IFD for Emission Classes G3E, G2B and G1B

The following notations are made: K_{MG} is the slope of the phase modulator, the phase deviation and $\Delta\varphi = K_{MG} \cdot X_m$ is expressed as product between the phase modulator slope K_{MG} and the modulation signal amplitude X_m, and $f(t)$ is the normalized modulation signal with zero mean value (the same quantities as in Section 2.2.2). Thus, the phase-modulated signals can be expressed as in Euation (15):

$$s_{MG}(t) = A_p\cos[\omega_p t + \varphi(t)] = A_p\cos[\omega_p t + K_{MG}x(t)] = A_p\cos[\omega_p t + \Delta\varphi \cdot f(t)] = A_p\cos[\omega_p t + K_{MG}X_m f(t)] \quad (15)$$

Because the actual information is contained only in the received signal phase $\varphi(t)$, before demodulation the signal is passed through an amplitude limiter. Therefore, at the demodulator input the limited signal power P_{SMG} is expressed with Equation (16):

$$P_{S_{MG}} = \overline{s_{MG}^2(t)} = A_p^2/2, \quad (16)$$

and the noise power is the same as in relationship (3).

The shape of the noise-affected signal can be described by Equation (17):

$$s_{MG}(t) = A_p \cos\left[\omega_p t + \varphi(t) + \varphi_z(t)\right] = A_p \cos\left[\omega_p t + \Phi\right], \tag{17}$$

where $\phi_z(t)$ is the noise component in the total phase of the signal Φ. The demodulated signal will be proportional to $\Phi = \varphi(t) + \varphi_z(t)$. Thus, the signal and noise power, respectively, will depend on the phase modulator slope K_{MG} and/or the phase demodulator slope K_{DG}, as seen in Equations (18) and (19):

$$P_{SoutG3E} = K_{DG}^2 \cdot K_{MG}^2 \overline{f^2(t)}, \tag{18}$$

$$P_{ZoutG3E} = \frac{K_{DG}^2}{A_p^2} \cdot \frac{1}{2\pi} \int\limits_{\omega_c-\pi B}^{\omega_c+\pi B} N_0 d\omega = 2\frac{K_{DG}^2}{A_p^2} N_0 B. \tag{19}$$

The result for G3E and G2E is an IFD expressed by Equation (20):

$$\rho_{G3E} = \frac{P_{Sout}/P_{Zout}}{P_{Sinp}/P_{Zinp}} = \frac{K_{DG}^2 K_{MG}^2 \overline{f^2(t)}}{2\left(K_{DG}^2/A_p^2\right) N_0 B} \cdot \frac{2N_0 B}{A_p^2} = K_{MG}^2 \cdot \overline{f^2(t)}, \tag{20}$$

2.2.5. Determining the IFD for Emission Classes F1B and F2B

Denoting K_{MF} the modulator frequency conversion constant, and $\Delta\omega = K_{MF}X_m$—the angular frequency deviation, the general form of these frequency-modulated signals is described by the following Equation (21):

$$s_{MF}(t) = A_p \cos\left[\omega_p t + \varphi(t)\right] = A_p \cos\left[\omega_p t + \Delta\omega \int_0^t f(\tau)d\tau\right] \Delta\omega \int_0^t f(\tau)d\tau, \tag{21}$$

In the case of F1B and F2B emission classes, the actual information is transmitted by varying the frequency of the carrier signal. Therefore, an amplitude limiter is inserted before the demodulator. The power of the input signal will be given by the relation (16) and that of the input noise by the relation (3).

The demodulated signal will be proportional to the time variation of the signal's total phase $d\Phi/dt$. If we note with K_{DF} the demodulator constant or efficiency, the respective powers of the output signal and the output noise are given by Equations (22) and (23):

$$P_{SoutF1B} = \overline{\left[\Delta\omega_M f(t)\right]^2} = \Delta\omega_M^2 \overline{f^2(t)} = K_{DF}^2 K_{MF}^2 X_m^2 \overline{f^2(t)}, \tag{22}$$

$$P_{ZoutF1B} = \frac{K_D^2}{A_p^2} \cdot \frac{1}{2\pi} \int\limits_{\omega_c-\pi B}^{\omega_c+\pi B} \omega^2 \cdot N_0 d\omega = \frac{2}{3} \frac{K_D^2 \cdot N_0 B^3}{A_p^2}. \tag{23}$$

The result for the improvement factor is expressed by relationship (24):

$$\rho_{F1B} = \frac{K_{DF}^2 K_{MF}^2 X_m^2 \overline{f^2(t)}}{(2/3)\cdot\left(K_{DF}^2 \cdot N_0 B^3/A_p^2\right)} \cdot \frac{2N_0 B}{A_p^2} = 3 \cdot \frac{K_{MF}^2 X_m^2}{B^2} \cdot \overline{f^2(t)} = 3 \cdot D^2 \cdot \overline{f^2(t)}, \tag{24}$$

where D is the deviation ratio.

2.2.6. Comparison of IFD (with the Use of Mathcad)

Figure 3 illustrates the IFD for all the modulated signals discussed above, depending on the modulation index m/slope K/deviation D, and the normalized modulation signal power $\overline{f^2(t)}$. Therefore, the blue dash-dotted line represents ρ_{a3x}—the IFD of a Double Sideband Amplitude-Modulated signal according to A3X emission class, computed with Equation (9) and depending on the modulation index m; the continuous red line represents

ρ_{j3e}—the IFD of a Single-Sideband Amplitude-Modulated signal according to J3E emission class and computed with Equation (14); the red dotted line represents ρ_{g3e}—the IFD of a phase-modulated signal according to G3E emission class, depending on the phase modulator shape K_{MG} and computed with Equation (20); and finally, the black dashed line represents ρ_{f1b}—the IFD of a Frequency-Modulated signal according to F1B emission class, depending on de frequency deviation ratio D, and computed with Equation (24). Three values of the normalized modulation signal power $\overline{f^2(t)}$ were considered 0.3, 0.5, and 0.8, respectively, and thus three graphs were drawn.

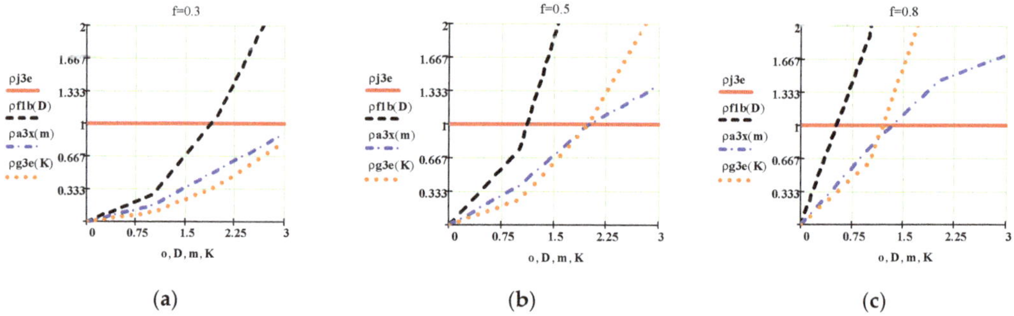

Figure 3. IFD values plotted based on the relationships (9), (14), (20) and (24) for different values of the normalized modulation signal power $\overline{f^2(t)}$: (**a**) 0.3; (**b**) 0.5; (**c**) 0.8.

By analysing Equations (9), (14), (20) and (24) implemented in the Mathcad software and the graphs drawn in Figure 3, we can state that:

- IFDs are constant and do not depend on the parameters of the modulating signal or of the modulation stage only for the analysed emission classes J3E and J2B;
- For small values of the standard modulator signal power, of the slope of the phase modulator for G3E and of the deviation ratio for F1B (default for small occupied bandwidths), IFD for emission classes J3E and J2B are clearly superior to all other classes. Additionally, in this case, the superiority of IFD of A3X class over those of G3E, G2B and G1B is found;
- A constant IFD of 1 for J3E and J2B could mean that we can recover the signal even if its power is equal to the noise power; thus, it is not depending on the standard modulating signal power, which gives these emission classes a special stability to disturbances;
- For the emission classes G3E, G2B G1B, and especially for F1B, F2B, IFD classes, the obtained values are much higher than those of the other classes, but only for high values of the normalized modulation signal power, and of the deviation ratio or the phase modulator slope, respectively, which rise as the modulated signal band increases.

2.3. *Verification of Results Using the Matlab—Simulink Simulation Program*

To verify the analytically obtained results, and starting from the equivalent noise demodulation schematic diagram presented in Figure 2, a block diagram of four types of demodulators was designed and implemented in the Matlab-Simulink programming environment. The diagram is shown Figure 4 and it describes the following demodulators: Double Sideband Amplitude Modulation (DSB AM) demodulator representing the A3X class, Single-Sideband Amplitude Modulation (SSB AM) demodulator representing the J3E class, frequency modulation (FM) demodulator respresenting the F1B class, and phase modulation (PM) demodulator representing the G3E class. Gaussian white noise was introduced through the "AWGN Channel" blocks.

Figure 5 illustrates five signals, listed from top to bottom: the modulation signal x(t) created by the waveform generator, and the signals obtained at the output of the following demodulators: DSB AM (A3X class), SSB AM (J3E class), FM (F1B class), and PM (G3E class). The scales of the signal graphs are different and illustrated on the left side of each oscilloscope channel. All five signals were determined for various values of the SNR and the phase deviation $\Delta\varphi$ (in the case of the PM demodulator).

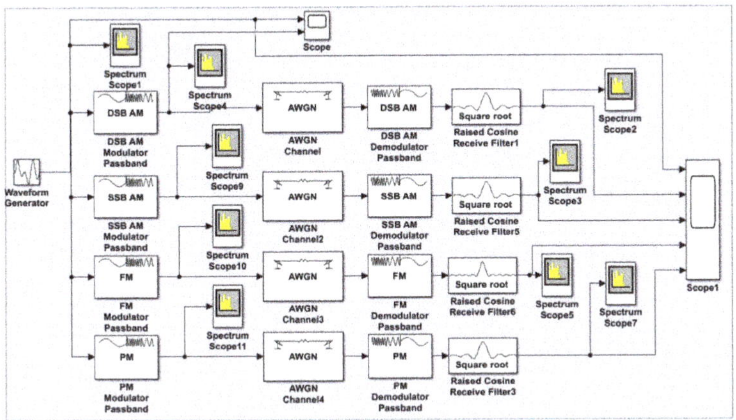

Figure 4. Schematic diagram used in the Matlab-Simulink software for verifying the stability to disturbances of the main emission classes used in search-and-rescue communications at sea.

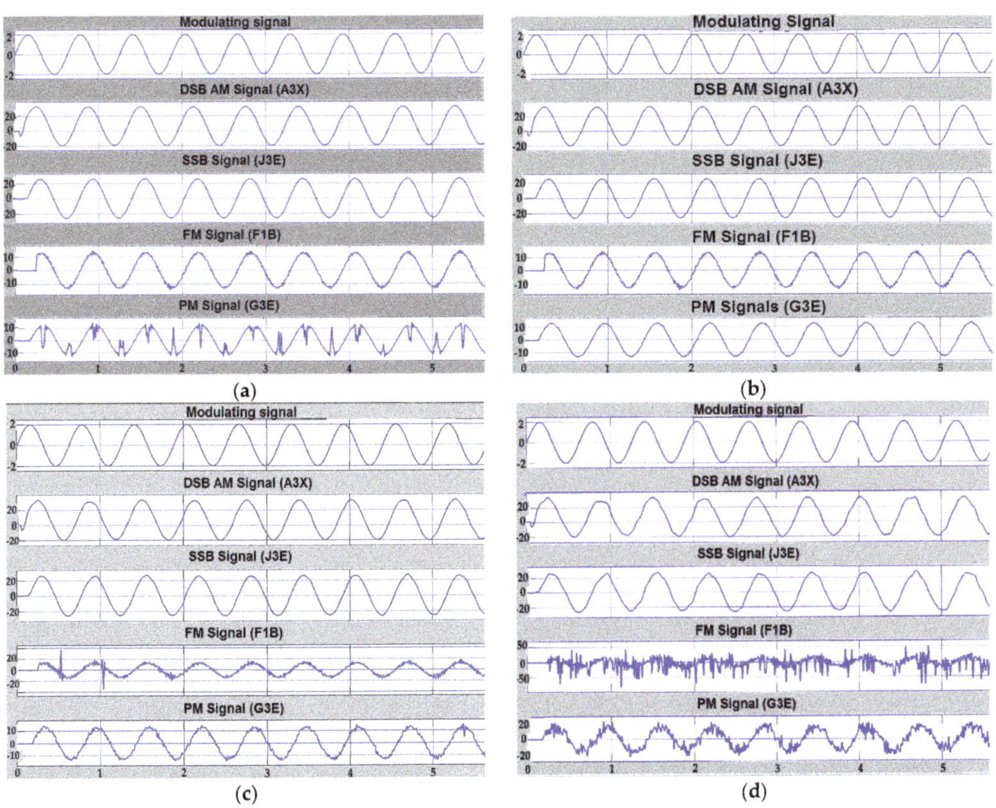

Figure 5. *Cont.*

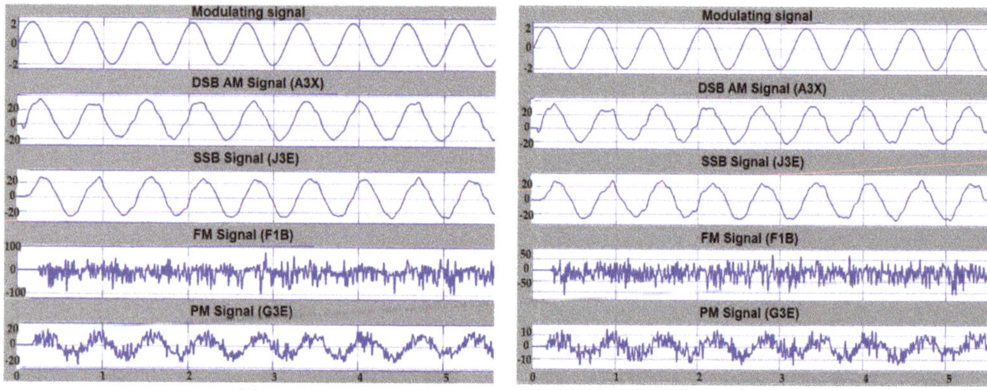

(e) (f)

Figure 5. Simulation results using the diagram in Figure 4, for different values of the noise signal ratio (SNR) at the demodulator input port: (**a**) SNR = 20 dB, for MP $\Delta\varphi = \pi/2$; (**b**) SNR = 20 dB, for MP $\Delta\varphi = \pi/3$; (**c**) SNR = 10 dB, for MP $\Delta\varphi = \pi/3$; (**d**) SNR = 3 dB, for MP $\Delta\varphi = \pi/3$; (**e**) SNR = 0 dB, for MP $\Delta\varphi = \pi/3$; (**f**) SNR = −3 dB, for MP $\Delta\varphi = \pi/3$.

3. Results and Discussion

Comparison of the results in Figure 5a,b provides an insight into the performance of the PM demodulator depending on the phase deviation value $\Delta\varphi = \pi/2$ in (a) as opposed to $\Delta\varphi = \pi/3$ in (b), for the same SNR = 20 dB. The following subfigures (c) to (f) maintain the same phase deviation $\Delta\varphi = \pi/3$ that yielded proper noise stability results of the PM demodulator, while varying the SNR, and in this case, an adequate comparison between various demodulator types can be carried out. Therefore:

- Emission classes that use phase modulation have been shown to be more resistant to interference when a $\pi/3$ phase deviation has been used, which is much more resistant than those that use frequency modulation.
- For emission classes using frequency modulation, at SNR values below 10 dB (considered to be the full improvement threshold) the actual signal shape can no longer be recognized—the effect of signal captured by noise occurs.
- Emission classes using DSB AM showed similar resistance to disturbance using SSB AM, so that, for both, the useful signal can be recognized even in the conditions of an SNR of −3 dB.

When comparing the simulation results in Figure 5 to those obtained analytically and illustrated in Figure 3, it is noted that the demodulating systems ranking from the perspective of noise stability is similar only for the situation of a high SNR = 20 dB (Figure 5b), and for small values (below 1) of the corresponding dependence parameters m, K, D in Figure 3. In this case, the J3E system (SSB AM) yields the highest stability in terms of both simulation results and IFD value. However, there is a notable difference for the DSB AM system (A3X class) compared to the FM system (F1B class): although the simulation in Figure 5 indicates a low performance of the F1B class system compared to the A3X one, the analytical calculations give high IFD values for the F1B, regardless of the normalized modulator signal power, the modulation index m, or the frequency deviation D. G3E (PM) systems yield the lowest performance in both cases.

Table 1 presents a comparative situation between the results obtained by the two methods adopted and the conclusions formulated in the specialized technical literature and the bibliographic references.

Table 1. Comparison of analog system stability to disturbance results obtained from both research method and from technical literature.

SNR	Emission Classes Used in Distress Communications	Stability to Disturbances		
		IFD Conclusions	Simulink Conclusions	Conclusions from the Technical Literature/References [21–24]
>10 dB	A3X	Depending on the modulation index m and the normalized modulation signal power $\overline{f^2(t)}$. Performances for m < 1 are below F1B, J3E but above G3E.	Very good. Similar performance as J3E and higher than F1B and G3E.	Good. According to [23] the poorest performance. According to [24] poorer performance than J3E.
	F1B	Depending on the deviation ratio D, and the normalized modulation signal power $\overline{f^2(t)}$. For low values of D and $\overline{f^2(t)}$ performing below J3E.	Very good.	Very good. Acording [21–23] the best performance.
	J3E	IFD = 1—good	Very good.	Good.
	G3E	Depending on the phase modulator slope K_{MG} and the normalized modulation signal power $\overline{f^2(t)}$. It has performance over J3E and A3X only for high values of K_{MG} and $\overline{f^2(t)}$.	Depends on the phase deviation; very good for phase deviation = $\pi/3$.	Very good. According to [21,23,24] better performance than J3E and A3X.
3–10 dB	A3X	N/A—we considered that the amplitude of the useful signal is higher in relation to the noise for most of the time	Very good	Good only for case of coherent detection, unacceptable for system with envelope detection.
	F1B	N/A—we considered that the amplitude of the useful signal is higher in relation to the noise for most of the time	Unacceptable	Very low below the SNR threshold of 10 dB [21]. Can be improved with preemphasis/deemphasis systems [22–24]. According to [24] better than A3X and J3E.
	J3E	IFD = 1—good for case of coherent detection.	Very good.	Good for case of coherent detection.
	G3E	N/A—we considered that the amplitude of the useful signal is higher in relation to the noise for most of the time.	Acceptable for phase deviation = $\pi/3$ (can be improved using additional processing methods).	Good. According to [24] better than F1B for most practical signals.
0–3 dB	A3X	Similar to case 3—10 dB.	Good	Unacceptable
	F1B	Similar to case 3—10 dB.	Unacceptable	Unacceptable [21].
	J3E	Similar to case 3—10 dB.	Good	Good for case of coherent detection.
	G3E	Similar to case 3—10 dB.	Bad, can be improved with additional processing	According to [24] better than F1B.

Table 1. *Cont.*

SNR	Emission Classes Used in Distress Communications	Stability to Disturbances		
		IFD Conclusions	Simulink Conclusions	Conclusions from the Technical Literature/References [21–24]
0–(−3) dB	A3X	Similar to case 3—10 dB.	Good	Unacceptable
	F1B	Similar to case 3—10 dB.	Unacceptable	Unacceptable
	J3E	Similar to case 3—10 dB.	Good	Unacceptable
	G3E	Similar to case 3—10 dB.	Unacceptable	Unacceptable

4. Conclusions

The paper presents a comparative analysis of analog (de)modulation systems employed in maritime communications for search-and-rescue operations. The noise stability of the following systems is investigated: DSB AM (A3X class), SSB AM (J3E class), FM (F1B class), and PM (G3E class). The analysis is carried out through analytical calculation of the improvement factor of the demodulator (IFD), and by implementing the modulator-demodulator systems in the Matlab-Simulink simulation environment. The research method employed in the two-phased analysis is scientifically accurate and thorough. For particular value ranges of the modulation parameters, the two phases of the analysis yield similar results. However, one drawback of the research method employed in the analysis is that it needs to be supplemented by experimental validation. Based on the results discussion in the previous section, several conclusions regarding the deployment of analog communication system in SAR operations communications can be formulated.

For radio communications allocated to MF/HF and VHF maritime bands, systems using emission classes related to linear modulations (in amplitude) prove to be more resistant to disturbances than systems using emission classes related to phase and frequency modulations. This is true for all radio communications in environments affected by noise and in which we cannot provide a proper signal-to-noise ratio (over 10 dB).

Keeping the J3E emission class for the maritime MF/HF range, both in ordinary communications and especially in distress communications, for the next at least 10 years is a perfectly justified thing due to:

- The constructive simplicity of the radio stations that use it;
- Resistance to disturbances;
- Very small occupied bandwidth, (default and lower transmission power);
- Efficient use of the electromagnetic spectrum;
- Propagation of HF waves at very long distances and on multiple paths, which leads to the summation at the input terminals of the receiver of both the multiple signals from the transmitter being worked with and from other transmitters;
- The multitude of users of this range and the very high requirement of allocated frequencies/channels.

The advantages listed above make the emission classes that use SSB modulation particularly attractive for terrestrial radiocommunications or fiber optic communications [25–29].

Taking into consideration that in ETSI EN 300 338-1 V1.4.2 (2017-11) standard it is provided that the integrated equipment used for communications in the MF/HF range must be able to operate and in J2B mode, it is very possible that in the future it will be introduced the obligation to transmit distress alerts via DSC in the MF/HF ranges in J2B mode.

The use of phase modulation for distress communications is possible only in the VHF range where with the assurance of a higher bandwidth (16 kHz) better performance of this modulation can be obtained and with the imposition of restrictions/communication disciplines (during communications any other type of communication that may interfere with them is prohibited).

Because the impact of the WRC extends over long periods of time, between 10 and 20 years after the conference, the emission classes analysed in this paper will certainly be used for distress maritime communications for at least the next 10 years [30]. Moreover, because one of the basic principles of SOLAS is the transmission of distress alerts by at least two independent means, the introduction of digital modulations in naval radio communications will not lead to the disappearance of communications using analog modulations, especially when the risk of cyberattacks is on the rise.

Author Contributions: Conceptualization, V.S.; methodology, V.S.; software, V.S.; validation, V.S. and G.R.; formal analysis, V.S.; investigation, V.S.; resources, V.S. and G.R.; data curation, G.R.; writing—original draft preparation, G.R.; writing—review and editing, M.G.; visualization, V.S.; supervision, M.G.; project administration, M.G.; funding acquisition, M.G. All authors have read and agreed to the published version of the manuscript.

Funding: This research was funded by the doctoral school of 'Dunarea de Jos' University Galati, Romania.

Informed Consent Statement: Not applicable.

Conflicts of Interest: The authors declare no conflict of interest.

References

1. *Australian Global Maritime Distress and Safety System (GMDSS) Handbook 2018*, 11th ed.; Australian Maritime Safety Authority (AMSA): Braddon, Australia, 2018.
2. *International Convention for the Safety of Life at Sea (SOLAS)*; International Maritime Organization (IMO): London, UK, 1974; Available online: https://www.imo.org/en/About/Conventions/Pages/International-Convention-for-the-Safety-of-Life-at-Sea-(SOLAS) (accessed on 10 March 2021).
3. *Brief History of IMO*; International Maritime Organization (IMO): London, UK, 2019; Available online: https://www.imo.org/en/About/HistoryOfIMO/Pages/Default.aspx (accessed on 10 March 2021).
4. Modern Maritime Communications. In Proceedings of the 29th World Radiocommunications Seminar, Online, 30 November–11 December 2020; Available online: https://www.itu.int/en/ITU-R/seminars/wrs/2020/ (accessed on 10 March 2021).
5. *SOLAS (Safety of Life at Sea) New or Upcoming Requirements*; Lloid's Register: London, UK, 2021; Available online: https://www.lr.org/en/solas/ (accessed on 24 April 2021).
6. Raulefs, R.; Wang, W. Radio Spectrum Scarcity for Maritime Communication and Navigation Systems. In Proceedings of the 2021 IEEE 4th International Conference on Electronic Information and Communication Technology (ICEICT), Xi'an, China, 18–20 August 2021; pp. 588–591. [CrossRef]
7. Gradiant. *The Digitalisation of Maritime Communications*; Gradiant: Vigo, Spain, 2019.
8. John, O.; Reinmann, M. Increasing Quality of Maritime Communication through Intelligent Speech Recognition and Radio Direction Finding. In Proceedings of the European Navigation Conference ENC 2020, Dresden, Germany, 23–24 November 2020.
9. A Comprehensive Guide to Maritime Cybersecurity, ©2021 Mission Secure. Available online: https://www.missionsecure.com/maritime-security-perspectives-for-a-comprehensive-approach (accessed on 24 April 2021).
10. European Commission. Setting the World's Radio Spectrum Rules. In Proceedings of the World Radiocommunication Conference, Sharm el-Sheikh, Egypt, 28 October–22 November 2019; Available online: https://ec.europa.eu/digital-single-market/en/world-radiocommunication-conference-setting-worlds-radio-spectrum-rules (accessed on 10 March 2021).
11. ITU. *Radio Regulations Articles*, 2012 ed.; ITU: Geneva, Switzerland, 2012.
12. ITU. *Radio Regulations Appendices*, 2016 ed.; ITU: Geneva, Switzerland, 2016.
13. ITU. *Final Acts WRC-15 World Radiocommunication Conference (GENEVA, 2015)*; ITU: Geneva, Switzerland, 2016.
14. Sternula, *Maritime Authority*, Sternula ApS, Denmark. Available online: https://www.sternula.com/index.php/maritime-authorities/ (accessed on 24 April 2021).
15. European Union Agency for the Space Programme. *Search and Rescue (SAR)/Galileo Service*. Available online: https://www.euspa.europa.eu/european-space/galileo/services/search-and-rescue-sar-galileo-service (accessed on 17 May 2021).
16. European Standard. ETSI EN 300 338-1 V1.4.2 (2017-11)—Technical Characteristics and Methods of Measurement. Available online: https://www.etsi.org/deliver/etsi_en/300300_300399/30033801/01.04.02_60/en_30033801v010402p.pdf (accessed on 17 May 2021).
17. Maritime Radio Transmitters and Receivers in the Band 156–162.5 MHz. Spectrum Management and Telecommunications Radio Standards Specification Industry Canada. Available online: https://www.ic.gc.ca/eic/site/smt-gst.nsf/vwapj/RSS182.pdf/$FILE/RSS182.pdf (accessed on 17 May 2021).
18. *RSS-287—Emergency Position Indicating Radio Beacons (EPIRB), Emergency Locator Transmitters (ELT), Personal Locator Beacons (PLB), and Maritime Survivor Locator Devices (MSLD)*; Government of Canada: Ottawa, ON, Canada, 2017.

19. *Specification for COSPAS-SARSAT 406 MHz Distress Beacons C/S T.001—Issue 4—Rev. 3*; Cospas-Sarsat: Montreal, QC, Canada, 2018; Available online: http://www.cospas-sarsat.int/en/ (accessed on 17 May 2021).
20. *International Aeronautical and Maritime Search and Rescue Manual, Volume III, Mobile Facilities*; International Maritime Organization: London, UK; International Civil Aviation Organization: Montréal, QC, Canada, 2016.
21. Constantin, I.M.I. *Transmisiuni Analogice şi Digitale*; Editura Tehnică: Bucureşti, Romania, 1995.
22. Ziemer, R.E.; Tranter, W.H. *Principles of Communications*, 7th ed.; Wiley: Hoboken, NJ, USA, 2014; Available online: https://it-ebooks.info/ (accessed on 19 May 2021).
23. Carlson, A.; Crilly, P. *Communication Systeems—An Introduction to Signals and Noise*, 5th ed.; McGraw-Hill: New York, NY, USA, 2011.
24. Agbo, S.O.; Sadiku, M.N.O. *Principles of Modern Communication, 7-Systems Noise in Analog Communications*; Cambridge University Press: Cambridge, UK, 2019.
25. Safavi, H.; Hosseini, S.E. Optical SSB modulator/frequency shifter with ultralow spurious sidebands. *Appl. Opt.* **2020**, *59*, 7408–7418. [CrossRef] [PubMed]
26. Zho, B.; Nai, W.; Ge, Y.; Xing, Y.; Zhu, H. Performance Analysis on Single Sideband Modulated Radio-Over-Fiber System Carrying Amplitude Shift Keying Signal with Different Duty Cycle. In Proceedings of the 2018 IEEE 9th International Conference on Software Engineering and Service Science (ICSESS), Beijing, China, 23–25 November 2018; Available online: https://ieeexplore.ieee.org/document/8663904 (accessed on 17 May 2021).
27. Sampath, K.I.A.; Takano, K.; Maeda, J. Peak-to-Average Power Ratio Reduction of Carrier-Suppressed Optical SSB Modulation: Performance Comparison of Three Methods. *Photonics* **2021**, *8*, 67. [CrossRef]
28. Yaakob, S.; Mahmood, R.M.; Zan, Z.; Rashidi, C.B.M.; Mahmud, A.; Anas, S.B.A. Modulation Index and Phase Imbalance of Dual-Sideband Optical Carrier Suppression (DSB-OCS) in Optical Millimeter-Wave System. *Photonics* **2021**, *8*, 153. [CrossRef]
29. Ganjali, M.; Safavi, N.; Qashqaei, M.; Hosseini, S.E. Optical single-sideband polarization modulator based on Sagnac interferometers and its applications in radio-over-fiber systems. *J. Opt. Soc. Am. B* **2020**, *37*, 3268–3276. [CrossRef]
30. ITU. Frequency Plans for Terrestrial Non-Broadcasting Services Document WRS20/8-E. In Proceedings of the ITU World Radiocommunication Seminar 2020 (WRS-20), Online, 3 November 2020.

inventions

MDPI

Article

Insights of Sea Buckthorn Extract's Encapsulation by Coacervation Technique

Diana Roman, Nina Nicoleta Condurache (Lazăr), Iuliana Aprodu, Elena Enachi, Vasilica Barbu, Gabriela Elena Bahrim, Nicoleta Stănciuc and Gabriela Râpeanu *

Faculty of Food Science and Engineering, "Dunărea de Jos" University of Galați, Domnească Street 111, 800201 Galați, Romania; diana.roman@ugal.ro (D.R.); nina.condurache@ugal.ro (N.N.C.); iuliana.aprodu@ugal.ro (I.A.); elena.ionita@ugal.ro (E.E.); vasilica.barbu@ugal.ro (V.B.); gabriela.bahrim@ugal.ro (G.E.B.); nicoleta.stanciuc@ugal.ro (N.S.)

* Correspondence: gabriela.rapeanu@ugal.ro

Abstract: Sea buckthorn (*Hippophae rhamnoides* L.) represents a valuable source of biologically active compounds such as carotenoids and polyphenols. High amounts of these substances are found in its fruits, bark, and leaves. However, their bioavailability is limited and must be increased in order to benefit from the properties they exert. Therefore, the purpose of this study was to increase the stability and bioavailability of sea buckthorn fruit's bioactives. The sea buckthorn's bioactive compounds were extracted with a solvent combination between glacial acetic acid, acetone, and water on one side and water only on the other side. Afterward, the phytochemicals from the extracts were encapsulated using the coacervation technique, followed by freeze-drying in order to obtain stable powders. The powders were characterized in terms of antioxidant activity, total carotenoids, β-carotene, lycopene, total polyphenol, and total flavonoid content, color, structure, and morphology. The phytochemical stability of the powders and their antioxidant activity was assessed during 270 days of storage at 4 °C. Moreover, the bioavailability of phytochemicals was measured during in vitro simulated digestibility. Our findings provide insights to promote carotenoids and polyphenols from sea buckthorn as bioactive ingredients with multiple purposes.

Keywords: *Hippophae rhamnoides* L.; bioactive compounds; carotenoids; flavonoids; antioxidant activity; microencapsulation

Citation: Roman, D.; Condurache, N.N.; Aprodu, I.; Enachi, E.; Barbu, V.; Bahrim, G.E.; Stănciuc, N.; Râpeanu, G. Insights of Sea Buckthorn Extract's Encapsulation by Coacervation Technique. *Inventions* **2021**, *6*, 59. https://doi.org/10.3390/inventions6030059

Academic Editor: Monique Lacroix

Received: 10 August 2021
Accepted: 10 September 2021
Published: 14 September 2021

1. Introduction

Sea buckthorn (Figure 1) represents an ancient versatile plant that has been used for centuries in the daily life of people for a variety of purposes (animal fodder, treatment for several diseases, firewood, or even decorative elements), according to Olas [1].

Figure 1. Sea buckthorn fruit [2].

Its fruits, leaves, seeds, and oils are sources of many biologically active substances, such as flavonoids, carotenoids, organic acids, unsaturated fatty acids, vitamins, proteins, amino acids, and minerals [3,4]. The average composition of sea buckthorn biologically active compounds is presented in Table 1, according to Ciesarova et al. [4].

Table 1. Composition of sea buckthorn.

Components	Fruits	Pulp Oil	Leaves
Vitamin C (g/kg fw)	0.2–30	-	-
Total Flavonoids (g/kg fw)	1.6–10.0	-	-
Quercetin (mg/kg dw)	40.0–375.0	-	332–1381
Kaempherol (mg/kg dw)	102.0	-	2.8–4.1
Isorhamnetin (mg/kg dw)	103.0–964.0	-	270.0
Gallic acid (mg/kg dw)	1.0–15.9	-	19.7–79.1
p-coumaric acid (mg/kg dw)	1.4–22.3	-	8.4–13.4
Salicylic acid (mg/kg dw)	21.0–47.5	-	-
Total carotenoids (mg/kg fw)	15.0–185.0	303–21,400	35.0–42.0
Lycopene (% from total carotenoids)	1.4–4.3	nd	nd
B-carotene (% from total carotenoids)	4.6–26.0	5.5	29.0–48.0
Linolenic acid (% fatty acids)	-	0.4–10.0	-
Linoleic acid (% fatty acids)	-	4.4–15.0	-
Oleic acid (% fatty acids)	-	10.0–26.2	-

nd—not detected; fw—fresh weight; dw—dry weight.

These substances and their free radicals scavenging activity endow sea buckthorn with health benefits. There are various studies suggesting that the consumption of sea buckthorn bioactives can promote health effects, such as neuroprotective effects, inhibition of cardiac cell destruction, anticoagulant effects, protection of the liver from oxidative stress, antiulcer, anticancer, antimicrobial effects, etc. [1,4,5].

The major compounds of scientific interest from sea buckthorn fruits are the carotenoids and flavonoids. Carotenoids represent lipophilic pigments with tetraterpenoid structure and colors including yellow, orange, and red. The most important biological activity associated with the carotenoids is provitamin A [6]. Flavonoids are secondary sea buckthorn metabolites and are a part of the polyphenols class, also having yellow and orange colors. Due to their colors and biological activities that can promote health effects, these compounds are suitable for use in food, nutraceutics, cosmeceutics, and pharmaceutics industries. However, the use of carotenoids and flavonoids as functional ingredients is limited due to their chemical instability, poor bioavailability, and bioaccessibility. Therefore, alternatives should be developed to increase the stability and bioavailability of these valuable bioactive compounds. The microencapsulation technique represents a valuable solution to these problems.

Nowadays, many encapsulation techniques are available, each with its advantages and disadvantages. The encapsulation method choice depends on both the properties of the bioactive compounds and encapsulating material, but also the desired final product [7]. Moreover, the encapsulation time, cost, and steps required, the volume of production, market requirements, and current legislation must be taken into account [7].

Coacervation is a superior technique to other microencapsulation techniques due to the advantages provided [8]. Depending on the amounts of polymers applied in the process, the coacervation technique could be simple or complex. The complex coacervation is based on the electrostatic attraction between the oppositely charged molecules. This attraction produces a complex with two distinct phases. One represents a polymer-rich phase (also named coacervate) and the other one is the solution solvent [9]. The biopolymer's interaction is influenced by the biopolymer's type, the pH, the ionic strength, the polymer's concentration, and the ratio between the biopolymers [10]. Natural biopolymers, such as proteins and polysaccharides, are widely used to create a variety of delivery systems to protect the biologically active compounds from environmental factors. The encapsulation of the biologically active compounds leads to the production of functional ingredients that can then be used in a variety of ways [11].

This study aimed to extract and microencapsulate by coacervation the biologically active compounds from sea buckthorn fruits. Biopolymer composite matrices were formed by whey protein isolate (WPI) in combination with carboxymethyl cellulose (CMC). Two resulting powders, coded P1 and P2, were characterized for their phytochemical profile and encapsulation efficiency, with emphasis on carotenoids, polyphenols, antioxidant activity, and in vitro digestibility. Further, the structure and morphology of the powders were analyzed using the confocal laser scanning microscopy technique (CLSM). The obtained results could provide evidence concerning how targeted biologically active compounds from sea buckthorn can be used in developing functional ingredients for multiple applications. The powders that we have obtained exert biological activities demonstrated in this study and these make them suitable for multiple applications as natural antioxidants. In addition to the role of natural antioxidants, our powders can also act as natural dyes. They are suitable for the food industry, textile, nutraceuticals, cosmetics, and pharmaceutics industries.

2. Materials and Methods

2.1. Chemicals

Hexane, acetone, glacial acetic acid, ethanol and methanol HPLC-grade, sodium carbonate (Na_2CO_3), 2,2′-azino-bis (3-ethylbenzothiazoline-6-sulfonic acid) diammonium salt (ABTS), 2,2-diphenyl1-picrylhydrazyl (DPPH), Folin–Ciocalteu reagent, gallic acid solution, hydrochloric acid (HCl), sodium nitrite ($NaNO_2$), aluminum chloride ($AlCl_3$), sodium hydroxide (NaOH), Trolox solution, whey protein isolate (WPI), and carboxymethyl cellulose (CMC) were purchased from Sigma Aldrich (Steinheim, Germany).

2.2. Materials

Sea buckthorn fruits were purchased from a local market named "December 30 market" in Galati, Romania, September 2018. The sea buckthorn berries were washed, frozen, and then lyophilized at −42 °C with a pressure of 0.1 mBar for 48 h (Alpha 1-4 LD Plus, Christ, Osterode am Harz, Germany). Dried sea buckthorn fruits were stored in plastic bags with a lid at room temperature until extraction.

2.3. Extraction of the Biologically Active Compounds from Sea Buckthorn

The biologically active compounds from sea buckthorn fruits were extracted with different solvent mixtures, using the ultrasound-assisted extraction method (UAE). Polar solvents like acetone, glacial acetic acid, and water were used. Two separate extractions with different solvent mixtures were performed using the same lyophilized matrix. On one hand, 5 g of sea buckthorn was mixed with 20 mL of glacial acetic acid/acetone/water (0.35/20/80 ratio), with the extract encoded E1. On the other hand, 5 g of sea buckthorn was mixed with 20 mL of water, with the extract encoded E2. Both extractions took place using an ultrasonic water bath (Smart MRC LTD, Holon, Israel) for 20 min at 40 °C and a frequency of 40 kHz. Further, the extracts were centrifuged for 10 min at 5000 rpm and 4 °C. The supernatant was collected and concentrated to dryness under reduced pressure at 40 °C (AVC 2-18, Christ, Osterode am Harz, Germany). Finally, the obtained extracts were phytochemically analysed by solubilization in the extraction solvent.

2.4. Phytochemical Profile of the Sea Buckthorn Extracts

The phytochemical profile of both extracts was analysed in terms of lipophilic compound and hydrophilic compound contents, using spectrophotometric methods. The lipophilic compounds content quantified in our extracts were the lycopene (LC), β-carotene (βC), and total carotenoids (TC). The hydrophilic compounds analysed were flavonoids (TFC) and polyphenols (TPC). For both extracts, the antioxidant activity against ABTS and DPPH free radicals was measured.

2.4.1. Total Carotenoid Content

The lycopene, β-carotene, and total carotenoid contents were measured and calculated using the spectrophotometric method described by Rodriguez-Amaya and Kimura [12]. Briefly, the extracts were solubilized in the same extraction solvent and the absorbance of the mixture was read at λ = 470 nm for total carotenoids, λ = 450 nm for β-carotene, and λ = 503 nm for lycopene. The results were expressed as mg/g of dry weight (dw).

2.4.2. Total Polyphenol Content (TPC)

The colorimetric Folin–Ciocâlteu method, described by Turturica et al. [13], was used for the TPC quantification. Briefly, 0.20 mL of extract solution were mixed with 15.8 mL of deionized water and 1 mL of Folin–Ciocâlteu reagent. After 10 min of rest, 3 mL of Na_2CO_3 20% was added to the mixture and vigorously shaken. The resulting mixture was kept for 60 min in the dark at room temperature. Finally, the absorbance of the mixture was measured at 765 nm. The results were expressed as mg Gallic acid equivalents (GAE)/g dw, using a calibration curve.

2.4.3. Total Flavonoid Content (TFC)

The total flavonoid content was determined using a spectrophotometric method based on the reaction between the aluminum chloride ($AlCl_3$) and phenolic compounds, described by Turturica et al. [13]. Briefly, 0.25 mL of extract solution was mixed with 2 mL of distilled water and 0.075 mL of $NaNO_2$ 5% solution. After 5 min of rest, 0.15 mL of $AlCl_3$ 10% solution were added, and the mixture was allowed to rest for another 6 min. Finally, 0.5 mL of NaOH (1 M) solution was added and the absorbance of the mixture was immediately measured at 510 nm. The results were expressed as mg catechin equivalents (CE)/g dw, using a calibration curve.

2.4.4. Antioxidant Activity

The antioxidant activities of the extracts were tested against ABTS [14] and DPPH free radical [13], the results being expressed as radical scavenging activity percentage (%).

For the DPPH scavenging activity, 3.9 mL of free radical solution 0.1 M in methanol were mixed with of 0.20 mL of extract solution (Af) and kept for 30 min at room temperature in the dark. Finally, the absorbance was recorded at λ = 515 nm against a blank (methanol instead of extract—A0).

For the ABTS scavenging activity, 1 mL of free radical stock solution 7 mM (in ethanol, mixed with 2.45 mM potassium persulfate) were mixed with 0.10 mL of extract solution (Af). The absorbance of the solution was measured at 734 nm after 1–6 min against a blank (ethanol instead of extract—A0).

The radical scavenging activity percentage (%) was calculated as follows:

$$\text{Radical scavenging activity percentage (\%)} = \frac{\text{A0} - \text{Af}}{\text{A0}} \times 100 \tag{1}$$

2.4.5. Chromatographic Analysis of the Carotenoids

The separation, identification, and quantification of the carotenoids from sea buckthorn extracts was performed using High-performance liquid chromatography (HPLC) as described by Mihalcea et al. [15] with slight modifications. Briefly, a Thermo Finnigan Surveyor HPLC system with Xcalibur software (Finnigan Surveyor LC, Thermo Scientific, Waltham, MA, USA) was used. The carotenoids were analyzed at 450 bn and 30 °C on a Lichrosorb RP-18 (5 lm) Hibar RT 125-4 column. The elution solvents used were 90% acetonitrile (A) and 100% ethyl acetate (B). Thus, 0.02 mL of extract was injected into the column and the flow was maintained at 0.500 mL/min. The identification and separation of the compounds were achieved based on their standard curves using ethyl acetate as solvent.

2.5. Microencapsulation of the Bioactive Compounds from Sea Buckthorn and Powder Characterization

Both extracts were encapsulated in a matrix formed by 2% WPI and 2% CMC, using the complex coacervation method, as described by Mihalcea et al. [15]. The wall materials were dissolved in distilled water in a 1:1 ratio (*w*/*w*) on a magnetic stirrer at 300 rpm and 25 °C, until fully hydrated. Prior to the addition, 5 g of the concentrated extract were mixed with 5 mL of sunflower seed oil and subjected to ultrasound for 1 h at 40 kHz. To begin the coacervation process, the pH of the solutions was adjusted to 3.75 with 1 N HCl solution, under constant mechanical stirring at 300 rpm. Subsequently, the reaction mixture was allowed to separate in a funnel. The separated phases were collected and stored at −20 °C for 24 h. Subsequently, the frozen coacervates were lyophilized (Alpha 1-4 LD plus, CHRIST, Osterode am Harz, Germany) at 42 °C, under a pressure of 10 Pa for 48 h. Finally, the powders were collected, packed in metallic bags, and stored in the refrigerator at 4 °C until further analysis.

2.5.1. Encapsulation Efficiency and Powder Characterization

To determine the carotenoids encapsulation efficiency, the slightly modified method of Mihalcea et al. [15] was used. For the total carotenoids, 100 mg of encapsulated powder was dissolved in 6 mL of 10% NaCl: methanol (1:1 ratio), followed by 30 min of rest to break the microcapsules. A volume of 30 mL of hexane was further added. The samples were treated with ultrasounds for 40 min at 50 °C and then centrifuged at 6000 rpm for 10 min. The TC of the supernatant were analysed. The LC, βC, TFC, TPC, and antioxidant activity were also quantified from the same supernatant, using the methods described above to characterize the powders.

The same procedure was used for the surface carotenoids. Approximately 100 mg of powder was weighed, and 30 mL of n-Hexane was added. This mixture was vortexed for 2 min and then centrifuged at 6000 rpm for 10 min at 10 °C. Finally, the resulting supernatant was analysed in terms of SC.

$$EE\ (\%) = (TC - SC)/TC \times 100 \tag{2}$$

EE—encapsulation efficiency, %;
TC—total carotenoids, mg/g dw;
SC—surface carotenoids, mg/g dw.

The encapsulated powders were evaluated regarding the encapsulation efficiency during storage at 4 °C in hermetically closed glass tubes with light protection after 180 days and 270 days of storage.

2.5.2. Colorimetric Analysis

For the color analysis of powders, a CR 410 Chroma Meter (Konica Minolta, Hino, Tokyo Japan) was used to appreciate the selected coordinates such as:
L*—clarity (=0 black)/(=100 white);
a*—shade of green (<0)/red (>0);
b*—shade of blue (<0)/yellow (>0).

2.5.3. Confocal Laser Scanning Microscopy (CLSM)

The CLSM technique has been used to observe the morphology and structure of the powders at a high resolution [15]. A Zeiss confocal laser system (LSM 710, (Carl Zeiss, Oberkohen, Germany) with a diode laser (405 nm), Ar-laser (458, 488, 514 nm), DPSS laser (diode-pumped solid-state e 561 nm), and HeNe-laser (633 nm) were used. The powders were stained with Red Congo (40 μM) and the distribution of the sea buckthorn bioactives into the matrices was observed using a Zeiss Axio Observer Z1 inverted microscope with a 40× apochromatic objective (numerical aperture 1.4). Moreover, for the analysis, the FS49, FS38, and FS15 filters (Carl Zeiss, Oberkohen, Germany) were used. Finally, ZEN 2012

SP1 software (black edition, (Carl Zeiss, Oberkohen, Germany) was used to analyse the 3D images.

2.5.4. In Vitro Digestion

The method described by Oancea et al. [16] has been used to evaluate the in vitro digestibility of the carotenoids from the obtained powders. The digestion was simulated using simulated gastric fluids (SGF) and intestinal fluids (SIF). Briefly, 200 mg of the powders mixed with 20 mL SGF (20 mg pepsin in 0.1 M HCl, pH 2.0) were incubated at 37 °C using an orbital shaking incubator (Medline Scientific, Oxon, UK) at 150 rpm, for 2 h. Afterward, in addition to the 15 mL of SGF which remained after digestion, 20 mL of SIF (40 mg pancreatin in 0.9 M baking soda, pH 7.8) were added and incubated for another 2 h at 150 rpm. Aliquots (1 mL) of each sample were taken every 30 min, diluted with 2 mL of n-hexane, and the absorbance was read at 450, 470, and 503 nm, respectively. Samples were prepared in duplicate (n = 2).

2.6. Statistical Analysis

The statistical analysis of the data was performed using Minitab 17 and the One-way ANOVA test. The experimental data were expressed as the mean of duplicate ± standard deviation.

3. Results and Discussion

3.1. Evaluation of the Phytochemical Profile of the Sea Buckthorn Extracts

The bioactive compounds from the freeze-dried sea buckthorn were extracted with different mixtures of solvents using the UAE method. Polar solvents, such as acetone and water, were used. Two extracts were obtained and the phytochemical contents of both are presented in Table 2.

Table 2. Phytochemical profile of the sea buckthorn extracts.

Phytochemical Content/Extracts	E1	E2
Lycopene content (mg LC/g dw)	1.69 ± 0.01 [a]	0.87 ± 0.01 [b]
β-carotene content (mg βC/g dw)	2.79 ± 0.02 [a]	1.38 ± 0.02 [b]
Total carotenoid content (mg TC/g dw)	3.30 ± 0.03 [a]	1.60 ± 0.01 [b]
Total flavonoid content (mg CE/g dw)	310.06 ± 6.01 [a]	127.80 ± 2.41 [b]
Total polyphenol content (mg GAE/g dw)	1023.50 ± 5.51 [a]	368.12 ± 5.37 [b]
ABTS antioxidant activity (% inhibition)	4.36 ± 0.01 [a]	3.81 ± 0.02 [b]
DPPH antioxidant activity (% inhibition)	92.00 ± 0.42 [a]	61.01 ± 0.42 [b]

For each phytochemical and extract tested, the values from the same row that do not share a lowercase letter are statistically different at $p < 0.05$ based on the Tukey method and 95% confidence.

The addition of acetone and glacial acetic acid into water led to a significant increase in the phytochemical content from sea buckthorn ($p < 0.05$) for both lipophilic and hydrophilic compounds (Table 2). This may be due to the stronger breaking of the cell walls. Thus, E1 presented a 1.69 ± 0.01 mg LC/g dw and E2 0.87 ± 0.01 mg LC/g dw lycopene content. In terms of β-carotene, E1 showed a 2.79 ± 0.02 mg βC/g dw and E2 1.38 ± 0.02 mg βC/g dw. Our results obtained are comparable to other studies. Ursache et al. [17] reported 38.34 ± 5.71 mg βC/g dw in the sea buckthorn extract.

Regarding the hydrophilic compounds, 310.06 ± 6.01 mg CE/g dw were found in E1 and 127.80 ± 2.41 mg CE/g dw in E2. The TPC of the extracts was 1023.50 ± 5.51 mg GAE/g dw and 368.12 ± 5.37 mg GAE/g dw for E1 and E2, respectively. Lower polyphenol contents in the sea buckthorn were reported by researchers like Korekar et al. [18], with values ranging from 9.64 to 107.04 mg GAE/g dw.

The antioxidant activity of the evaluated extracts was also different, the best results being obtained with acetic acid, acetone, and water for both free radicals. Thus, E1 presented a 4.36 ± 0.01 % inhibition against the ABTS free radical and 92.00 ± 0.42 % inhibition against DPPH. For E2, significantly lower inhibition results were highlighted ($p < 0.05$)

such as 3.81 ± 0.02 % inhibition against ABTS and 61.01 ± 0.42 % inhibition against DPPH. Ursache et al. [17] reported lower DPPH inhibition with values of 33.7 ± 0.29 % for sea buckthorn extracts.

The extraction solvent choice represents a decisive step in the extraction of biologically active compounds from sea buckthorn, from both contents of the bioactive compounds and antioxidant activity point of view. In our study, the combination of acetic acid, acetone, and water provided the extract with the highest content of biologically active compounds and antioxidant activity.

In order to characterize the carotenoid profile from the extracts obtained, the analysis of the chromatographic profile was performed by HPLC (Figure 2).

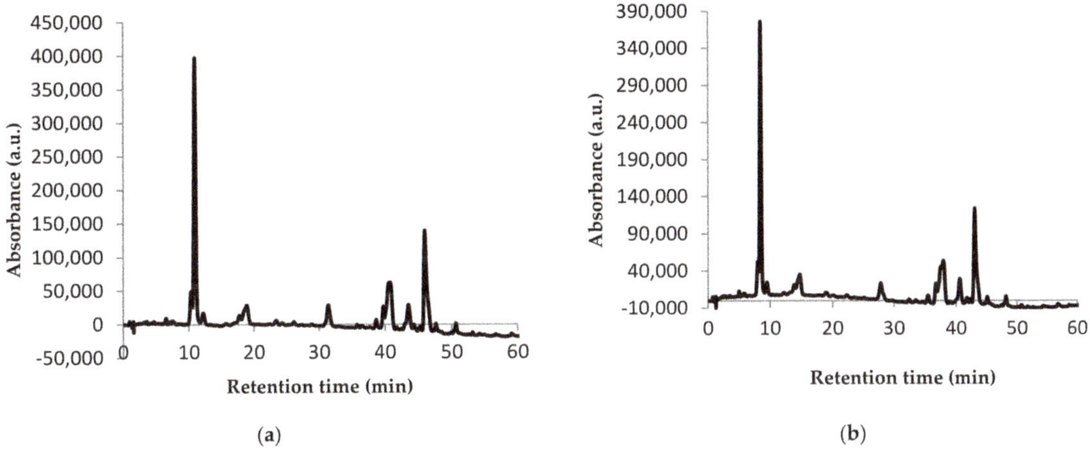

Figure 2. Chromatographic profile of carotenoids from sea buckthorn E1 (**a**) and E2 (**b**): peak 1—astaxanthin; peak 2—zeaxanthin; peak 3—β-cryptoxanthin; peak 4—lycopene; peak 5—β-carotene.

The HPLC technique allowed the identification of five compounds in both extracts (Figure 2). The difference between the chromatographic profiles of the two extracts consists in the amount of compounds quantified. Thus, E1 presented 0.81 mg/g dw astaxanthin, 5.11 mg/g dw zeaxanthin, 0.43 mg/g dw β-cryptoxanthin, 0.96 mg/g dw lycopene, and 2.71 mg/g dw β-carotene (Figure 2a). For E2, almost half of the amounts of E1 were quantified as follows: 0.56 mg/g dw astaxanthin, 4.74 mg/g dw zeaxanthin, 0.55 mg/g dw β-cryptoxanthin, 0.98 mg/g dw lycopene, and 1.31 mg/g dw β-carotene (Figure 2b). Our results are in agreement with Mihalcea et al. [15] and Ursache et al. [17], who also identified these compounds in sea buckthorn extracts. However, while Ursache et al. [17] reported zeaxanthin as the major compound from sea buckthorn extracts as we did, Mihalcea et al. [15] reported β-carotene and zeaxanthin palmitate as major compounds.

3.2. Evaluation of the Phytochemical Profile of the Powders

In this paper, the bioactive compounds from sea buckthorn were encapsulated by complex coacervation, using whey proteins and CMC. The choice of wall materials is justified by the fact that they are effective carriers for bioactive compounds, fats, oils, fatty acids, and flavors [19].

For both powders, the same amount of 0.5 g extract was used. For a better compound dispersion and taking into account their lipophilic profile, the extracts were mixed with sunflower oil. The characterization of the microencapsulated sea buckthorn extract was performed by quantifying the LC, βC, TC, TFC, TPC, and antioxidant activity. The results obtained from these experiments are shown in Table 3.

Table 3. Phytochemical profile of the powders obtained by the microencapsulation of the sea buckthorn extracts.

Phytochemical Content/Powders	P1	P2
Encapsulation efficiency (%)	52.20 ± 1.60 [a]	87.23 ± 0.05 [b]
Lycopene content (mg LC/g dw)	1.22 ± 0.01 [a]	0.86 ± 0.02 [b]
β-carotene content (mg βC/g dw)	0.63 ± 0.03 [a]	0.98 ± 0.05 [b]
Total carotenoid content (mg TC/g dw)	2.13 ± 0.03 [a]	1.14 ± 0.04 [b]
Total flavonoid content (mg CE/g dw)	219.86 ± 5.71 [a]	154.68 ± 10.36 [b]
Total polyphenol content (mg GAE/g dw)	529.84 ± 10.54 [a]	319.14 ± 7.37 [b]
ABTS antioxidant activity (% inhibition)	6.67 ± 0.38 [a]	7.09 ± 0.01 [b]
DPPH antioxidant activity (% inhibition)	93.05 ± 0.77 [a]	74.90 ± 0.99 [b]

For each phytochemical and powder tested, the values from the same row that do not share a lowercase letter are statistically different at $p < 0.05$ based on the Tukey method and 95% confidence.

The encapsulation efficiency is an important parameter that provides information about the delivery system. In our study, the extracts were encapsulated in CMC and WPI using the coacervation method, followed by freeze-drying to obtain stable powders. The encapsulation efficiency of P1 was significantly lower compared to P2 ($p < 0.05$). Our results are comparable to those reported by other researchers. Ursache et al. [17] reported an encapsulation efficiency of 56.16 ± 1.24% after the microencapsulation of sea buckthorn extract in WPI and acacia gum (1:1).

However, P1 presented the highest LC, TC, TFC, TPC, and DPPH antioxidant activity. The β-carotene content, on the other hand, was higher in P2, as the antioxidant activity against ABTS. Our results are in agreement with Ursache et al. [17] who reported values of 2.82 ± 0.17 mg/g dw for total carotenoids content after the microencapsulation of the sea buckthorn extract in a WPI and acacia gum matrix (1:1). Laos et al. [20] reported 11.12 mg/100 g of β-carotene content in sea buckthorn juice microspheres obtained by ionotropic gelation. Interesting results were also published by Neagu et al. [21]. They encapsulated the sea buckthorn extract obtained with supercritical fluids in WPI and casein using different encapsulation methods. They obtained values in the range of 302.98 ± 2.30 to 352.90 ± 1.02 mg/g dw for the total carotenoid contents. The results from our study demonstrate that the use of WPI and CMC as wall materials is suitable for the sea buckthorn phytochemical encapsulation.

The encapsulation efficiency over time can predict the stability of the compounds inside the encapsulation matrix. Table 4 shows the variation of encapsulation efficiency during 270 days of storage at 4 °C in the dark.

Table 4. Initial encapsulation efficiency of the powders obtained by the microencapsulation of the sea buckthorn extracts and its stability during 270 days of storage.

Powders	Initial Encapsulation Efficiency, (%)	Encapsulation Efficiency after 180 Days, (%)	Encapsulation Efficiency after 270 Days, (%)
P1	52.20 ± 1.60 [a]	49.20 ± 0.02 [b]	48.88 ± 0.01 [b]
P2	87.23 ± 0.05 [a]	80.45 ± 0.03 [b]	76.43 ± 0.01 [c]

For each storage time and powder tested, the values from the same row that do not share a lowercase letter are statistically different at $p < 0.05$ based on the Tukey method and 95% confidence.

A slight decrease over time in the encapsulation efficiencies can be observed for both powders (Table 4). Thus, the encapsulation efficiency of P1 is 7% lower after 270 days of storage than the initial encapsulation efficiency. P2 showed a 12% decrease in the encapsulation efficiency after 270 days of storage.

3.3. Colorimetric Analysis of the Powders and Its Stability

The results of colorimetric analysis conducted with the powders are reported in Table 5. The lightness (L*), the redness (a*), and the yellowness (b*) were the parameters analysed during 270 days of storage.

Table 5. Colorimetric parameters of the powders obtained by the microencapsulation of the sea buckthorn extracts and its stability during 270 days of storage.

Storage Period, Days	P1 Colorimetric Parameters			P2 Colorimetric Parameters		
	L*	a*	b*	L*	a*	b*
0	80.90 ± 2.11 [a]	7.26 ± 0.011 [a]	32.39 ± 1.50 [a]	86.37 ± 0.12 [a]	7.67 ± 0.21 [a]	38.33 ± 1.91 [a]
180	89.41 ± 5.25 [b]	8.18 ± 0.21 [b]	30.20 ± 0.49 [b]	90.80 ± 3.41 [b]	9.11 ± 0.34 [b]	36.70 ± 0.44 [b]
270	109.61 ± 6.20 [c]	8.21 ± 0.57 [b]	29.13 ± 2.11 [c]	95.21 ± 1.21 [c]	10.62 ± 1.60 [c]	29.60 ± 0.73 [c]

For each storage time, colorimetric parameter, and powder tested, the values from the same column that do not share a lowercase letter are statistically different at $p < 0.05$ based on the Tukey method and 95% confidence. L*—lightness, a*—redness, b*—yellowness.

The b* values presented in Table 5 suggest a high initial yellowness for both powders due to the carotenoids and flavonoids from the extract, but higher for P2. Our results are in agreement with those of Ursache et al. [17], who also reported a high yellowness for the powders obtained by the microencapsulation of the sea buckthorn extract in WPI and acacia gum. The color measurements over time show a significantly decrease for the yellowness parameter ($p < 0.05$), which can be correlated with the encapsulation efficiency variation over the same time. This may be due to the release of the compounds from the matrices which led to degradation and lower color intensities.

3.4. The Confocal Laser Scanning Microscopy (CLSM)

The biochemical complexity of the sea buckthorn fruits has been proven by numerous studies. The remarkable content of carotenoids (β-carotene, lycopene, lutein, zeaxanthin), vitamins (A, B, C, E, K), flavonoids (quercetin, kaempferol, catechin, isorhamnetin, etc.), and mono- and polyunsaturated fatty acids recommend this fruit as a food-medicine [17]. Depending on the variety, the climatic conditions, the time of harvest but also on the extraction method used, the proportion of biologically active compounds can differ considerably [22]. The confocal analysis of the two samples (Figures 3 and 4) aimed at clarifying the morphological and structural aspect of the powders depending on the extraction method. For the excitation of the samples, the following lasers were used: Ar laser (458, 488, and 514 nm) and DPSS (561 nm pumped solid-state diodes) because the data from the literature indicate for the carotenoids' absorption range the wavelengths of 448, 476, and 505 nm [23,24]. The studies by Llansola-Portoles et al. [25], conducted on an in vitro carotenoid assembly, recommend the excitation at 532 nm to obtain the full set of transient absorption signals. The complex aggregates of carotenoids and flavonoids microencapsulated in the polymer matrices showed an autofluorescence in a fairly wide range 530–630 nm, and sometimes it could be observed that the spectrum shifts to red (685–750 nm) upon the aggregation [23,25]. The carbohydrate biopolymers that have been introduced into the microencapsulating matrices displayed the maximum absorbance between 300 and 400 nm which have been excited by the diode laser (405 nm).

The images acquired by point-by-point laser scanning (Figure 3) revealed a scalariform appearance of the microcapsules. The fine biofilms (in blue or green depending on the ratio between the WPI and the polysaccharide biopolymer) were formed. Within the biopolymer matrix, several plant pigments can be seen (in yellow-orange) microspherosomes (1–2μm) that are anchored. The interaction between the vegetal pigments from sea buckthorn fruits with the polysaccharide compounds from the microencapsules did not cause modifications on the absorbance pattern of the spectrum. The strong linkages are usually established between carotenoids and polyphenols, and these preserve the UV absorption capability of the pigments in time [26].

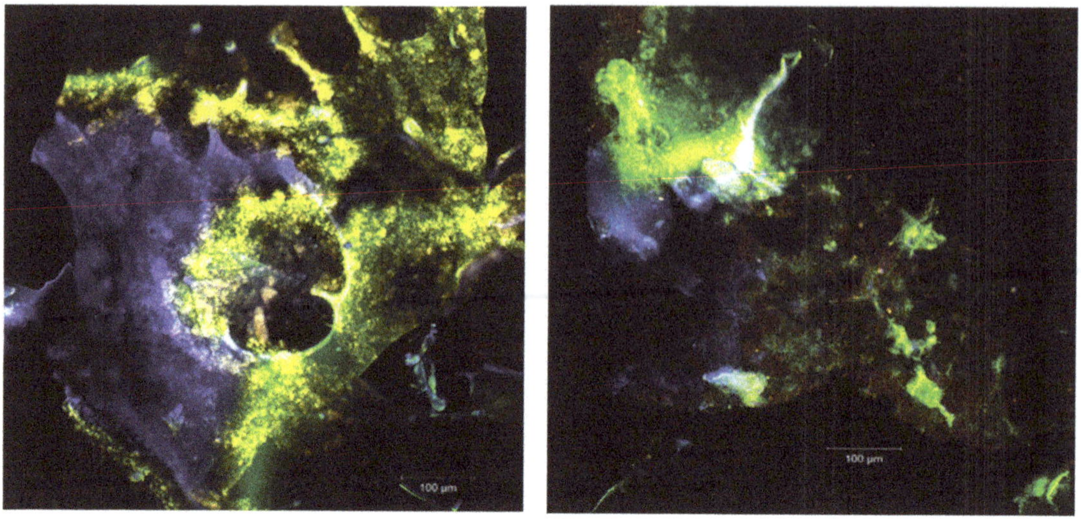

Figure 3. The confocal laser scanning microscopy (CLSM) images of the unstained native powders.

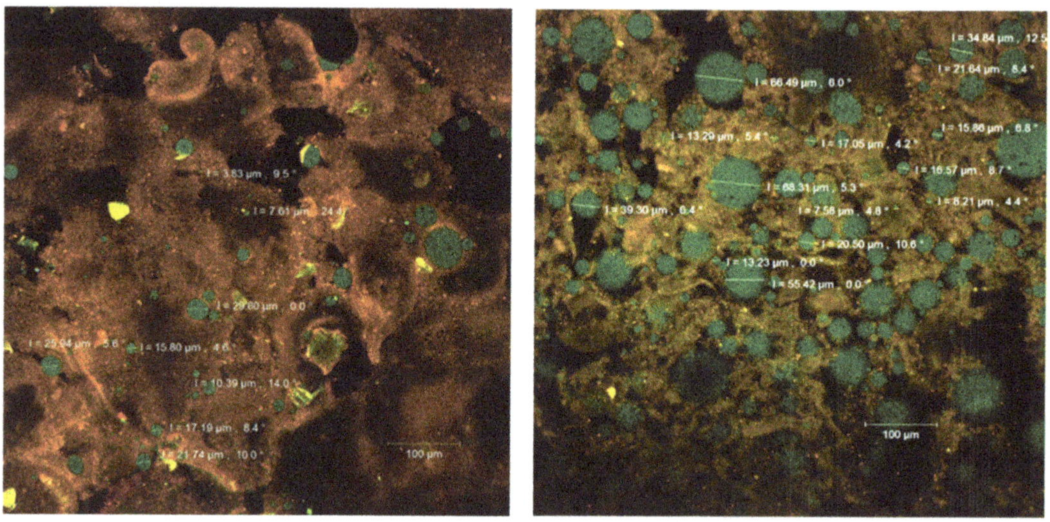

Figure 4. The confocal laser scanning microscopy (CLSM) images of the fluorophore dyed powders.

The confocal analysis of the samples after the fluorescent staining with Red Congo revealed the biologically active compounds from sea buckthorn fruit extracts in the form of spherosomes (in green) with different diameters between 3.83 and 215.78 µm. The Congo red dye binds the intact β-d-glucans from the biopolymer matrices [27] as well as the peptides from the WPI, marking them in orange (Figure 4). Small spherosomes (<25 µm) are visualized in P1, while in P2 they are of medium size. P2 displayed a large number of spherosomes of medium size, well individualized, with a tendency to group in clusters and well stabilized in the WPI and CMC matrix. Similar results were previously obtained by Mihalcea et al. [15], who use WPI and acacia gum as the encapsulation material, and by Neagu et al. [21], who compared the native sea buckthorn extract with the cross-linked one mediated by transglutaminase.

3.5. In Vitro Digestion

The in vitro digestibility of the encapsulated powders was also studied. Figures 5 and 6 represent the release percentage of TC, βC, and LC from both powders at the gastric and intestinal level.

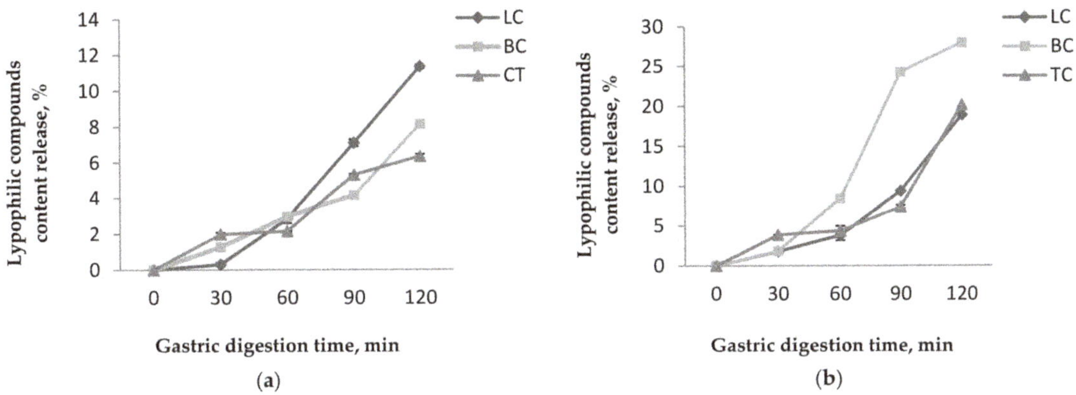

Figure 5. Lypophilic compounds content release from P1 (**a**) and P2 (**b**) during the in vitro gastric digestion.

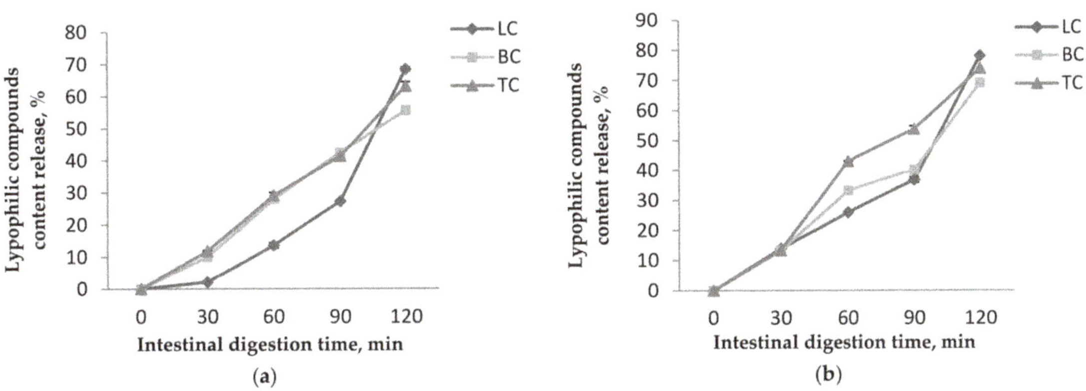

Figure 6. Lypophilic compounds content release from P1 (**a**) and P2 (**b**) during the in vitro intestinal digestion.

The results obtained for the in vitro digestibility in simulated gastric conditions showed that the coating materials presented a protective effect on the lipophilic compounds, especially in P1. However, a slight release of the lipophilic compounds was observed for both powders. Among the investigated lipophilic compounds, in P1, the lycopene content presented the highest release from all compounds up to 11.33 ± 0.11% after 120 min of digestion (Figure 5a). The β-carotene and total carotenoids released up to 8.12 ± 0.1% and 6.33 ± 0.12%, respectively after 120 min. In P2, the β-carotene presented the highest release from all compounds up to 27.97 ± 0.26% after 120 min (Figure 5b). The lycopene and total carotenoids were released up to 18.90 ± 0.14% and 20.17 ± 0.20% after 120 min. For all that, P2 showed the highest release percentages for all lipophilic compounds.

In simulated intestinal conditions, the results show that the maximum amount of lipophilic compounds is released during 120 min of digestion and are recorded for P2 (Figure 6). Thus, in P1 release, values of 55.59 ± 0.99%, 63.14 ± 1.3%, and 68.36 ± 0.62% were achieved for βC, TC, and LC, respectively, after 120 min (Figure 6a). P2 presented higher release values, i.e., 68.98 ± 0.41 % of the βC, 73.86 ± 1.03% of the TC, and 78.01 ± 0.91% of the LC released after 120 min of digestion (Figure 6b). Our results are in agreement with other studies. In research conducted by Neagu et al. [21], a 10% release of the carotenoids in gastric juice was reported for the microencapsulated sea buckthorn extract in WPI and casein matrices. They also reported that, in simulated intestinal juice, the release increased significantly up to 65%.

4. Conclusions

This study aimed to increase the stability and bioavailability of sea buckthorn fruit's bioactives. For this purpose, the biologically active compounds from *Hippophae rhamnoides* L. fruits have been extracted using the ultrasound-assisted extraction method. Different polar solvents have been used to extract as many bioactive compounds as possible. On one side, a combination between glacial acetic acid, acetone, and water was used (E1), with water only on another side (E2). By adding acetic acid and acetone into the water, the highest phytochemical content was obtained. The chromatographic profile of both extracts revealed the presence of five compounds, with zeaxanthin being the major one. Further, to increase the bioactive's stability and bioavailability, both extracts were encapsulated in a matrix composed of CMC and WPI. The coacervation technique was used, followed by freeze-drying. Different analyses have been made to test the functionality of the ingredients. The phytochemicals characterization revealed that both powders presented high phytochemicals content. Although the powder with E2 presented the highest encapsulation efficiency, the one with E1 showed the highest phytochemical content and encapsulation efficiency stability. P2 highlighted the highest amount of yellowness, being correlated with the high content of phytochemicals. The CLSM images revealed that the coacervation led to the scalariform appearance of the microcapsules. The bioactives from sea buckthorn are observable as yellow-orange microspherosomes. Small spherosomes were visualized in P1, while in P2, they were of medium size. Both powders had a substantially higher release of the biologically active substances in the simulated intestinal juice, but the best transmission of the target substances has P2. This research suggests that, despite the differences between the powders, both could be used in food applications as targeted delivery vehicles for carotenoids and polyphenols as bioactive ingredients.

Author Contributions: Conceptualization, D.R., N.N.C. and G.R.; methodology, D.R.; software, E.E. and V.B.; validation, E.E, V.B. and I.A.; formal analysis, D.R.; investigation, D.R.; resources, G.E.B.; data curation G.R. and N.S.; writing—original draft preparation D.R. and N.N.C.; writing—review and editing, G.R. and N.N.C.; visualization, N.S.; supervision, G.E.B. All authors have read and agreed to the published version of the manuscript.

Funding: This research received no external funding.

Institutional Review Board Statement: Not applicable.

Informed Consent Statement: Not applicable.

Data Availability Statement: The data that support the findings of this study are available from the corresponding author (G.R.) upon reasonable request.

Acknowledgments: The results of this work have been presented to the 9th edition of the Scientific Conference organized by the Doctoral Schools of "Dunărea de Jos" University of Galati (http://www.cssd-udjg.ugal.ro/ (accessed on 5 August 2021)), that was held on 10–11 June 2021 in Galati, Romania.

Conflicts of Interest: The authors declare no conflict of interest.

Abbreviations

ABTS	2,2′-azino-bis (3-ethylbenzothiazoline-6-sulfonic acid) diammonium salt
$AlCl_3$	aluminum chloride
CE	Catechin Equivalents
CLSM	Confocal Laser Scanning Microscopy

CMC	Carboxymethil cellulose
DPPH	2,2-diphenyl1-picrylhydrazyl
dw	dry weight
fr	fresh weight
GAE	Gallic Acid Equivalents
HCl	hydrochloric acid
HPLC	High Performance Liquid Chromatography
LC	Lycopene Content
Na_2CO_3	sodium carbonate
NaCl	sodiul chloride
$NaNO_2$	sodium nitrite
NaOH	sodium hydroxide
nd	not detected
P	Powder
rpm	rotations per minute
SGF	Simulated Gastric Fluid
SIF	Simulated Intestinal Fluid
TC	Total Carotenoid Content
TFC	Total Flavonoid Content
TPC	Total Polyphenol Content
UAE	Ultrasounds Assisted Extraction
E	Extract
WPI	Whey Protein Isolate
βC	Beta carotene Content

References

1. Olas, B. Sea buckthorn as a source of important bioactive compounds in cardiovascular diseases. *Food Chem. Toxicol.* **2016**, *97*, 199–204. [CrossRef] [PubMed]
2. Shutterstock. Available online: https://www.shutterstock.com/ro/image-photo/sea-buckthorn-fresh-ripe-berries-leaves-604664045 (accessed on 13 September 2021).
3. de Freitas Santos, P.D.; Rubio, F.T.V.; da Silva, M.P.; Pinho, L.S.; Favaro-Trindade, C.S. Microencapsulation of carotenoid-rich materials: A review. *Food Res. Int.* **2021**, *147*, 110571. [CrossRef] [PubMed]
4. Ciesarová, Z.; Murkovic, M.; Cejpek, K.; Kreps, K.; Tobolková, B.; Koplík, R.; Belajová, E.; Kukurová, K.; Daško, L.; Panovská, Z.; et al. Why is sea buckthorn (*Hippophae rhamnoides* L.) so exceptional? A review. *Food Res. Int.* **2020**, *133*, 109170. [CrossRef] [PubMed]
5. Xu, Y.; Kaur, M.; Dhillon, R.S.; Tappia, P.S.; Dhalla, N.S. Health benefits of sea buckthorn for the prevention of cardiovascular diseases. *J. Funct. Foods* **2011**, *3*, 2–12. [CrossRef]
6. Saini, R.K.; Keum, Y.S. Carotenoid extraction methods: A review of recent developments. *Food Chem.* **2018**, *240*, 90–103. [CrossRef]
7. Celli, G.B.; Ghanem, A.; Brooks, M.S.L. Bioactive Encapsulated Powders for Functional Foods—A Review of Methods and Current Limitations. *Food Bioproc. Tech.* **2015**, *8*, 1825–1837. [CrossRef]
8. Taneja, A.; Harjinder, S. Challenges for the delivery of long-chain n-3 fatty acids in functional foods. *Annu. Rev. Food Sci. Technol.* **2012**, *3*, 105–123. [CrossRef]
9. de Souza Simões, L.; Madalena, D.A.; Pinheiro, A.C.; Teixeira, J.A.; Vicente, A.A.; Ramos, O.L. Micro- and nano bio-based delivery systems for food applications: In vitro behavior. *Adv. Colloid Interface Sci.* **2017**, *243*, 23–45. [CrossRef]
10. Turgeon, S.L.; Schmitt, C.; Sanchez, C. Protein-polysaccharide complexes and coacervates. *Curr. Opin. Colloid Interface Sci.* **2007**, *12*, 63–70. [CrossRef]
11. Ye, Q.; Georges, N.; Selomulya, C. Microencapsulation of active ingredients in functional foods: From research stage to commercial food products. *Trends Food Sci. Technol.* **2018**, *78*, 167–179. [CrossRef]
12. Rodriguez-Amaya, D.B.; Kimura, M. *HarvestPlus Handbook for Carotenoid Analysis*; HarvestPlus: Washington, DC, USA, 2004; pp. 8–19.
13. Turturică, M.; Râpeanu, G.; Stănciuc, N.; Bahrim, G. Fluorescence spectroscopy investigation on pH and heat changes of cherries anthocyanin extracts. *J. Biotechnol.* **2015**, *208*, S68. [CrossRef]
14. Re, R.; Pellegrini, N.; Proteggente, A.; Pannala, A.; Yang, M.; Rice-Evans, C. Antioxidant activity applying an improved ABTS radical cation decolorization assay. *Free Radic. Biol. Med.* **1999**, *26*, 1231–1237. [CrossRef]
15. Mihalcea, L.; Turturica, M.; Barbu, V.; Ionita, E.; Patrascu, L.; Cotarlet, M.; Dumitrascu, L.; Aprodu, I.; Rapeanu, G.; Stanciuc, N. Transglutaminase mediated microencapsulation of sea buckthorn supercritical CO_2 extract in whey protein isolate and valorization in highly value added food products. *Food Chem.* **2018**, *262*, 30–38. [CrossRef]
16. Oancea, A.M.; Hasan, M.; Vasile, A.M.; Barbu, V.; Enachi, E.; Bahrim, G.; Rapeanu, G.; Silvi, S.; Stanciuc, N. Functional evaluation of microencapsulated anthocyanins from sour cherries skins extract in whey proteins isolate. *LWT* **2018**, *95*, 129–134. [CrossRef]

17. Ursache, F.M.; Ghinea, I.O.; Turturica, M.; Aprodu, I.; Rapeanu, G.; Stanciuc, N. Phytochemicals content and antioxidant properties of sea buckthorn (*Hippophae rhamnoides* L.) as affected by heat treatment–Quantitative spectroscopic and kinetic approaches. *Food Chem.* **2017**, *233*, 442–449. [CrossRef] [PubMed]
18. Korekar, G.; Dolkar, P.; Singh, H.; Srivastava, R.B.; Stobdan, T. Variability and the genotypic effect on antioxidant activity, total phenolics, carotenoids and ascorbic acid content in seventeen natural population of Seabuckthorn (*Hippophae rhamnoides* L.) from trans-Himalaya. *LWT* **2014**, *55*, 157–162. [CrossRef]
19. Zhang, R.; Belwal, T.; Li, L.; Lin, X.; Xu, Y.; Luo, Z. Recent advances in polysaccharides stabilized emulsions for encapsulation and delivery of bioactive food ingredients: A review. *Carbohydr. Polym.* **2020**, *242*, 219495666. [CrossRef] [PubMed]
20. Laos, K.; Lõugas, T.; Mändmets, A.; Vokk, R. Encapsulation of β-carotene from sea buckthorn (*Hippophaë rhamnoides* L.) juice in furcellaran beads. *Innov. Food Sci. Emerg. Technol.* **2007**, *3*, 395–398. [CrossRef]
21. Neagu, C.; Mihalcea, L.; Enachi, E.; Barbu, V.; Borda, D.; Bahrim, G.E.; Stanciunc, N. Cross-linked microencapsulation of CO_2 supercritical extracted oleoresins from sea buckthorn: Evidence of targeted functionality and stability. *Molecules* **2020**, *10*, 2442. [CrossRef] [PubMed]
22. Topală, C.M.; Mazilu, I.C.; Vulpe, M.; Vîjan, L.E. Quality study of fruits and extracts from six romanian sea buckthorn varieties. *Curr. Trends Nat. Sci.* **2020**, *9*, 273–283. [CrossRef]
23. Katoh, T.; Nagashima, U.; Mimuro, M. Fluorescence properties of the allenic carotenoid fucoxanthin: Implication for energy transfer in photosynthetic pigment systems. *Photosyn. Res.* **1991**, *27*, 221–226. [CrossRef]
24. Marinova, D.; Ribarova, F. HPLC determination of carotenoids in Bulgarian berries. *J. Food Compost. Anal.* **2007**, *20*, 370–374. [CrossRef]
25. Llansola-Portoles, M.J.; Redeckas, K.; Streckaité, S.; Ilioaia, C.; Pascal, A.A.; Telfer, A.; Vengris, M.; Valkunasde, L.; Robert, B. Lycopene crystalloids exhibit singlet exciton fission in tomatoes. *Phys. Chem. Chem. Phys.* **2018**, *20*, 8640–8646. [CrossRef]
26. Salgado, L.T.; Tomazetto, R.; Cinelli, L.P.; Farina, M.; Amado Filho, G.M. The influence of brown algae alginates on phenolic compounds capability of ultraviolet radiation absorption in vitro. *Braz. J. Oceanogr.* **2007**, *55*, 145–154. [CrossRef]
27. McDonald, J.E.; Rooks, D.J.; McCarthy, A.J. Chapter nineteen-Methods for the Isolation of Cellulose-Degrading Microorganisms. In *Methods in Enzymology*; Gilbert, H.J., Ed.; Academic Press: London, UK, 2012; Volume 510, pp. 349–374. [CrossRef]

inventions

MDPI

Article

Determination of Atorvastatin with Voltammetric Sensors Based on Nanomaterials

Ramona Oana Gunache (Roșca), Alexandra Virginia Bounegru and Constantin Apetrei *

Department of Chemistry, Physics and Environment, Faculty of Sciences and Environment, "Dunarea de Jos" University of Galati, 47 Domneasca Street, 800008 Galati, Romania; oana.gunache@ugal.ro (R.O.G.); alexandra.meresescu@ugal.ro (A.V.B.)
* Correspondence: apetreic@ugal.ro; Tel.: +40-727-580-914

Abstract: This paper presents an accurate and fast electrochemical method for atorvastatin determination in pharmaceutical products. Two screen-printed sensors, one—carbon based (SPCE) and one based on carbon nanotubes and gold nanoparticles (AuNP-CNT/SPCE) were used during the electrochemical analyses. At all experimental stages, cyclic voltammetry was employed, both for the characterization of the sensors and their electrochemical behavior, and for quantitative determinations. AuNP-CNT/SPCE has showed an extended active area, higher intensity peaks, better reversibility and lower background current than the unmodified sensor. For atorvastatin quantification, a calibration curve has been developed within the 1.2–606.25 μM concentration range. A linearity relation between the current of the anodic peak and concentration has been obtained in the range 1.2–53.33 μM for both sensors. With the AuNP-CNT/SPCE sensor, low values of limit of detection, LOD (1.92×10^{-7} M) and limit of quantification, LOQ (6.39×10^{-7} M) have been obtained, which demonstrates the feasibility of the method of determining atorvastatin from real samples. Atorvastatin amount has been successfully determined from pharmaceutical products using AuNP-CNT/SPCE. The results were similar to the manufacturer's specifications regarding the dosage per tablet and to the concentrations obtained by applying the FTIR spectrometric method.

Keywords: atorvastatin; sensor; cyclic voltammetry; carbon nanotubes; gold nanoparticles

Citation: Gunache (Roșca), R.O.; Bounegru, A.V.; Apetrei, C. Determination of Atorvastatin with Voltammetric Sensors Based on Nanomaterials. *Inventions* **2021**, *6*, 57. https://doi.org/10.3390/inventions6030057

Academic Editors: Eugen Rusu and Gabriela Rapeanu

Received: 20 July 2021
Accepted: 11 August 2021
Published: 12 August 2021

1. Introduction

Hyperlipidemia is a chronic condition manifested through the increase of the triglyceride levels in blood. The most effective and widely used medication for the treatment of hyperlipidemia is statins. Atorvastatin falls in this category, being an inhibitor of HMG-CoA reductase (an enzyme indispensable for cholesterol biosynthesis in liver) with significant anti-hypercholesterolemic effects [1]. Structurally, atorvastatin is an aromatic amide, a monofluorobenzene derivative, a dihydroxy monocarboxylic acid and a pyrrole derivative. Atorvastatin is also a derivative of heptanoic acid [2] (Scheme 1).

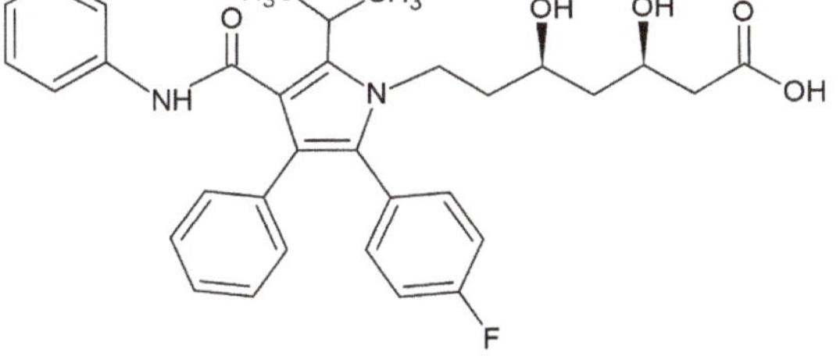

Scheme 1. Chemical structure of atorvastatin.

This compound reduces the cholesterol quantity in blood, lowering the risk of cardiovascular diseases and cerebrovascular accidents. It is generally administered in the form of atorvastatin calcium trihydrate, in doses of 10 to 80 mg per day. As any drug, atorvastatin may have some side effects, such as muscular or hepatic problems or increased glycemia [3]. Consequently, monitoring the atorvastatin concentration in pharmaceutical formulation may have an extremely significant role in adjusting the administered dose [4].

Up until now, researchers have applied more techniques for the analytical determination of atorvastatin both in drugs and biological samples, such as: liquid chromatography [5,6], high performance liquid chromatography [7,8], spectrophotometry [9,10], but also electrochemical techniques [11–13]. Even though the chromatographic and spectrophotometric methods provide precise results, they require using expensive instruments and devices, and additional stages of sample pretreatment, which increase both the cost and time of the analysis. On the other hand, the electrochemical methods require the use of electrodes modified with various nanomaterials, which present multiple advantages, such as short response time, selectivity, sensitivity, precision and, above all, the possibility of miniaturizing [14–18].

Among nanomaterials, carbon nanotubes (CNT) are frequently used for the modification of screen-printed sensors, owing to their mechanical properties, remarkable electrical conductivity, optimal surface, high chemical resistance and stability [19,20].

In addition, in the last few years, metallic nanoparticles have drawn the scientific community's attention with their chemical and physical particularities, which make them suitable for electrochemical sensors modification [21,22]. Special attention was granted to gold nanoparticles (AuNP) because of their good biological compatibility, excellent conductibility and high surface-volume ratio. Modifying the sensor surface with AuNP has increased the value of electrochemical tests [23,24]. The combination of CNT with AuNP with a view to modifying the screen-printed sensors may positively influence the detection process, favoring the transfer of electrons and selectivity [25].

This study aims at evaluating the electrochemical behavior of two screen-printed sensors, the former, carbon-based (SPCE) and the latter, based on gold-particles modified carbon nanotubes (AuNP-CNT/SPCE) and the qualitative and quantitative determination of atorvastatin in a pharmaceutical product. The AuNP-CNT/SPCE is different as design, materials and detection technique comparing with other sensors reported in literature [26]. The electrochemical method applied for detection will be cyclic voltammetry. The quantitative results obtained with the AuNP-CNT/SPCE sensor using voltammetry will be subsequently compared with the ones obtained using the FTIR method.

The novelty of this study is the development of a novel electroanalytical method based on AuNP-CNT/SPCE sensor for the sensitive and accurate determination of atorvastatin in real samples.

2. Materials and Methods

2.1. Reactives and Solutions

All reactives have been used without additional purification, as they presented analytical purity. The studies on the characterization of electrodes and optimization of the experimental parameters were conducted in potassium chloride 10^{-1} M and potassium ferrocyanide 10^{-3} M, the compounds having been purchased from Sigma—Aldrich (St. Louis, MO, USA). All solutions were prepared with MilliQ ultrapure water (resistivity 18.2 MΩ $\times$ cm) obtained from a Milli-Q Simplicity® water purification system. For the atorvastatin detection studies, real samples, purchased from local pharmacies, were used.

2.2. Electrochemical Measurements

All electrochemical determinations using cyclic voltammetry were conducted using an EG&G potentiostat/galvanostat (Princeton Applied Research, Oak Ridge, TN, USA) controlled by a Windows-operated computer with an ECHEM software, used for characterizing the electrodes signals and for atorvastatin electroanalysis. A classical three-electrode system was employed. The working electrode is the carbon screen-printed electrode (SPCE) or the one modified with carbon nanotubes and gold particles (AuNP-CNT/SPCE) from Metrohm-Dropsens. The reference electrode was the silver/silver chloride electrode, and the counter electrode was a platinum wire. The working electrode unmodified and modified with nanomaterials have a surface of 0.1256 cm^2. The electrochemical measurements are repeatable, the sensors could be used more than 100 measurements without losing their sensitivity. All potentials indicated in this study refer to the Ag/AgCl reference electrode.

The electrochemical method employed in all experimental analyses was cyclic voltammetry. In this study, for the stabilization of the sensors' electrochemical signal, 6 successive cycles with a 0.1 V/s scan rate were recorded using cyclic voltammetry. After that, 2 successive cycles at the same scan rate were also recorded, and the second cycle was saved and it is considered the stable signal of the sensor in the tested sample solution.

The FTIR spectra were obtained using a Bruker ALPHA FT-IR spectrometer (BrukerOptik GmbH, Ettlingen, Germany) which uses the OPUS software (BrukekrOptik GmbH, Ettlingen, Germany) within the 4000–500 cm^{-1} range (32 scans, resolution 4 cm^{-1}) in the attenuated total reflection (ATR) mode. Between measurements, the ATR ZnSe crystal was cleaned with ultrapure water and isopropanol. The spectra were recorded towards the air as background.

2.3. *Analysis of Phramaceutical Samples*

The sensors' applicability was studied by atorvastatin detection analysis in PRF (i.e., medical prescription withheld in the pharmacy) pharmaceutical products (Sortis 20 mg, Sortis 10 mg). Sortis is one of the medicaments known as statins, which act towards lipids (fats) regulation in the body, being used for lowering the concentrations of cholesterol and triglycerides in blood.

The stock solution of atorvastatin in the form of atorvastatin calcium was obtained from pharmaceutical product Sortis. Sortis 20 mg is a filmed tablet whose core contains atorvastatin calcium trihydrate, alongside other excipients, such as lactose monohydrate (54.50 mg/cp) and benzoic acid (0.00008 mg/cp), calcium carbonate, microcrystalline cellulose, sodium croscarmellose, polysorbate 80, hydroxypropyl cellulose, magnesium stearate.

For the preparation of the stock solution, more tablets of pharmaceutical product were triturated and dissolved in water. Prior to trituration, the coat on the tablet surface had been removed by rinsing with double distilled water. After ultrasonication, the respective mixture was filtered, and the effluent collected. Then, the water was evaporated, obtaining a solid powder, which was further characterized using FTIR spectrometry. The FTIR spectrum was similar to the one reported in the literature for atorvastatin calcium [23]. The required quantity was dissolved in KCl solution of 0.1 M concentration, resulting a stock solution of atorvastatin with 6×10^{-4} M concentration. The solution was ultrasonicated using the Elma S10H Elmasonic device for dissolution and homogenization. The stock solution obtained was appropriately diluted and then introduced into the electrochemical cell and the three electrodes (working, reference and counter electrode) were connected.

In the validation studies tablets of Sortis 20 mg and Sortis 10 mg were used. The pharmaceutical samples were analyzed in triplicate.

3. Results and Discussion

3.1. *Study of the Electrochemical Behavior of Sensors in KCl Solution and Potassium Ferrocyanide/KCl Solution*

The purpose of the preliminary analyses was the evaluation of the electrochemical behavior of two screen-printed sensors, the former, screen-printed carbon electrode (SPCE) and the latter, based on carbon modified with carbon nanotubes and gold nanoparticles (AuNP-CNT/SPCE).

During the first stage of analyses, an electrochemically inactive KCl 0.1 M was used with a view to evaluate the active surface of the sensors. The potential range was optimized, and the stable signal was obtained in the range from −0.4 to +0.7 V. No peaks were highlighted in the recorded cyclic voltammograms, which could have suggested a possible contamination of the sensor surface. The sensors present low background currents, which guarantees their suitability for further experiments.

In the next stage, the SPCE and AuNP-CNT/SPCE electrodes were immersed in a solution with electrochemical activity, obtained from potassium ferrocyanide 10^{-3} M and KCl 10^{-1} M, with a view to analyzing the electrochemical behavior of the sensors. The cyclic voltammograms were recorded at a scan rate of 0.1 V/s, in the potential range between −0.4 V and +0.7 V. This range was proven stable and optimal for potassium ferrocyanide/KCl solution too, having been applied in other studies [27].

Figure 1 presents the cyclic voltammograms of the SPCE and AuNP-CNT/SPCE sensors in potassium ferrocyanide 10^{-3} M-KCl 10^{-1} M solution recorded at a 0.1 V/s scan rate.

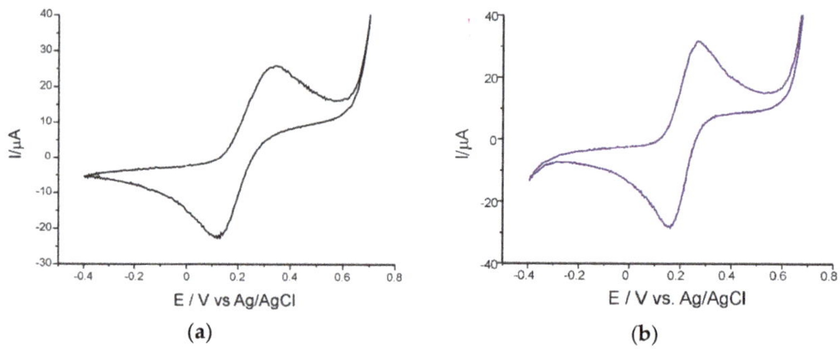

Figure 1. Cyclic voltammograms of the SPCE (**a**) and AuNP-CNT/SPCE (**b**) electrodes in 10^{-3} M of $K_4[Fe(CN)_6]$ and 10^{-1} M KCl solution. Scan rate 0.1 V/s.

As seen in Figure 1, in both cases, the redox of the potassium ferrocyanide ion which takes place at the sensor surface is obvious, being observed an anodic and a cathodic peak.

The parameters of interest obtained, and the ones calculated following the voltammetric analysis are presented in Table 1.

Table 1. Electrochemical parameters obtained from cyclic voltammograms of the sensors immersed in $K_4[Fe(CN)_6]$ 10^{-3} M-KCl 10^{-1} M solution.

Sensor	E_{pa} (V)	E_{pc} (V)	$E^{1/2}$ (V)	E (V)	I_a (μA)	I_c (μA)	I_c/I_a
SPCE	0.334	0.118	0.226	0.216	25.95	−21.94	0.84
AuNP-CNT/SPCE	0.272	0.156	0.214	0.116	31.31	−28.02	0.89

Abbreviations: E_{pa}—potential of the anodic peak; E_{pc}—potential of the cathodic peak; I_a—current of the anodic peak; I_c—current of the cathodic peak; $E^{1/2}$—half wave potential, $E = E_{pa} - E_{pc}$.

In the case of AuNP-CNT/SPCE, the peaks are better defined, of higher intensity and better reversibility (E and $E^{1/2}$ have lower values). By assessing the values of the principal parameters in the table, one can contend that the redox process of the ferrocyanide ion is quasi-irreversible in both cases, and that the sensors present good sensitivity.

In order to calculate the active area of the sensors, the cyclic voltammograms have been recorded in 10^{-3} M potassium ferrocyanide—10^{-1} M potassium chloride solution at various scan rates between the 0.05 and 1.0 V/s. The results obtained in the case of AuNP-CNT/SPCE are displayed in Figure 2.

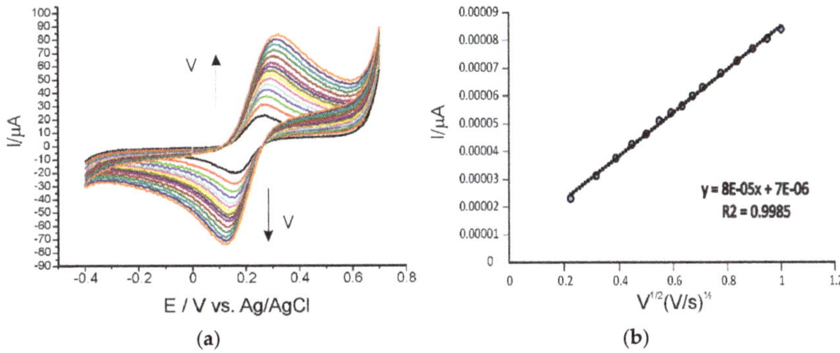

Figure 2. (**a**) Cyclic voltammograms of the AuNP-CNT/SPCE electrode in solution 10^{-3} M $K_4[Fe(CN)_6]$ -10^{-1} M KCl at various scan rates (0.05–1.0 V/s) (**b**) Dependence I_{pa} vs. square root of the scan rate.

After recording the cyclic voltammograms, one notes that the intensity of currents corresponding to the redox process of potassium ferrocyanide increases with the scan rate. By developing the linear regression between the intensity of the anodic current and scan rate, respectively, square root of the scan rate, one can determine the stage determining the rate of the oxidation process.

Figure 2b shows good linearity between I_{pa} and square root of the scan rate, with a determination coefficient (R^2) of 0.9985. This result demonstrates that the electrochemical process at the level of the working electrode is controlled by the diffusion of the electroactive species, which concords with the literature [28]. The same type of linearity was also obtained by SPCE.

The active area of the electrodes was calculated by using the linear dependences between I_{pa} and $v^{1/2}$ and applying the Randles-Sevcik equation [27,29].

$$I_{pa} = 268,600 \times n^{3/2} \times A \times D^{1/2} \times Cv^{1/2} \quad (1)$$

where: I_{pa} = current of the anodic peak (A); n = number of electrons transferred in the redox process; A = electrode area (cm^2); D = diffusion coefficient (cm^2/s); C = concentration (mol/cm^3); v = scan rate (V/s).

The diffusion coefficient of the ferrocyanide ion is considered known: $D = 7.26 \times 10^{-6}\ cm^2 \times s^{-1}$ [27].

Table 2 presents the values obtained for the area of active surface and the roughness factor of the two working electrodes.

Table 2. Geometric area, active surface area and roughness factor for SPCE and AuNP-CNT/SPCE.

Electrode	Geometric Area (cm^2)	Active Area (cm^2)	Roughness Factor
SPCE	0.1256	0.2836	2.25
AuNP-CNT/SPCE		0.9842	7.84

AuNP-CNT/SPCE displays an active surface approximately 8 times larger than its geometric area and approx. 4 times larger than the active surface of the SPCE. The high sensitivity of the AuNP-CNT/SPCE modified sensor can be explained by the presence of the carbon nanotubes and gold particles which, owing to their shape, size and structure, present good electrical and mechanical properties, favoring the fast transfer of electrons [30,31], necessary in the case of electrochemical sensors. In addition, the association of carbon nanotubes and gold nanoparticles results in good conductibility and electrocatalytic capacity, thus increasing the sensor's selectivity, sensitivity and reproducibility [32,33].

3.2. Study of the Electrochemical Behavior of the Sensors in Atorvastatin Solition (Support Electrolyte Solution KCl 10^{-1} M)

During the following stage, the response of the sensors in an atorvastatin solution of 3×10^{-5} M and KCl 10^{-1} M was analyzed. The purpose of the analysis is to investigate the oxidation-reduction process of atorvastatin, using the sensors selected for this study. Furthermore, using cyclic voltammetry the sensitivity and selectivity of the sensors for the detection of atorvastatin were studied.

For stabilizing the sensors' responses, 6 cycles in the potential range between −0.4 V and +1.3 V were necessary. The results presented were obtained after the stabilization of the voltammetric signal.

Figure 3 presents the cyclic voltammograms of the sensors in atorvastatin 3×10^{-5} M solution (electrolyte KCl 10^{-1} M solution).

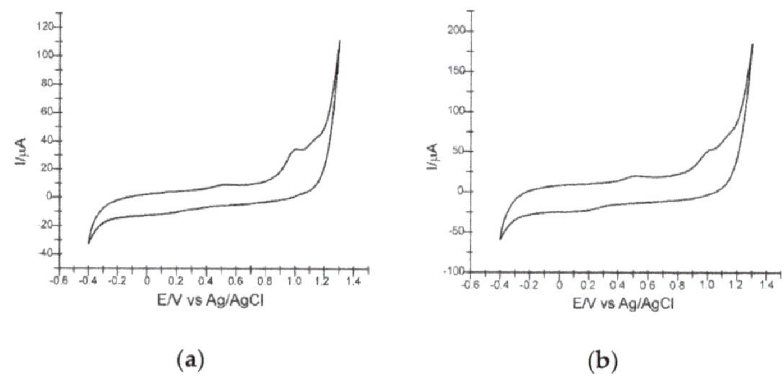

Figure 3. Cyclic voltammograms of the sensors: (**a**) SPCE and (**b**) AuNP-CNT/SPCE in atorvastatin 3×10^{-5} M solution. Scan rate 0.1 V/s.

Using both electrodes, one notes similar electrochemical behavior. The recorded cyclic voltammograms highlight two anodic and one cathodic peaks. The second anodic peak (E_{pa} = 1.004 V, I_{pa} = 34.208 µA for SPCE and E_{pa} = 1.004 V, I_{pa} = 51.82 µA for AuNP-CNT/SPCE) is related to the oxidation process of atorvastatin. The oxidation process is irreversible, as no cathodic peak associated to the oxidation of atorvastatin is observed.

In the initial phase of the oxidation process, atorvastatin loses one electron to form a cationic radical, which, in the following stage, loses one proton and one electron, a quaternary Schiff base being form. The resulting quaternary Schiff base suffers a rapid hydrolysis, two compounds appearing: the 3,5-dihydroxi-7-oxo-heptanoic acid and the fenylamide of acid 5-(4-Fluorophenyl)-2-isopropyl-4-phenyl-1H-pyrrole-3-carboxylic. This mechanism has been suggested by many researchers [10,31,32]. The scheme of the oxidation process of atorvastatin is presented in Figure 4.

Therefore, the electrochemical process entails the transfer of two electrons and two protons concomitantly detected by the screen-printed sensors. This electrochemical behavior of atorvastatin obtained with screen-printed sensors is similar to the one reported in the literature [10,33].

It is noted that the anodic peak related to the oxidation of atorvastatin occurs at the same potential (at a 0.1 V/s scan rate) for both sensors, but the intensity of current is higher in the case of AuNP-CNT/SPCE. This difference is given by the modifications of the surface of the screen-printed sensor. It is known that carbon nanotubes have good mechanical, electrical and thermic properties [17] and their association with the gold nanoparticles facilitates, in the case of atorvastatin oxidation, a well-defined electrochemical response, with higher intensity and lower background current.

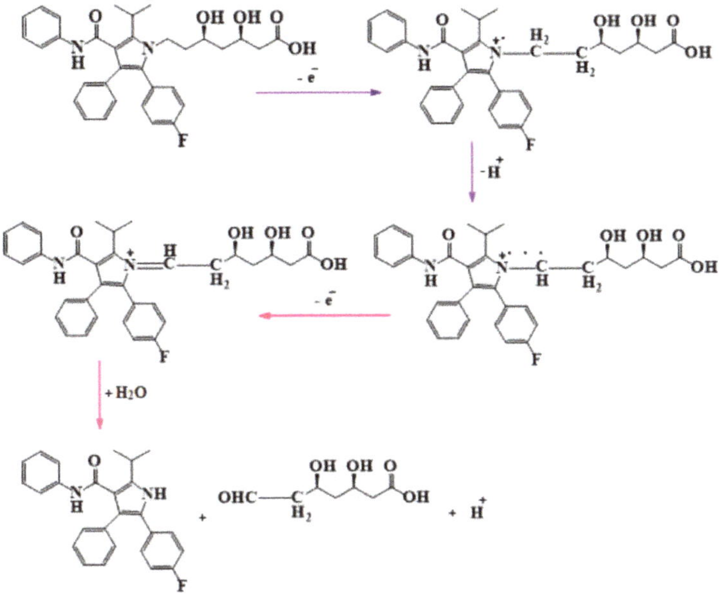

Figure 4. Mechanism proposed for electrochemical oxidation of atorvastatin. Adapted from [34].

According to the literature, rosuvastatin, for example, shows a slight oxidation at a different potential, usually with a higher value (+1.26 V) [35] (+1.39 V) [36], which is why, in a simultaneous detection analysis, the signals would not interfere and the selectivity of the sensor would prove to be optimal. More studies will be needed in this direction.

3.3. *Influence of the Scan Rate on the Sensors' Responses*

This section presents the scan rate's influence over the intensity and potential of the anodic peak corresponding to atorvastatin oxidation, using, in turn, SPCE and AuNP-CNT/SPCE. Scan rates varied within a range from 0.1 and 1.0 V/s, modifying the scan rate gradually with 0.1 V/s at each recording.

Applying gradual increases of scan rates the determining stage of the electrochemical process was achieved.

The intensity of the anodic peak increases, and the potential reaches higher values with the increase of the scan rate (Figure 5). This behavior is explained by the irreversible nature of the oxidation process, as confirmed by other studies [33]. The linear dependence of the anodic current to the scan rate and to the scan rate square root was also evaluated in order to determine whether the process is controlled by diffusion or adsorption.

Assessing the two linear dependences, one notes that the better linearity and a determination factor (R^2) closer to 1 was obtained between the intensity of the anodic current and scan rate, as evident from Figure 6. This trend was observed for both sensors.

The linear regression equations for the anodic current intensity versus scan rate were as follows:

$$\text{Ip}(\mu\text{A}) = 132.16\text{v}(\text{V/s}) + 25.357,\ \text{R}^2 = 0.9956\ (\text{for SPCE})$$

$$\text{Ip}(\mu\text{A}) = 239.24\ \text{v}\left(\frac{\text{V}}{\text{s}}\right) + 36.527,\ \text{R}^2 = 0.995\ (\text{for AuNP} - \text{CNT/SPCE})$$

In addition, the equations for the anodic current intensity logarithm versus the scan rate logarithm were also obtained:

$$\log\ \text{Ip}(\mu\text{A}) = 0.6627\ \log\ \text{v}(\text{V/s}) + 2.1786,\ \text{R}^2 = 0.9974\ (\text{for SPCE})$$

$$\log \text{Ip}(\mu\text{A}) = 0.7198 \log \text{v}(\text{V/s}) + 2.4262,\ \text{R}^2 = 0.9988\ (\text{for AuNP} - \text{CNT/SPCE})$$

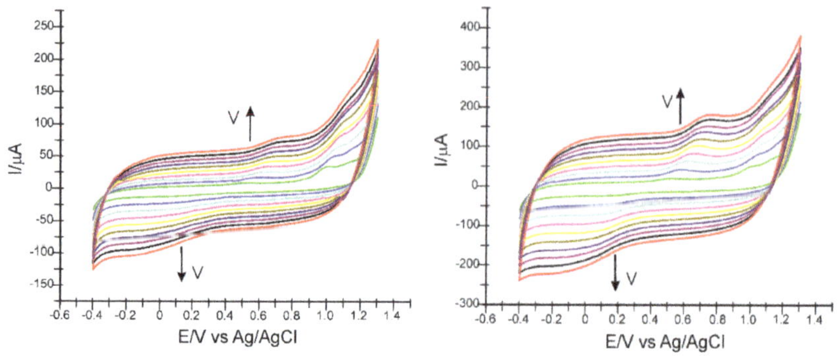

(**a**) (**b**)

Figure 5. Cyclic voltammograms recorded by (**a**) SPCE and (**b**) AuNP-CNT/SPCE in atorvastatin 3×10^{-5} M solution (electrolyte KCl 10^{-1} M solution) at scan rates ranging from 0.1 V/s to 1.0 V/s.

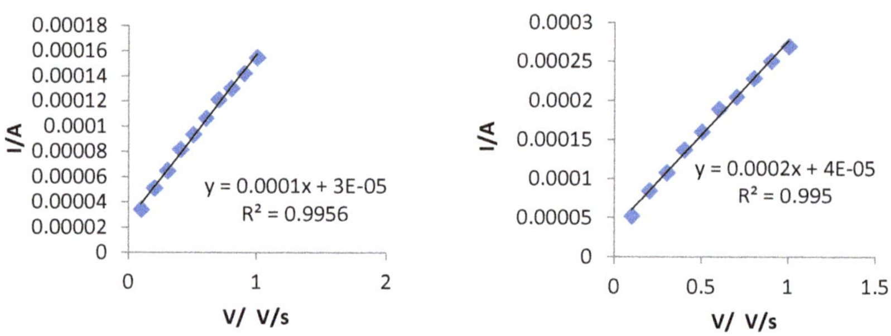

(**a**) (**b**)

Figure 6. Linear dependence of the anodic current dependent on scan rate for (**a**) SPCE and (**b**) AuNP-CNT/SPCE.

The values of the slopes are close to the theoretical value (0.5) which corresponds to an ideal reaction of the species in solution [34]. Consequently, the oxidation mechanism of atorvastatin which takes place at the SPCE and AuNP-CNT/SPCE surface is an adsorption-controlled process [13,35].

However, the graphic between the anodic peak logarithm and the square root of scan rate logarithm shows a slope of 0.7589 and R^2 = 0.9973 for SPCE, respectively, 0.6938 and R^2 = 0.9988 for AuNP-CNT/SPCE, also indicating the presence of a diffusion process.

$$\log \text{Ip}(\mu\text{A}) = 0.7525 \log \text{v}^{1/2}\ (\text{V/s}) + 1.64,\ \text{R}^2 = 0.9973\ (\text{for SPCE})$$

$$\log \text{Ip}(\mu\text{A}) = 0.6938 \log \text{v}^{\frac{1}{2}} \left(\frac{\text{V}}{\text{s}}\right) + 1.6834,\ \text{R}^2 = 0.9988\ (\text{for AuNP} - \text{CNT/SPCE})$$

In order to investigate more thoroughly the atorvastatin electrochemical process which takes places at the electrodes surface, 5 cycles were recorded by cyclic voltammetry. The anodic peak current corresponding to atorvastatin displayed a significant decrease during the successive recordings (the results are not presented) in the first and the second scanning cycle, while the other three cycles remained almost unchanged, the signal being stable. These results confirm that the process that takes place at the surface is adsorption-controlled, but the diffusion process also plays an important role. Such electrochemical behavior of atorvastatin has also been reported in other studies [33].

In order to calculate the concentration of the electroactive species adsorbed on the active surface of the electrode, Laviron equation was employed [37].

$$i_p = \frac{n^2F^2\Gamma Av}{4RT} \quad (2)$$

where: Γ—surface concentration of the electroactive species, mol × cm^{-2}; i_p—current corresponding to the peak, A; A—electrode supraface, cm^2; n—number of electrons transferred during the redox processes; F—Faraday constant, 96,485 C × mol^{-1}; R—molar gas constant, 8.314 J/mol K; T—absolute temperature, K; 298 K.

According to the atorvastatin oxidation process, the number of transferred electrons is 2 [10,33].

Table 3 displays the Γ values obtained from Equation (2).

Table 3. Equation of the linear regression I_p vs. v, R^2 and Γ corresponding to the two sensors.

Electrode	Equation	R^2	(mol × cm^{-2})
SPCE	$I_p = 1.32 \times 10^{-4}$ v + 2.53×10^{-5}	0.9956	1.11×10^{-11}
AuNP-CNT/SPCE	$I_p = 2.39 \times 10^{-4}$ v + 3.65×10^{-5}	0.9950	4.94×10^{-10}

The values obtained for the degree of covering the surface with electroactive species conform with the values reported by other studies [33].

It can be asserted that AuNP-CNT/SPCE has superior electroanalytical properties in comparison with SPCE at atorvastatin detection. The presence of carbon nanotubes and gold nanoparticles ensures better conductibility and selectivity of the sensor in real samples.

3.4. Calibration Curve and Detection Limit

In optimized experimental conditions, the SPCE and AuNP-CNT/SPCE electrodes were used for the voltammetric determination of atorvastatin. A calibration curve was developed to determine the sensor response linearity range towards concentrations and to calculate the detection and quantification limit. The results are displayed in Figure 7. For the calibration curve, an atorvastation stock solution was used (6×10^{-4} M), out of which clearly defined volumes were added to the electrolyte solution of KCl 10^{-1} M. The concentration range under focus was 1.2–606.25 μM. A good linearity was obtained in the 1.2–53.33 μM range for both sensors, as shown in Figure 7.

With the help of the linear calibration curve equation and standard deviation, the values of detection limit (LOD = 3σ/m) and quantification (LOQ = 10σ/m) [38] were calculated (Table 4)

Table 4. The equations of linear fitting I_{pa} vs. c, R^2, LOD and LOQ values for the two sensors for atorvastatin detection.

Sensor	Equation of the Linear Calibration Curve	R^2	LOD (μM)	LOQ (μM)
SPCE	I_{pa} = 0.0383 c + 2.2746	0.9716	6.75×10^{-7}	2.25×10^{-6}
AuNP-CNT/SPCE	I_{pa} = 0.0252 c + 10.762	0.9892	1.92×10^{-7}	6.39×10^{-7}

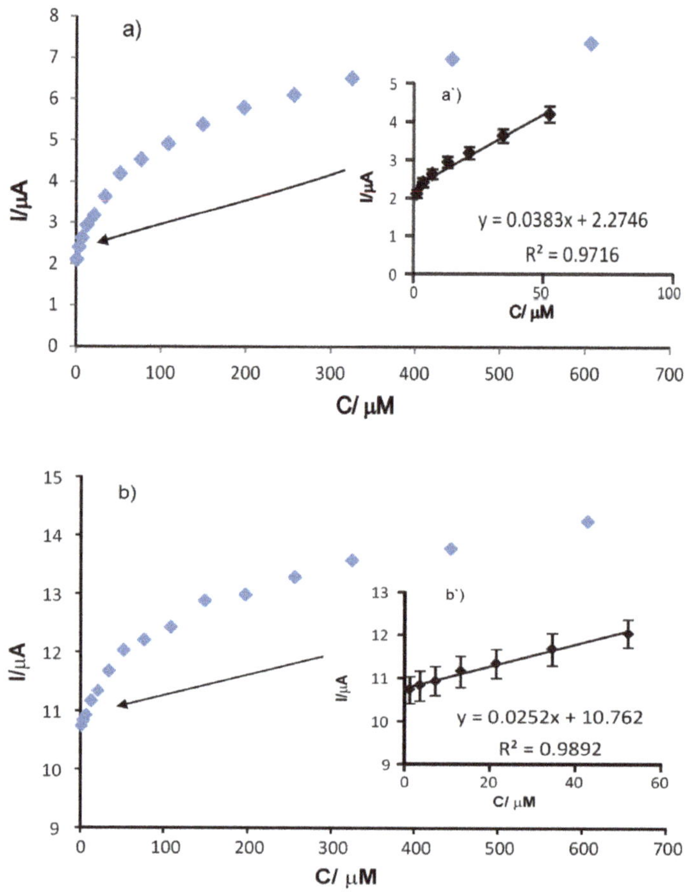

Figure 7. The linear fitting between I_{pa} and the atorvastatin concentration for (**a**) SPCE and (**b**) AuNP-CNT/SPCE in the 1.2–606.25 μM concentration range. The inserted figures present the linear dependence between I_{pa} and c for (a') SPCE and (b') AuNP-CNT/SPCE in the 1.2–53.33 μM concentration range. Error bars are represented for a standard error of 5%.

SPCE/CNT-AuNP presents low LOD and LOQ values, which indicates high sensitivity. Consequently, the voltammetric method is precise and feasible for qualitative and quantitative atorvastatin determination. The values obtained are even below the ones reported in the literature (Table 5), which proves that SPCE/CNT-AuNP has very good electroanalytical performance for atorvastatin detection.

This study has emphasized the excellent electrochemical properties of the SPCE/CNT-AuNP sensor, which are the result of the association of carbon nanotubes and gold nanoparticles. These superior properties have been marked out through an extended area of active surface and low detection limit and quantification values. Therefore, AuNP-CNT/SPCE has been used for the determination of atorvastatin in real samples.

Table 5. Sensitive materials, detection technique, linear range and LOD of some sensors used for the detection of atorvastatin.

Sensitive Materials	Detection Technique	Linear Range (μM)	LOD (μM)	Real Sample	Ref.
PbTe NPs/SPE [1]	cronAmp	1–70	0.05	Tablet urine	[34]
GCE [2]	DPV SWV	1–50	0.2 0–3	Tablet	[12]
Fe_3O_4@PPy/MWCNTs/GE [3]	SWASV	0.0314–201	0.0230	Tablet and human serum	[38]
AgNWs/Au/GCE [4]	CV DPV	10×10^{-6}–1×10^{-3}	1×10^{-3}	Tablet	[14]
EPPGE [5]	DPV SWV	0.0960–5.60 0.0960–7.74	21.0 23.0	Tablet	[13]
PPY/CNTs/GCE [6]	LSV	0.005–1.0	0.0015	Tablet	[39]
ZnO/NS/CPE [7]	DPV SWV	1.00–12.0 0.01–0.1	7×10^{-4} 1.2×10^{-4}	Tablet and urine	[11]
PPY-GNs/GCE [8]	DPV	20.0–200	1.20	Tablet	[40]
VACNT-GO electrode [9]	DPAdSV	0.09–3.81	9.4×10^{-3}	Urine and human serum	[41]
CPE in the presence of CTAB [10]	DPV	0.05–10	4.08×10^{-3}	Tablet and urine	[42]

[1] Lead telluride nanoparticles-modified graphite screen-printed electrodes; [2] glassy carbon electrode; [3] graphite electrode (GE) modified with polypyrrole-coated Fe_3O_4 nanohybrid by core-shell structure (Fe_3O_4@PPy NPs) and multiwall carbon nanotubes (MWCNTs); [4] silver nanowires/gold-modified glassy carbon electrode; [5] edge-plane pyrolytic graphite electrode; [6] Polypyrrole/Carbon nanotube/Glassy carbon electrode; [7] zinc oxide nanoparticles and nano-silica carbon paste electrode; [8] polypyrrole functionalized graphene sheets/glassy carbon electrode; [9] vertically aligned carbon nanotube/graphene oxide; [10] carbon paste electrode (CPE) in the presence of an enhancing agent, cetyltrimethyl ammonium bromide (CTAB).

For this study, the real samples were pharmaceutical products in the form of tablets. As can be seen in the Table 5, many authors have determined atorvastatin from various types of biological samples. Biological samples involve several stages of pretreatment and storage [43], involving higher costs and a longer time. Moreover, the detection from biological samples implies the existence of interferences that can be difficult to control. Metabolites of other compounds in urine or plasma may significantly impair or alter the sensor signal. The future step of our research will be the detection of statins in biological samples.

3.5. Atorvastatin Determination in the Pharmaceutical Products

In the next experimental step, the aim is to demonstrate the sensitivity and accuracy of the AuNP-CNT/SPCE sensor for the qualitative and quantitative determination of the analyte of interest in a pharmaceutical products. For the determination of atorvastatin in the pharmaceutical products, AuNP-CNT/SPCE was used, applying cyclic voltammetry. The potential range was between −0.4 V and +1.3 and the scan rate was of 0.1 V/s. The cyclic voltammogram points out the peak corresponding to the presence of atorvastatin in the pharmaceutical products. Taking into consideration the anodic peak intensity, the quantity of product analyzed and the linear calibration equation, the quantity of atorvastatin in the pharmaceutical products were calculated.

The results obtained conform to the values reported by manufacturers, which demonstrates the precision of the AuNP-CNT/SPCE sensor (Table 6).

Table 6. Atorvastatin concentrations in Sortis obtained by voltammetry and, respectively, by FTIR.

Product	Atorvastatin Quantity Reported by the Manufacturer (mg/cp)	Atorvastatin Quantity Experimentally Determined	
		CV (mg/cp) method	FTIR (mg/cp) method
Sortis 20 mg	20	20 ± 1	21 ± 2
Sortis 10 mg	10	10 ± 0.5	10.5 ± 0.5

The voltammetric method was validated using the FTIR spectrometric method. The FTIR method presents a considerable potential in medicine quality control, as it is capable of identifying the chemical structure and the identity of the active substance in various samples. For this test, a solid standard of atorvastatin-KBr in 10 mg/g concentration was prepared, and the linear calibration curve was determined based on the absorbers of the peak at a wave length of 1316 cm^{-1}, specific to the stretching vibration of the C-N group [44].

Figure 8 displays the FTIR spectrum of the sample obtained from the pharmaceutical product Sortis 20 mg.

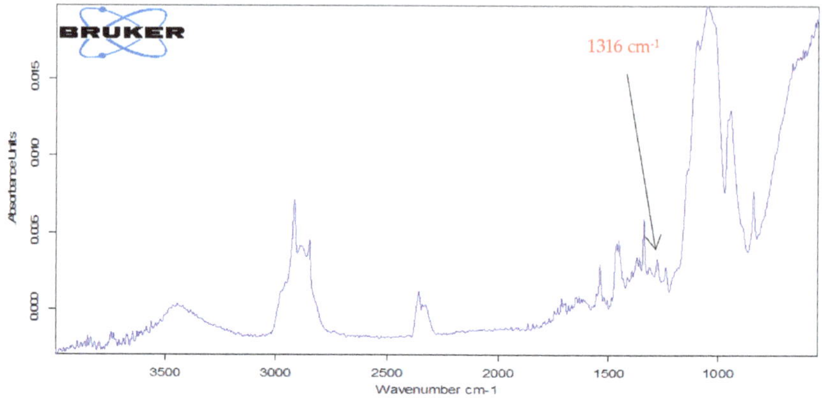

Figure 8. FTIR spectrum of Sortis 20 mg.

All samples were analyzed in triplicate. The results are outlined in Table 6.

It is noted that the values obtained are similar for both methods. In addition, the atorvastatin concentrations obtained in this study correspond to the concentration reported by manufacturer. These results demonstrate the accuracy and precision of the voltammetric method, as well as the sensitivity and selectivity of the AuNP-CNT/SPCE sensor.

3.6. Recovery Test

The analytical recovery was determined using cyclic voltammetry, after the addition of clearly determined quantities of pharmaceutical product solutions, more precisely, 3×10^{-6}, 5×10^{-6} and 7×10^{-6} M.

Analytical recoveries ranged between 99.4 and 100.9% (Table 7) at approx. 1% RSD (relative standard deviation). The values of the analytical recovery obtained are related to the presence of some excipients, which may interfere with the electrochemical response. The results obtained are positive, and the sensor may be used in pharmaceutical practice.

Table 7. Recovery test at atorvastatin detection.

Atorvastatin Concentration ($\times 10^6$ M)	Atorvastatin Quantity ($\times 10^6$ M)	Recovery (%) $\pm$ RSD (n = 5)
3	3.03	100.5 $\pm$ 0.9
5	4.97	99.4 $\pm$ 0.8
7	7.06	100.9 $\pm$ 0.9

4. Conclusions

This study has characterized the SPCE and AuNP-CNT/SPCE sensors with the purpose of determining the atorvastatin in pharmaceutical products. AuNP-CNT/SPCE displayed superior electrochemical behavior in comparison with SPCE as early as during the preliminary experiments. With a wider active surface area, AuNP-CNT/SPCE has recorded more intense and better defined peaks in the atorvastatin solution, the process being mainly controlled by the adsorption of the species at the electrode surface. For the quantification of atorvastatin, a calibration curve was determined and the calculated LOD and LOQ values proved higher sensitivity for AuNP-CNT/SPCE. In the quantitative analysis, the atorvastatin concentrations obtained with AuNP-CNT/SPCE are very close to the ones obtained by spectrometry, and also to the values reported by manufacturer. It can be thus concluded that the method employed is simple, versatile, precise and entails low costs. In addition, AuNP-CNT/SPCE is suitable for the screening analysis of complex samples that contain atorvastatin.

The present study could be a first step towards the development of a sensitive, portable, single-use device for the detection of atorvastatin in biological samples. Such a device could be used at home or in the hospital for the benefit of the patient. By monitoring the level of atorvastatin in the blood, the dose administered daily can be adjusted, and thus, liver function would be protected. Furthermore, further research could address the simultaneous detection of several statins (rosuvastatin or simvastatin), results that could make important contributions to drug control.

Author Contributions: Conceptualization, C.A. and R.O.G.; methodology, C.A.; validation, R.O.G. and A.V.B.; formal analysis, R.O.G.; investigation, R.O.G. and A.V.B.; resources, C.A.; writing—original draft preparation, R.O.G. and A.V.B.; writing—review and editing, C.A.; supervision, C.A. All authors have read and agreed to the published version of the manuscript.

Funding: The contribution of author Ramona Oana Gunache (Roșca) was supported by the project entitled "Excelența academică valori antreprenoriale—sistem de burse pentru asigurarea oportunităților de formare dezvoltare a competențelor antreprenoriale ale doctoranzilor post doctoran-zilor—ANTREPRENORDOC", Contract number 36355/23.05.2019 POCU/380/6/13—SMIS code: 123847.

Institutional Review Board Statement: Not applicable.

Informed Consent Statement: Not applicable.

Data Availability Statement: The authors confirm that the data supporting the findings of this study are available within the article.

Acknowledgments: The results of this work have been presented to the 9th edition of the Scientific Conference organized by the Doctoral Schools of "Dunărea de Jos" University of Galati (SCDS-UDJG) http://www.cssd-udjg.ugal.ro/, accessed on 10 June 2021 that will be held on 10 and 11 June 2021, in Galati, Romania.

Conflicts of Interest: The authors declare no conflict of interest.

References

1. Tsoukas, M.A.; Mantzoros, C.S. Lipodystrophy Syndromes. In *Endocrinology: Adult and Pediatric*; Elsevier: New York, NY, USA, 2016; pp. 648–661.e5, ISBN 978-0-323-18907-1.
2. Atorvastatin (CHEBI:39548). Available online: https://www.ebi.ac.uk/chebi/searchId.do?chebiId=CHEBI:39548 (accessed on 31 July 2021).
3. Ramkumar, S.; Raghunath, A.; Raghunath, S. Statin Therapy: Review of Safety and Potential Side Effects. *Acta Cardiol. Sin.* **2016**, *32*, 631–639. [CrossRef] [PubMed]
4. Sewright, K.A.; Clarkson, P.M.; Thompson, P.D. Statin Myopathy: Incidence, Risk Factors, and Pathophysiology. *Curr. Atheroscler. Rep.* **2007**, *9*, 389–396. [CrossRef]
5. Martins, A.F.; Frank, C.D.S.; Altissimo, J.; de Oliveira, J.A.; da Silva, D.S.; Reichert, J.F.; Souza, D.M. Determination of Statin Drugs in Hospital Effluent with Dispersive Liquid–Liquid Microextraction and Quantification by Liquid Chromatography. *J. Environ. Sci. Health Part A* **2017**, *52*, 939–945. [CrossRef] [PubMed]
6. Danafar, H.; Hamidi, M. Method Validation of Amlodipine and Atorvastatin by Liquid Chromatography–Mass Spectrometry (LC–MS) Method in Human Plasma. *Cogent Med.* **2016**, *3*, 1129790. [CrossRef]
7. Emhofer, L.; Himmelsbach, M.; Buchberger, W.; Klampfl, C.W. High-Performance Liquid Chromatography Drift-Tube Ion-Mobility Quadrupole Time-of-Flight/Mass Spectrometry for the Identity Confirmation and Characterization of Metabolites from Three Statins (Lipid-Lowering Drugs) in the Model Plant Cress (Lepidium Sativum) after Uptake from Water. *J. Chromatogr. A* **2019**, *1592*, 122–132. [CrossRef]
8. Dastkhoon, M.; Ghaedi, M.; Asfaram, A.; Arabi, M.; Ostovan, A.; Goudarzi, A. Cu@SnS/SnO_2 Nanoparticles as Novel Sorbent for Dispersive Micro Solid Phase Extraction of Atorvastatin in Human Plasma and Urine Samples by High-Performance Liquid Chromatography with UV Detection: Application of Central Composite Design (CCD). *Ultrason. Sonochem.* **2017**, *36*, 42–49. [CrossRef]
9. Bioanalytical Method Development and Validation of Atorvastatin in Human Plasma by Using UV-Visibile Spectrophotometry—ProQuest. Available online: https://www.proquest.com/openview/ffa8b2baa35d2ac4fbec030eeda904e2/1?pq-origsite=gscholar&cbl=54977 (accessed on 15 July 2021).
10. Yilmaz, B.; Kaban, S. UV and First Derivative Spectrophotometric Methods for the Estimation of Atorvastatin in Pharmaceutical Preparations. *J. Adv. Pharm. Res.* **2018**, *2*, 89–94. [CrossRef]

11. Bukkitgar, S.D.; Shetti, N.P.; Kulkarni, R.M. Construction of Nanoparticles Composite Sensor for Atorvastatin and Its Determination in Pharmaceutical and Urine Samples. *Sens. Actuators B Chem.* **2018**, *255*, 1462–1470. [CrossRef]
12. Yilmaz, B.; Kaban, S. Electrochemical Behavior of Atorvastatin at Glassy Carbon Electrode and Its Direct Determination in Pharmaceutical Preparations by Square Wave and Differential Pulse Voltammetry. *Indian J. Pharm. Sci.* **2016**, *78*, 360–367. [CrossRef]
13. Fazlolahzadeh, O.; Rouhollahi, A.; Hadi, M. Electroanalytical Determination of Atorvastatin in Pharmaceutical Formulations Based on Edge-Plane Pyrolytic Graphite Electrode. *Anal. Bioanal. Electrochem.* **2016**, *8*, 566–577.
14. Naseri, A.; Hormozi-Nezhad, M.R.; Shahrokhian, S.; Asadian, E. Silver Nanowires Immobilized on Gold-Modified Glassy Carbon Electrode for Electrochemical Quantification of Atorvastatin. *J. Electroanal. Chem.* **2020**, *876*, 114540. [CrossRef]
15. Kumar, S.; Bukkitgar, S.D.; Singh, S.; Pratibha; Singh, V.; Reddy, K.R.; Shetti, N.P.; Venkata Reddy, C.; Sadhu, V.; Naveen, S. Electrochemical Sensors and Biosensors Based on Graphene Functionalized with Metal Oxide Nanostructures for Healthcare Applications. *ChemistrySelect* **2019**, *4*, 5322–5337. [CrossRef]
16. Dakshayini, B.S.; Reddy, K.R.; Mishra, A.; Shetti, N.P.; Malode, S.J.; Basu, S.; Naveen, S.; Raghu, A.V. Role of Conducting Polymer and Metal Oxide-Based Hybrids for Applications in Ampereometric Sensors and Biosensors. *Microchem. J.* **2019**, *147*, 7–24. [CrossRef]
17. Kuzin, Y.; Kappo, D.; Porfireva, A.; Shurpik, D.; Stoikov, I.; Evtugyn, G.; Hianik, T. Electrochemical DNA Sensor Based on Carbon Black—Poly(Neutral Red) Composite for Detection of Oxidative DNA Damage. *Sensors* **2018**, *18*, 3489. [CrossRef]
18. Shamgsumova, R.V.; Shurpik, D.N.; Evtugyn, V.G.; Stoikov, I.I.; Evtugyn, G.A. Electrochemical Determination of Malathion on an Acetylcholinesterase-Modified Glassy Carbon Electrode. *Anal. Lett.* **2018**, *51*, 1911–1926. [CrossRef]
19. Iijima, S. Helical Microtubules of Graphitic Carbon. *Nature* **1991**, *354*, 56–58. [CrossRef]
20. Apetrei, C.; Apetrei, I.M. Biosensor Based on Tyrosinase Immobilized on a Single-Walled Carbon Nanotube-Modified Glassy Carbon Electrode for Detection of Epinephrine. *IJN* **2013**, 4391. [CrossRef] [PubMed]
21. Claussen, J.C.; Kumar, A.; Jaroch, D.B.; Khawaja, M.H.; Hibbard, A.B.; Porterfield, D.M.; Fisher, T.S. Nanostructuring Platinum Nanoparticles on Multilayered Graphene Petal Nanosheets for Electrochemical Biosensing. *Adv. Funct. Mater.* **2012**, *22*, 3399–3405. [CrossRef]
22. Apetrei, I.M.; Apetrei, C. A Modified Nanostructured Graphene-Gold Nanoparticle Carbon Screen-Printed Electrode for the Sensitive Voltammetric Detection of Rutin. *Measurement* **2018**, *114*, 37–43. [CrossRef]
23. Cheng, W.; Dong, S.; Wang, E. Gold Nanoparticles as Fine Tuners of Electrochemical Properties of the Electrode/Solution Interface. *Langmuir* **2002**, *18*, 9947–9952. [CrossRef]
24. Guo, S.; Wang, E. Synthesis and Electrochemical Applications of Gold Nanoparticles. *Anal. Chim. Acta* **2007**, *598*, 181–192. [CrossRef]
25. Alim, S.; Vejayan, J.; Yusoff, M.M.; Kafi, A.K.M. Recent Uses of Carbon Nanotubes & Gold Nanoparticles in Electrochemistry with Application in Biosensing: A Review. *Biosens. Bioelectron.* **2018**, *121*, 125–136. [CrossRef]
26. Serafín, V.; Martínez-García, G.; Agüí, L.; Yáñez-Sedeño, P.; Pingarrón, J.M. Multiplexed Determination of Human Growth Hormone and Prolactin at a Label Free Electrochemical Immunosensor Using Dual Carbon Nanotube–Screen Printed Electrodes Modified with Gold and PEDOT Nanoparticles. *Analyst* **2014**, *139*, 4556–4563. [CrossRef] [PubMed]
27. Apetrei, I.M.; Apetrei, C. Voltammetric Determination of Melatonin Using a Graphene-Based Sensor in Pharmaceutical Products. *Int. J. Nanomed.* **2016**, *11*, 1859–1866. [CrossRef]
28. Bounegru, A.; Apetrei, C. Voltammetric Sensors Based on Nanomaterials for Detection of Caffeic Acid in Food Supplements. *Chemosensors* **2020**, *8*, 41. [CrossRef]
29. Bard, A.J.; Faulkner, L.R. *Electrochemical Methods: Fundamentals and Applications*, 2nd ed.; Wiley: New York, NY, USA, 2001; ISBN 978-0-471-04372-0.
30. Bounegru, A.V.; Apetrei, C. Carbonaceous Nanomaterials Employed in the Development of Electrochemical Sensors Based on Screen-Printing Technique—A Review. *Catalysts* **2020**, *10*, 680. [CrossRef]
31. García-Carmona, L.; Moreno-Guzmán, M.; Sierra, T.; González, M.C.; Escarpa, A. Filtered Carbon Nanotubes-Based Electrodes for Rapid Sensing and Monitoring of L-Tyrosine in Plasma and Whole Blood Samples. *Sens. Actuators B Chem.* **2018**, *259*, 762–767. [CrossRef]
32. Bounegru, A.V.; Apetrei, C. Voltamperometric Sensors and Biosensors Based on Carbon Nanomaterials Used for Detecting Caffeic Acid—A Review. *IJMS* **2020**, *21*, 9275. [CrossRef]
33. Bounegru, A.V.; Apetrei, C. Development of a Novel Electrochemical Biosensor Based on Carbon Nanofibers–Gold Nanoparticles–Tyrosinase for the Detection of Ferulic Acid in Cosmetics. *Sensors* **2020**, *20*, 6724. [CrossRef]
34. Pourtaheri, E.; Taher, M.A.; Ali, G.A.; Agarwal, S.; Gupta, V.K. Low-Cost and Highly Sensitive Sensor for Determining Atorvastatin Using PbTe Nanoparticles-Modified Graphite Screen-Printed Electrode. *Int. J. Electrochem. Sci.* **2019**, *14*, 9622–9632. [CrossRef]
35. Silva, T.A.; Zanin, H.; Vicentini, F.C.; Corat, E.J.; Fatibello-Filho, O. Electrochemical Determination of Rosuvastatin Calcium in Pharmaceutical and Human Body Fluid Samples Using a Composite of Vertically Aligned Carbon Nanotubes and Graphene Oxide as the Electrode Material. *Sens. Actuators B Chem.* **2015**, *218*, 51–59. [CrossRef]
36. Silva, T.A.; Pereira, G.F.; Fatibello-Filho, O.; Eguiluz, K.I.B.; Salazar-Banda, G.R. Square-Wave Voltammetric Determination of Rosuvastatin Calcium in Pharmaceutical and Biological Fluid Samples Using a Cathodically Pretreated Boron-Doped Diamond Electrode. *Diam. Relat. Mater.* **2015**, *58*, 103–109. [CrossRef]

37. Apetrei, I.M.; Rodriguez-Mendez, M.L.; Apetrei, C.; de Saja, J.A. Fish Freshness Monitoring Using an E-Tongue Based on Polypyrrole Modified Screen-Printed Electrodes. *IEEE Sens. J.* **2013**, *13*, 2548–2554. [CrossRef]
38. Tavousi, A.; Ahmadi, E.; Mohammadi-Behzad, L.; Riahifar, V.; Maghemi, F. Sensitive Electrochemical Sensor Using Polypyrrole-Coated Fe_3O_4 Core-Shell Nanoparticles/Multiwall Carbon Nanotubes Modified Graphite Electrode for Atorvastatin Analysis. *Microchem. J.* **2020**, *158*, 105159. [CrossRef]
39. Kamalzadeh, Z.; Shahrokhian, S. Electrochemical Determination of Atorvastatin on Nano-Scaled Polypyrrole Film. *Bioelectrochemistry* **2014**, *98*, 1–10. [CrossRef] [PubMed]
40. Wei, L. Electrochemical Sensor for Ultrasensitive Determination of Atorvastatin Based on Polypyrrole Functionalized Graphene Sheets. *Int. J. Electrochem. Sci.* **2018**, 10173–10180. [CrossRef]
41. Silva, T.A.; Zanin, H.; Vicentini, F.C.; Corat, E.J.; Fatibello-Filho, O. Differential Pulse Adsorptive Stripping Voltammetric Determination of Nanomolar Levels of Atorvastatin Calcium in Pharmaceutical and Biological Samples Using a Vertically Aligned Carbon Nanotube/Graphene Oxide Electrode. *Analyst* **2014**, *139*, 2832. [CrossRef]
42. Abbar, J.C.; Nandibewoor, S.T. Voltammetric Oxidation and Determination of Atorvastatin Based on the Enhancement Effect of Cetyltrimethyl Ammonium Bromide at a Carbon Paste Electrode. *Colloids Surf. B Biointerfaces* **2013**, *106*, 158–164. [CrossRef] [PubMed]
43. Eskiköy, D.; Durmuş, Z.; Kiliç, E. Electrochemical Oxidation of Atorvastatin and Its Adsorptive Stripping Determination in Pharmaceutical Dosage Forms and Biological Fluids. *Collect. Czech. Chem. Commun.* **2011**, *76*, 1633–1649. [CrossRef]
44. Shahraeini, S.S.; Akbari, J.; Saeedi, M.; Morteza-Semnani, K.; Abootorabi, S.; Dehghanpoor, M.; Rostamkalaei, S.S.; Nokhodchi, A. Atorvastatin Solid Lipid Nanoparticles as a Promising Approach for Dermal Delivery and an Anti-Inflammatory Agent. *AAPS PharmSciTech* **2020**, *21*, 263. [CrossRef]

inventions

MDPI

Article

Development of a Novel Sensor Based on Polypyrrole Doped with Potassium Hexacyanoferrate (II) for Detection of L-Tryptophan in Pharmaceutics

Ancuța Dinu and Constantin Apetrei *

Department of Chemistry, Phyiscs and Environment, Faculty of Science and Environment, "Dunărea de Jos" University of Galati, 47 Domnească Street, RO-800008 Galati, Romania; ancuta.dinu@ugal.ro
* Correspondence: apetreic@ugal.ro; Tel.: +40-727-580-914

Abstract: This study describes the development of a new sensor with applicability in the determination and quantification of yjr essential amino acid (AA) L-tryptophan (L-TRP) from pharmaceutical products. The proposed sensor is based on a carbon screen-printed electrode (SPCE) modified with the conductor polymer polypyrrole (PPy) doped with potassium hexacyanoferrate (II) (FeCN). For the modification of the SPCE with the PPy doped with FeCN, the chronoamperometry (CA) method was used. For the study of the electrochemical behavior and the sensitive properties of the sensor when detecting L-TRP, the cyclic voltammetry (CV) method was used. This developed electrode has shown a high sensibility, a low detection limit (LOD) of up to 1.05×10^{-7} M, a quantification limit (LOQ) equal to 3.51×10^{-7} M and a wide linearity range between 3.3×10^{-7} M and 1.06×10^{-5} M. The analytical performances of the device were studied for the detection of AA L-TRP from pharmaceutical products, obtaining excellent results. The validation of the electroanalytical method was performed by using the standard method with good results.

Keywords: L-tryptophan; polypyrrole; sensor; amino acid; cyclic voltammetry; chronoamperometry

Citation: Dinu, A.; Apetrei, C. Development of a Novel Sensor Based on Polypyrrole Doped with Potassium Hexacyanoferrate (II) for Detection of L-Tryptophan in Pharmaceutics. *Inventions* **2021**, *6*, 56. https://doi.org/10.3390/inventions6030056

Academic Editors: Eugen Rusu and Gabriela Rapeanu

Received: 15 July 2021
Accepted: 21 July 2021
Published: 28 July 2021

1. Introduction

Tryptophan (TRP), with the molecular formula $C_{11}H_{11}N_2O_2$, is a well-known substance, being part of the essential AA category that cannot be produced naturally by the human body [1]. This chemical compound is hardly found in its natural form, which makes it expensive. TRP is the precursor of the neurotransmitter serotonin, the neurohormone melatonin and niacin (or the PP vitamin) [2]. TRP works as a chemical precursor for 5-hydroxytryptophan (5-HTP), as well as a coenzyme for the nicotinamide adenine dinucleotide (NAD) and nicotinamide adenine dinucleotide phosphate (NADP) [3]. In Figure 1, the pathway for the biosynthesis of TRP is presented.

Being a chiral molecule, TRP is found under two isomers—a natural form (L-TRP) and a synthetic one (D-TRP)—with both forms being almost identical in their physical-chemical properties, a fact which leads to difficulties in distinguishing them [2,4]. In the literature, there are multiple methods explained on how to identify enantiomers [5–7]. The chemical structures of the two enantiomers can be seen in Figure 2.

TRP is most frequently found in food, such as chicken and turkey meat, peanut tofu, sesame seeds, eggs, sunflower seeds, pumpkin seeds, soy and chocolate, as well as in dairy products and pharmaceutical products available on the market with different concentrations of TRP [8]. The recommended dose of TRP by the World Health Organization is 4 mg/kg/day [9]. In the case of improper metabolism of TRP, hallucinations and disappointment can take place, which is why a deficiency or an excess of L-TRP could lead to problems that could affect the nervous system's health [10]. Additionally, phytoproducts or nutraceuticals containing TRP can be administered in the case of depression caused by the new coronavirus (SARS-CoV-2) [11]. For these reasons, L-TRP was selected to be detected as fast as possible from pharmaceutics by a new device: a compound which contributes to preventing neuropsychiatric disorders led by the lack of AA, such as depression [12], anxiety [13], insomnia [14], attention deficit hyperactivity disorder (ADHD) [15], the premenstrual syndrome [16] and Parkinson's disease [17].

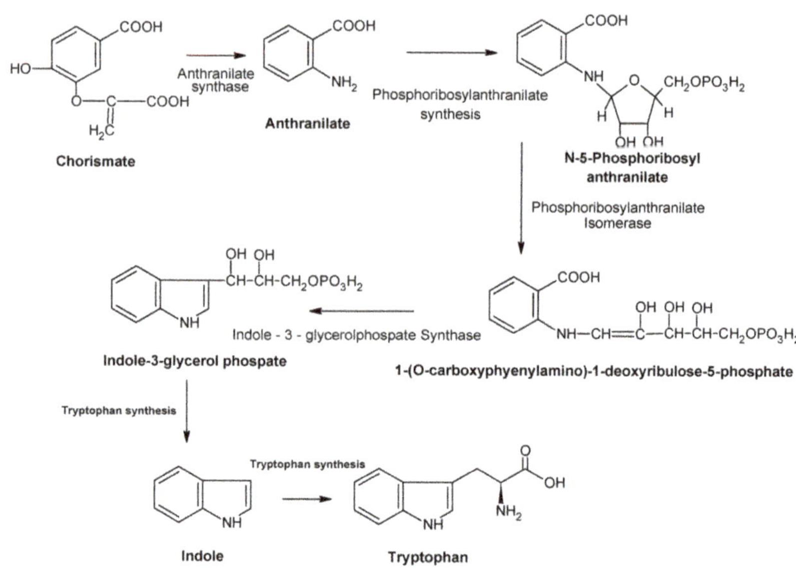

Figure 1. The pathway for the synthesis of TRP (adapted from [3]).

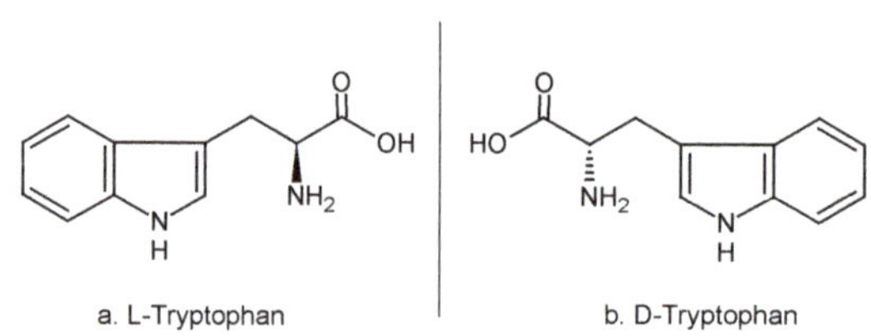

Figure 2. The chemical structure of the two enantiomers. (adapted from [5]).

The physicochemical properties of TRP are very important, being important features for accurate detection. Over time, a variety of methods for the detection of L-TRP and its derivatives were tested, as can be seen in Table 1.

Table 1. Physicochemical methods used for L-TRP detection.

The Detection Method	Reference
Chromatography	[18]
Spectroscopy	[19–22]
Colorimetry	[23]
High-performance liquid chromatography (HPLC)	[19,21,24,25]
Capillary electrophoresis	[26]
Chemiluminescence	[27]
Atomic force spectroscopy	[28]

These methods proved to be expensive and required time to obtain the results, facts that led to more versatile methods that were based on electroanalysis using sensors [29] and biosensors for the much more simple, sensible and selective detection of TRP from pharmaceutical, biological and food products, using electrochemical detection methods such as CV [30,31], differential pulse voltammetry (DPV) [20,30] and square wave voltammetry (SWV) [9,32].

The main sensors developed over the years, used for the determination of the L-TRP quantity, sensitive materials and electroanalytical performance characteristics, are presented in Table 2.

Table 2. The sensitive material, the detection limit and the linear range of the main sensors for the determination of L-TRP through electrochemical methods.

Electrode Materials	The Detection Technique	The Detection Limit	The Linear Range	Reference
Cu_2O-ERGO/GCE (nanocomposite of cuprous oxide and electrochemically reduced graphene oxide)	CV [1], SWV [2]	0.01×10^{-6} M	0.02–20 $\times 10^{-6}$ M	[9]
Nafion-MIP-MWCNTs@IL/GCE (molecularly imprinted copolymer of dual functional monomers and ionic liquid (1-butyl-3-methylimidazolium hexafluorophosphate) functionalized multi-walled carbon nanotubes)	LSV [3], DPV [4]	6×10^{-9} M	8×10^{-9}–26×10^{-6} M	[33]
PVP-GR/GCE (glassy carbon electrode modified with polyvinylpyrrolidonefunctionalized graphene/glassy carbon electrode)	CV	0.01×10^{-6} M	0.06–10.0 $\times 10^{-6}$ M and 10.0–100.0 $\times 10^{-6}$ M	[34]
GR/PEDOT:PSS/GCE (glassy carbon electrode modifed with exfoliated graphene and poly (3,4-ethylenedioxythiophene): poly (styrene sulfonate)	CV	0.015×10^{-6} M	0.1–100 $\times 10^{-6}$ M and 100–1000 $\times 10^{-6}$ M	[35]
PPy/FeCN/SPCE	CV, CA [5]	1.05×10^{-7} M	3.3×10^{-7} M–1.06×10^{-5} M	This work

[1] Cyclic voltammetry. [2] Square wave voltammetry. [3] Linear sweep voltammetry. [4] Differential pulse voltammetry. [5] Chronoamperometry.

In addition, the novel sensor developed in this study is presented in Table 2, developed using a conducting polymer doped with an electroactive compound (PPy and ferrocyanide). The device obtained a lower detection limit and a wider linearity range compared with the other sensors, proving more sensitive L-TRP detection.

The selected polymer for the modification of the sensor was the PPy so that this might present a series of characteristics and advantages that led to higher performance of the sensor, including improved stability and sensibility and high electrical conductivity, malleability and biocompatibility [36]. The doping of PPy by electropolymerization with the help of CA is the usual electrochemical synthesis method, used in improving properties such as stability, conductivity, fast electron transfers and permeability so that this type of sensor could be applied in multiple fields, such as chemistry, pharmaceutics, medicine, the food industry and biology [37,38]. In the literature, for the synthesis of PPy and its doping, certain electrochemical methods were used often, namely CA, CV and chronopotentiometry (CP) [37]. The doping agent of PPy selected for this study was FeCN, which offered to the sensor good sensitivity, reproducibility and stability.

The analytical performances of the sensor developed in this study were evaluated in model solutions and in solutions obtained through pharmaceutical products with distinct concentrations of L-TRP from different manufacturers, followed by validation with the standard method based on Fourier-transform infrared (FT-IR) spectroscopy.

2. Materials and Methods

2.1. Reagents and Solutions

L-TRP (≥98%), potassium chloride (≥99.0%), potassium hexacyanoferrate (II) trihydrate (≥99.5%) and pyrrole (98%) were purchased from Sigma-Aldrich (St. Louis, MO, USA).

All reagents were of analytical grade, and all of the solutions within this study were prepared in ultrapure water (18.3 MΩ × cm, Milli—Q Simplicity® Water Purification System from Millipore Corporation) (Bedford, MA, USA).

Potassium chloride (KCl) was used as an electrolyte support, the concentration being somewhere around 0.1 M in all of the analyzed solutions.

In order to validate the obtained results with the modified sensor, two pharmaceutical products that contained L-TRP—Sleep Optimizer SOLARAY (150 mg tryptophan) and Cebrium NEUROPHARMA (1.02 mg tryptophan)—were used. The preparation of the solutions was realized every day, and only freshly solutions were used in the electrochemical determinations.

2.2. Instrumentation

For the modification and characterization of the sensors, two potentiostats were used: a model 263A EG&G potentiostat/galvanostat (Princeton Applied Research, Oak Ridge, TN, USA) controlled by ECHEM software and a Biologic SP 150 potentiostat/galvanostat (Bio-Logic Science Instruments SAS, Seyssinet-Pariset, France) controlled by EC—Lab Express software. The first potentiostat was used for modification of the electrodes with PPy (polypyrrole) through the CA (chronoamperometry) method, and the second potentiostat was used for obtaining the electrochemical responses of the sensors using the CV (cyclic voltammetry) method.

The analysis of the obtained data from chronoamperograms and voltammograms was carried out using Origin (version 6.0) and Microsoft Office Excel (version 2007) software.

The electrochemical cell used had a capacity of 15 mL, and the electrode system was formed from the reference electrode and the counter electrode integrated into the sensor device (counter electrode: carbon; reference electrode: Ag/AgCl). The working electrode was the modified electrode with PPy. For modification, an SPCE (DRP-C110 working in a solution) was used, which was purchased from Dropsens (www.dropsens.com, accessed on 20 May 2021).

For the TRP study, the FT-IR method was used with the Bruker ALPHA FT-IR spectrophotometer (BrukerOptik GmbH, Ettlingen, Germany), controlled by OPUS software (BrukerOptik GmbH, Ettlingen, Germany). An Elmasonic S10H ultrasonic bath was used for the dissolution of the compounds and the homogenization of the solutions. In addition, analytical balance, volumetric flasks, pipettes and micropipettes were used for weighting the solid samples and for the preparation of the solutions.

2.3. Preparation of PPy/FeCN/SPCE

For the deposition of PPy on the SPCE, a solution of 0.1 M pyrrole and 0.1 M FeCN was prepared. Then, 15 mL of the solution was introduced in the electrochemical cell, and in the solution, the DRP-C110 electrode was immersed, making connections to the EG&G potentiostat. The deposition of the PPy thin film in the presence of the doping agent was realized with the help of the CA method, employing the following working parameters: a potential of 0.8 V and a deposition time of 90 s. The obtained chronoamperograms related to the electropolymerization processes are presented in Figure 3 in two forms: the current's dependence on the time (Figure 3A) and the dependence of the electric charge on the time (Figure 3B). These chronoamperograms were recorded for three different sensors developed in the same experimental conditions.

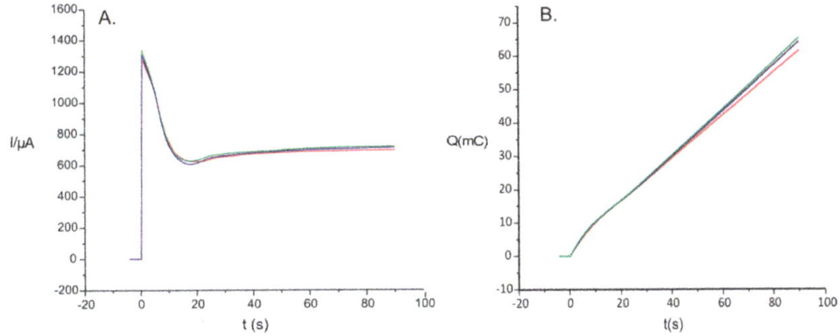

Figure 3. (**A**) Dependence of current (I) versus time (t) of curves registered in the electrosynthesis process of PPy /FeCN films at 0.8 V for 90 s. (**B**) Charge (Q) versus time (t) when PPy underwent electrosynthesis in the presence of FeCN for three replicate sensors.

The chronoamperograms presented in Figure 3A,B were compared for the purpose of determination of the reproducibility of electropolymerization. There were small differences, with a coefficient of variation of 2.9% between the chronoamperograms of the sensors prepared in the same conditions.

Additionally, from the chronoamperograms recorded during electropolymerization, the thickness of the PPy layer deposited on the surface of the carbon electrode was determined with the help of the ratio between the product of the oxidation load (Q), the molecular mass of the monomer (M = 67.0892 g/mol), the product of the number of electrons (n) involved with the Faraday constant (F = 96,485.33 C/mol), the active surface of the sensor (A) and the polymer's density (ρ_{PPy} = 1.5 g/cm^3) using the following equation [39]:

$$d = QM_w/nFA\rho \quad (1)$$

The value of d obtained for the PPy/FeCN sensor was of 0.203 μm, which demonstrates that the electrodeposition was facilitated in the presence of the doping agent FeCN, and the thickness of this film was optimal for the detection of L-TRP.

2.4. Samples Tested

Two pharmaceutical products that one can get without a medical prescription were tested with the PPy/FeCN sensor. Each one of them had different contents of the AA of L-TRP, and they came from different manufacturers, such as Cebrium (EVER NEURO PHARMA) (1.02 mg L-tryptophan per capsule) and Sleep Optimizer (SECOM) (150 mg L-tryptophan per capsule). Cebrium is a product that contains the following AAs besides L-TRP: glutamic acid, leucine, lysine, aspartic acid, arginine, phenylalanine, serine, threonine, valine, isoleucine, tyrosine, histidine and methionine.

Using the contents of the capsules mentioned earlier, they were prepared for different solutions with the purpose of analyzing three different concentrations for every sample, dissolved in 0.1 M KCl. The electrochemical analysis of the products was realized with the help of the CV method at a potential range between −1.0 V and +0.5 V at a scanning rate of 0.1 V × s^{-1}.

The standard FT-IR method was used for validating the obtained results of the electroanalytical method developed in this study. The IR spectra obtained for the two pharmaceutics with the Bruker ALPHA (BrukerOptik GmbH, Ettlingen, Germany) spectrophotometer ranged between 4000 and 500 cm^{-1}. The resolution of the recorded spectra was 4 cm^{-1} within 32 scans. The ZnSe crystals of this spectrophotometer were cleaned with ultra-pure water and isopropanol before adding the samples to be analyzed, eliminating possible impurities.

3. Results and Discussions

The sensor's analysis was performed with a Biologic SP 150 potentiostat/galvanostat by the CV electrochemical method in the potential range between −1.0 V and +0.5 V.

3.1. The Electrochemical Behavior of the Unmodified DRP-110 Electrode in 0.1 M KCl–10^{-3} M L-TRP Solution

In order to prove the device's efficiency developed in this study, few preliminary analyses were carried out. Before the modification, the SPCE was immersed in a solution of 0.1 M KCl–10^{-3} M L-TRP, and the cyclic voltamogram was recorded, comparing the results obtained from the unmodified sensor and the results from the modified sensor with PPy/FeCN. Using the electrochemical parameters mentioned above, the voltamogram obtained with the unmodified sensor at a scan rate of 0.1 V × s^{-1} is presented in Figure 4.

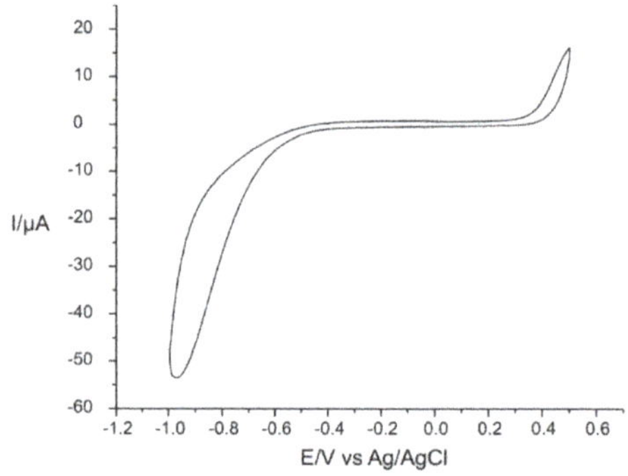

Figure 4. The electrochemical behavior of the unmodified sensor in a double solution of 0.1 M KCl and 10^{-3} M L-TRP at a scan rate of 0.1 V × s^{-1}.

It can be seen that the unmodified electrode did not show peaks in the studied potential range, and the background current was reduced. Therefore, the electrode's modification was necessary for L-TRP detection.

3.2. The Electrochemical Response of the Modified Electrode with PPy/FeCN in a 0.1 M KCl Solution and in a 0.1 M KCl–10^{-3} M L-TRP Double Solution

After the modification with PPy/FeCN, the electrochemical behavior of the sensor was initially analyzed in a 0.1 M KCl solution to observe the redox processes of PPy and the ferrocyanide ion included in the polymer matrix. In order to stabilize the sensor's signal in the electrolyte solution, six consecutive cycles were recorded. After this stage, the signal became stable. Using the same electrochemical parameters and going through the same stages with a scan rate of 0.1 V × s^{-1} and a potential range between −1.0 V and +0.5 V, the PPy/FeCN/SPCE sensor was analyzed in a double solution of 0.1 M KCl and 10^{-3} M L-TRP, where it was observed that the potentials and currents of the peaks, as well as their forms, were influenced by the presence of AA L-TRP through shifting the potentials of the peaks, increasing their currents, especially in the case of cathodic peaks. Figure 5 shows the stable signal of the modified sensor immersed in 0.1 M KCl overlayed with the stable signal of PPy/FeCN/SPCE immersed in a double solution of 0.1 M KCl and 10^{-3} M L-TRP.

Both in the first solution and in the second solution, we can see two pairs of peaks; the first pair (I) corresponded to the redox processes of the PPy polymer, and the second pair of peaks (II) was due to the redox processes of the doping ion immobilized in the polymer matrix (FeCN) [40]. The important difference between the two voltammograms was represented by the presence of AA L-TRP in the second solution, in which the electrode was immersed in the 10^{-3} M L-TRP solution, having as an electrolyte support 0.1 M KCl.

This difference can be seen in Table 3, which includes the potentials and peak currents for both redox systems obtained from the cyclic voltammograms recorded with the PPy/FeCN sensor in the two analyzed solutions.

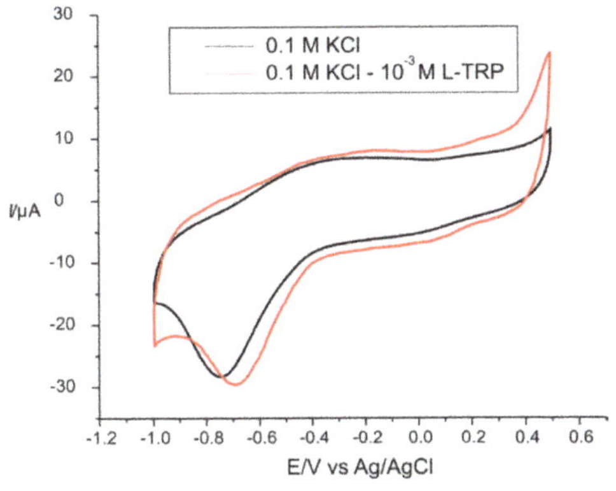

Figure 5. Electrochemical response of PPy/FeCN sensor immersed in a 0.1 M KCl solution (black line) and in a double solution 0.1 M KCl–10^{-3} M L-TRP (red line) at 0.1 V × s^{-1}.

Table 3. The potentials and currents of the peaks of PPy-modified sensors and immersed in a 0.1 M KCl solution and in a double solution 0.1 M KCl–10^{-3} M L-TRP at scan rate of 0.1 V × s^{-1}.

Sensor	Solution		Electrochemical Parameters					
			E_{pa} [1] (V)	E_{pc} [2] (V)	ΔE [3] (V)	I_{pa} [4] (μA)	I_{pc} [5] (μA)	I_{pc}/I_{pa}
PPy/FeCN/SPCE	0.1 M KCl	Redox system I	−0.27	−0.74	−1.01	6.76	−28.33	4.19
		Redox system II	0.16	0.03	0.13	7.07	−4.94	0.69
PPy/FeCN/SPCE	0.1 M KCl–10^{-3} M L-TRP	Redox system I	−0.31	−0.78	0.47	6.81	−31.24	4.58
		Redox system II	0.22	0.06	0.16	9.33	−7.70	0.82

[1] Potential of the anodic peak. [2] Potential of the cathodic peak. [3] ΔE = $E_{pa} - E_{pc}$. [4] Current of the anodic peak. [5] Current of the cathodic peak.

It was proven by the obtained results that the PPy/FeCN/SPCE sensor could be useful for the detection of L-TRP, similar to reports from other scientific works, with the mention that some characteristics of the electrode (such as the electrode's surface and the modifier material) and some electrochemical parameters (the potential field and the scan rate) were not the same.

The TRP showed redox activity, being oxidized to 2-amino-3- (5-oxo-3,5-dihydro-2H-indol-3-yl) propionic acid, with two protons and two electrons involved in the oxidation reaction [41]. The mechanism of the oxidation reaction is illustrated in Figure 6.

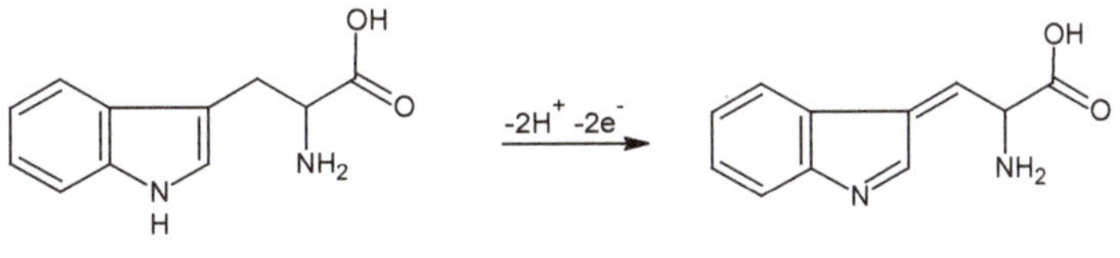

Figure 6. Process of the electrochemical oxidation of L-TRP (adapted from [35]).

3.3. The Influence of the Scan Rate on the Sensor Responses Immersed in a 0.1 M KCl and 10^{-3} M L-TRP Solution

The proposed sensor for L-TRP detection was immersed in the double solution of 10^{-3} M L-TRP and 0.1 M KCl, recording cyclic voltammograms at 10 different scan rates, and the results are shown in Figure 7A. The scan rates varied between 0.1 and 1.0 V × s^{-1}. Playing an important role for the electrochemical measurements, the scan rate influenced the redox processes. In the case of the intensity of the cathodic peak for the PPy/FeCN electrode, a linear dependence with the scan rate was observed. That way, the determining factor of rate could be established. From the linear relationship between the intensity of the cathodic peak and the scan rate (Figure 7B), it was concluded that the redox process from the level of the sensitive element was controled by the electron transfer from the electrochemical reaction. The linear fitting equation was $I_{pc} = 0.87268 \times v - 20.486$, and the coefficient of determination (R^2) was 0.9901, showing good quality for the linear regression model.

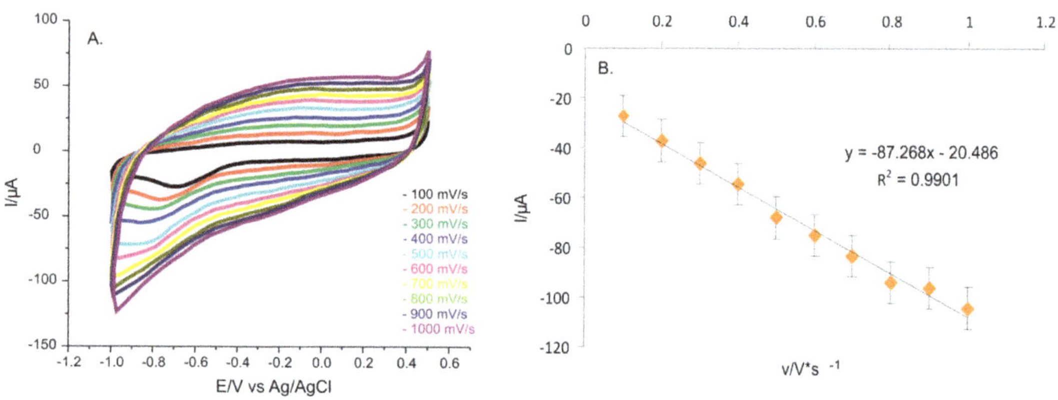

Figure 7. (**A**) CVs of PPy/FeCN/SPCE sensor immersed in a 0.1 M KCl and 10^{-3} M L-TRP solution at scan rates between 0.1 and 1.0 V × s^{-1}. (**B**) The plot of the linear dependence between I_{pc} and the scan rate.

Laviron's equation, $I_{pc} = n^2 F^2 \Gamma Av/4 RT$ (where *n* is the number of transfered electrons, F is Faraday's constant (F = 96 485 C mol^{-1}), A is the covered surface expressed in mol × cm^{-2}, v is the scan rate, R is the universal gas constant (R = 8.314 J mol^{-1} K^{-1}) and T is the absolute temperature (T = 298 K)), allowing the calculation of the degree of coverage on the electrode surface with active centers (Γ), that being 1.76×10^{-10} mol × cm^{-2}, considering the linear equation between the scan rate and the current of the most intense cathodic peak. This value was close to the results reported in the literature [42].

In conclusion, the modified elctrode with PPy/FeCN demonstrated a high sensitivity for L-TRP from the solution to be analyzed, the peaks being more intense with the increasing scan rate.

3.4. Influence of the Concentration on Responses of the Sensor Immersed in a 0.1 M KCl and 10^{-3} M L-TRP Solution: Calibration Curve

The concentration of the analyzed solution proved to be important for the electrochemical responses of the PPy/FeCN sensor, used in the present study solutions with different concentrations of L-TRP dissolved in a solution of 0.1 M KCl. The concentration range studied was between 3.33×10^{-7} M and 2.72×10^{-5} M, recording the cyclic voltamograms presented in Figure 8A with a scan rate of 0.1 V × s^{-1} and in the potential range between −1.0 V and +0.5 V. Figure 8B presents the most intense cathodic peak currents as a function of the L-TRP concentration. The calibration curve was obtained from the dependence of the cathodic current on the concentration (Figure 8C). The linearity range (Figure 8D) was observed to be between 3.3 × 10^{-7} M and 1.06 × 10^{-5} M, and the calibration equation and the calculated values of the LOD and LOQ are reported in Table 4.

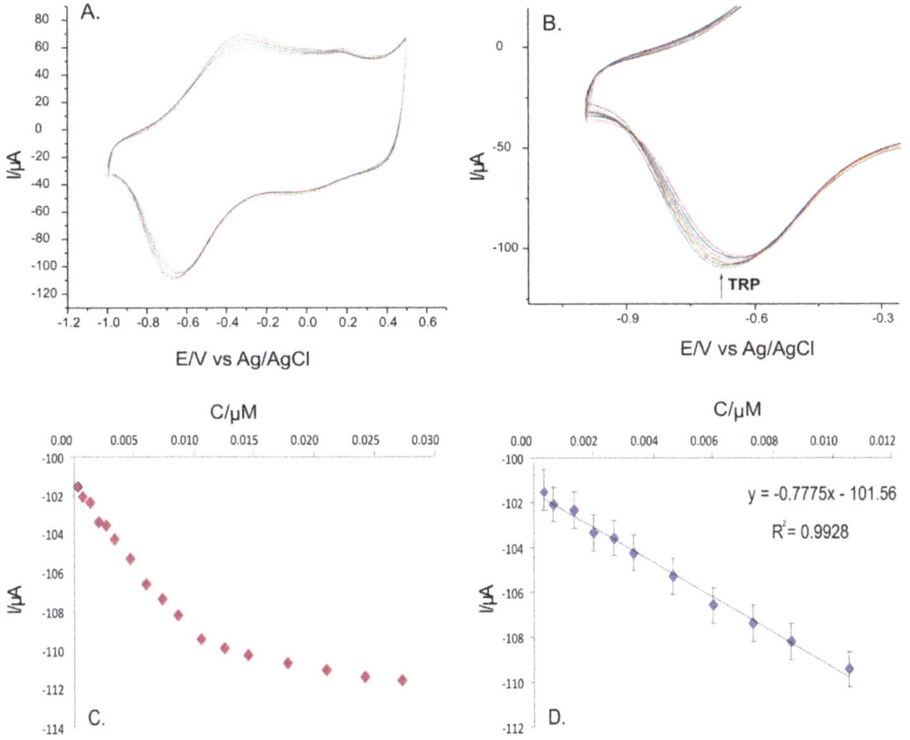

Figure 8. (**A**) Cyclic voltammograms of the PPy/FeCN sensor immersed in a double solution of 0.1 M KCl and 10^{-3} M L-TRP at different concentrations between 3.33 × 10^{-7} M and 2.72 × 10^{-5} M. (**B**) Zoomed in image on the catodic peak for which the calibration curve was developed. (**C**) Variation of the current of cathodic peak I with the TRP concentration. (**D**) Calibration curve in the range between 3.3 × 10^{-7} M and 1.06 × 10^{-5} M.

Table 4. LOD and LOQ obtained with the PPy/FeCN sensor detecting L-TRP.

Sensor	LOD [1] (M)	LOQ [2] (M)
PPy/FeCN-SPCE	1.05 × 10^{-7}	3.51 × 10^{-7}

[1] Limit of detection. [2] Limit of quantification.

The LOD and LOQ were calculated using the slope of the calibration equation and the standard deviation of the sensor response in a blank solution [38].

The detection limit obtained with the PPy/FeCN sensor was lower than the sensor's LODs presented in Table 2, making it possible to be used for the sensitive detection of L-TRP from pharmaceutical samples.

3.5. Method Precision, Stability and Reproducibility

Precision studies performed for the Ppy/FeCN sensor were performed both interday and intraday, based on solutions with L-TRP contents alongside the concentration of 5 × 10^{-6} M. The interday precision was evaluated on 4 distinct days, and the intraday precision was analyzed in 3 different moments of the day at an interval of 2 h. The relative standard deviation (RSD (%)) presented the following values: 4.2% interday and 3.8%

intraday. The good stability of the Ppy/FeCN sensor, both in the short and long term, was demonstrated by the CV method. In the short term, there were 30 consecutive scans recorded with the sensor developed in a double solution of 5×10^{-5} M L-TRP–0.1 M KCl, keeping the intensity of the peaks at 97.8% compared with the initial response. In the long term, there 96% stability out of the initial response was obtained after 5 days, with the RSD representing a value of 95%.

In addition, the sensor's reproductibility was created in a double solution of 5×10^{-5} M L-TRP–0.1 M KCl, preparing three different sensors. The RSD value for the cathodic peak observed in all 3 cases was 3.1%.

3.6. Validation of the Modified Sensor by Quantitative Determination of L TRP in Pharmaceutical Samples

A series of existing products of the pharmaceutical market contains the active compound L-TRP, which is the subject of the present study. Of these, Cebrium and Sleep Optimizer were tested for sensor validation using the FT-IR method, as well as comparing the electroanalytical results with those indicated by the manufacturers. The two pharmaceutical products have different compositions, different concentrations of L-TRP and come from different manufacturers (Cebrium: 1.02 mg L-TRP per capsule, Ever Neuro Pharma; Sleep Optimizer: 150 mg L-TRP per capsule, Solaray).

The L-TRP from the two products was quantified through two methods—CV and FT-IR—where CV was the method developed in this study, and FT-IR was the standard method.

Figure 9 presents the responses of the sensor immersed in solutions of L-TRP obtained from the analyzed pharmaceutical products. The estimated concentration of L-TRP in the solutions was 5×10^{-6} M. It was observed that the signals were in solutions made out of pharmaceutical products, which were influenced by the presence of other substances, especially in the anodic part of the cyclic voltamogram. L-TRP quantification was performed from the cathodic peak current using the calibration linear equation of I vs. c.

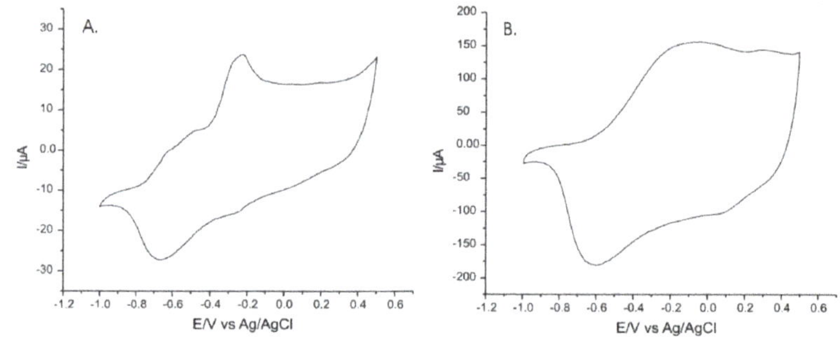

Figure 9. Voltammetric responses of the PPy/FeCN sensor in solutions of (**A**) Cebrium (EVER NEURO PHARMA) and (**B**) Sleep Optimizer (SOLARAY).

For the TRP quantification using the FT-IR method, we used the two solid standards formed of pure TRP and KBr of 1 mg/g and 150 mg/g concentrations, respectively, similar values with those of pharmaceutical products. The absorbance of the 1650 cm^{-1} peak characteristic of the vibration of the N-H group (bending) was used for quantification [43]. The results are presented in Table 5.

Table 5. Quantitative determination of L-TRP in pharmaceutical products.

Pharmaceutical Product	L-TRP Concentration Reported by Producer (mg)	Concentration of L-TRP	
		Method CV ([1]/mg)	Method FT-IR ([2]/mg)
Cebrium EVER NEURO PHARMA	1.02 mg	1.04 ± 0.05	1.06 ± 0.06
Sleep Optimizer SOLARAY	150 mg	150 ± 3	150 ± 4

[1] Cyclic voltammetry. [2] Fourier-transform infrared spectroscopy.

It was observed that the differences between the three sets of values were very small, demonstrating that the quantification method of L-TRP from pharmaceutical products with the sensor had a very good accuracy and could be used in laboratory practice.

4. Conclusions

The CA method proved to be efficient for the PPy doped with FeCN deposition through electropolymerization on an SPCE's surface. The developed sensor in this study, PPy/FeCN/SPCE, presents utility in detecting L-TRP both from model solutions and from pharmaceutical products, showing excellent electroanalytical results, with higher sensibility, precision and good stability. The fast response, low cost and the variety of the fields in which this new device could be applied for the L-TRP, namely the medicine, pharmaceutical, chemistry and food industries, are important advantages for placement in the commercial market, contributing to the control of pharmaceutical products, monitoring some effects caused by an L-TRP deficiency or excess and food quality control.

Author Contributions: Conceptualization, C.A. and A.D.; methodology, C.A.; validation, C.A. and A.D.; formal analysis, A.D.; investigation, C.A. and A.D.; data curation, C.A. and A.D.; writing—original draft preparation, A.D.; writing—review and editing, C.A.; supervision, C.A. All authors have read and agreed to the published version of the manuscript.

Funding: The contribution of the author Ancuta Dinu was supported by the project ANTREPRENOR-DOC, in the framework of the Human Resources Development Operational Programme 2014–2020, financed from the European Social Fund under the contract number 36355/23.05.2019 HRD OP /380/6/13–SMIS Code: 123847.

Institutional Review Board Statement: Not applicable.

Informed Consent Statement: Not applicable.

Data Availability Statement: The authors confirm that the data supporting the findings of this studyare available within the article.

Acknowledgments: The results of this work have been presented to the 9th edition of the Scientific Conference organized by the Doctoral Schools of "Dunărea de Jos" University of Galati (SCDS-UDJG) (http://www.cssd-udjg.ugal.ro/ (accessed on 10 July 2021)) [4] that will be held on 10 and 11 June 2021 in Galati, Romania.

Conflicts of Interest: The authors declare no conflict of interest. The funders had no role in the design of the study; in the collection, analyses, or interpretation of data; in the writing of the manuscript, or in the decision to publish the results.

References

1. Sadok, I.; Tyszczuk-Rotko, K.; Mroczka, R.; Staniszewska, M. Simultaneous voltammetric analysis of tryptophan and kynurenine in culture medium from human cancer cells. *Talanta* **2020**, *209*, 120574. [CrossRef]
2. Liu, X.; Luo, L.; Ding, Y.; Ye, D. Poly-glutamic acid modified carbon nanotube-doped carbon paste electrode for sensitive detection of L-tryptophan. *Bioelectrochemistry* **2011**, *82*, 38–45. [CrossRef] [PubMed]
3. Kagan, J.; Sharon, I.; Béjà, O.; Kuhn, J.C. The tryptophan pathway genes of the Sargasso Sea metagenome: New operon structures and the prevalence of non-operon organization. *Genome Biol.* **2008**, *9*, R20. [CrossRef] [PubMed]

4. Dinu, A.; Apetrei, C. A Review on Electrochemical Sensors and Biosensors Used in Phenylalanine Electroanalysis. *Sensors* **2020**, *20*, 2496. [CrossRef] [PubMed]
5. Yu, L.-Y.; Liu, Q.; Wu, X.-W.; Jiang, X.-Y.; Yu, J.-G.; Chen, X.-Q. Chiral electrochemical recognition of tryptophan enantiomers at a multi-walled carbon nanotube–chitosan composite modified glassy carbon electrode. *RSC Adv.* **2015**, *5*, 98020–98025. [CrossRef]
6. Colombo, S.; Coluccini, C.; Caricato, M.; Gargiulli, C.; Gattuso, G.; Pasini, D. Shape selectivity in the synthesis of chiral macrocyclic amides. *Tetrahedron* **2010**, *66*, 4206–4211. [CrossRef]
7. Caricato, M.; Olmo, A.; Gargiulli, C.; Gattuso, G.; Pasini, D. A 'clicked' macrocyclic probe incorporating Binol as the signalling unit for the chiroptical sensing of anions. *Tetrahedron* **2012**, *68*, 7861–7866. [CrossRef]
8. Kałużna-Czaplińska, J.; Gątarek, P.; Chirumbolo, S.; Chartrand, M.S.; Bjørklund, G. How important is tryptophan in human health? *Crit. Rev. Food Sci. Nutr.* **2019**, *59*, 72–88. [CrossRef] [PubMed]
9. He, Q.; Tian, Y.; Wu, Y.; Liu, J.; Li, G.; Deng, P.; Chen, D. Electrochemical Sensor for Rapid and Sensitive Detection of Tryptophan by a Cu_2O Nanoparticles-Coated Reduced Graphene Oxide Nanocomposite. *Biomolecules* **2019**, *9*, 176. [CrossRef]
10. Heine, W.; Radke, M.; Wutzke, K.-D. The significance of tryptophan in human nutrition. *Amino Acids* **1995**, *9*, 91–205. [CrossRef]
11. Eroğlu, I.; Eroğlu, B.Ç.; Güven, G.S. Altered tryptophan absorption and metabolism could underlie long-term symptoms in survivors of coronavirus disease 2019 (COVID-19). *Nutrition* **2021**, *90*, 111308. [CrossRef]
12. Shaw, K.A.; Turner, J.; Del Mar, C. Tryptophan and 5-Hydroxytryptophan for depression. *Cochrane Database Syst. Rev.* **2002**. [CrossRef]
13. Hudson, C.; Hudson, S.; MacKenzie, J. Protein-source tryptophan as an efficacious treatment for social anxiety disorder: A pilot studyThis article is one of a selection of papers published in this special issue (part 1 of 2) on the Safety and Efficacy of Natural Health Products. *Can. J. Physiol. Pharmacol.* **2007**, *85*, 928–932. [CrossRef] [PubMed]
14. Schneider-Helmert, D.; Spinweber, C.L. Evaluation of L-Tryptophan for Treatment of Insomnia: A Review. *Psychopharmacology* **1986**, *89*. [CrossRef] [PubMed]
15. Mette, C.; Zimmermann, M.; Grabemann, M.; Abdel-Hamid, M.; Uekermann, J.; Biskup, C.S.; Wiltfang, J.; Zepf, F.D.; Kis, B. The impact of acute tryptophan depletion on attentional performance in adult patients with ADHD. *Acta Psychiatr. Scand.* **2013**, *128*, 124–132. [CrossRef]
16. Menkes, D.B.; Coates, D.C.; Fawcett, J.P. Acute tryptophan depletion aggravates premenstrual syndrome. *J. Affect. Disord.* **1994**, *32*, 37–44. [CrossRef]
17. Lovelace, M.D.; Varney, B.; Sundaram, G.; Franco, N.F.; Ng, M.L.; Pai, S.; Lim, C.K.; Guillemin, G.J.; Brew, B.J. Current Evidence for a Role of the Kynurenine Pathway of Tryptophan Metabolism in Multiple Sclerosis. *Front. Immunol.* **2016**, *7*, 246. [CrossRef]
18. Liu, Y.; Tian, A.; Wang, X.; Qi, J.; Wang, F.; Ma, Y.; Ito, Y.; Wei, Y. Fabrication of chiral amino acid ionic liquid modified magnetic multifunctional nanospheres for centrifugal chiral chromatography separation of racemates. *J. Chromatogr. A* **2015**, *1400*, 40–46. [CrossRef] [PubMed]
19. Boulet, L.; Faure, P.; Flore, P.; Montérémal, J.; Ducros, V. Simultaneous determination of tryptophan and 8 metabolites in human plasma by liquid chromatography/tandem mass spectrometry. *J. Chromatogr. B* **2017**, *1054*, 36–43. [CrossRef]
20. Yıldız, C.; Bayraktepe, D.E.; Yazan, Z. Electrochemical low-level detection of l-tryptophan in human urine samples: Use of pencil graphite leads as electrodes for a fast and cost-effective voltammetric method. *Mon. Chem.* **2020**, *151*, 871–879. [CrossRef]
21. Whiley, L.; Nye, L.C.; Grant, I.; Andreas, N.J.; Chappell, K.E.; Sarafian, M.H.; Misra, R.; Plumb, R.S.; Lewis, M.R.; Nicholson, J.K.; et al. Ultrahigh-Performance Liquid Chromatography Tandem Mass Spectrometry with Electrospray Ionization Quantification of Tryptophan Metabolites and Markers of Gut Health in Serum and Plasma—Application to Clinical and Epidemiology Cohorts. *Anal. Chem.* **2019**, *91*, 5207–5216. [CrossRef]
22. Hawkins, G.; Zipkin, I.; Marshall, L. Determination of Uric Acid, Tyrosine, Tryptophan, and Protein in whole Human Parotid Saliva by Ultraviolet Absorption Spectrophotometry. *J. Dent. Res.* **1963**, *42*, 1015–1022. [CrossRef]
23. Fan, Y.; Liu, J.-H.; Lu, H.-T.; Zhang, Q. Electrochemistry and voltammetric determination of L-tryptophan and L-tyrosine using a glassy carbon electrode modified with a Nafion/TiO_2-graphene composite film. *Microchim. Acta* **2011**, *173*, 241–247. [CrossRef]
24. Zhao, J. Simultaneous determination of plasma creatinine, uric acid, kynurenine and tryptophan by high-performance liquid chromatography: Method validation and in application to the assessment of renal function: Simultaneous Determination of Cr, UA, KYN and TRP by HPLC. *Biomed. Chromatogr.* **2015**, *29*, 410–415. [CrossRef] [PubMed]
25. Ritota, M.; Manzi, P. Rapid Determination of Total Tryptophan in Yoghurt by Ultra High Performance Liquid Chromatography with Fluorescence Detection. *Molecules* **2020**, *25*, 5025. [CrossRef] [PubMed]
26. Zhao, M.; Zhou, M.-F.; Feng, H.; Cong, X.-X.; Wang, X.-L. Determination of Tryptophan, Glutathione, and Uric Acid in Human Whole Blood Extract by Capillary Electrophoresis with a One-Step Electrochemically Reduced Graphene Oxide Modified Microelectrode. *Chromatographia* **2016**, *79*, 911–918. [CrossRef]
27. Li, S.; Xing, M.; Wang, H.; Zhang, L.; Zhong, Y.; Chen, L. Determination of tryptophan and tyrosine by chemiluminescence based on a luminol–N-bromosuccinimide–ZnS quantum dots system. *RSC Adv.* **2015**, *5*, 59286–59291. [CrossRef]
28. Moscetti, I.; Cannistraro, S.; Bizzarri, A.R. Probing direct interaction of oncomiR-21-3p with the tumor suppressor p53 by fluorescence, FRET and atomic force spectroscopy. *Arch. Biochem. Biophys.* **2019**, *671*, 35–41. [CrossRef] [PubMed]
29. Viter, R.; Iatsunskyi, I. Metal Oxide Nanostructures in Sensing. In *Nanomaterials Design for Sensing Applications*; Elsevier: Amsterdam, The Netherlands, 2019; pp. 41–91. ISBN 978-0-12-814505-0.

30. Tığ, G.A. Development of electrochemical sensor for detection of ascorbic acid, dopamine, uric acid and L-tryptophan based on Ag nanoparticles and poly(L-arginine)-graphene oxide composite. *J. Electroanal. Chem.* **2017**, *807*, 19–28. [CrossRef]
31. Hashkavayi, A.B.; Raoof, J.B.; Park, K.S. Sensitive Electrochemical Detection of Tryptophan Using a Hemin/G-Quadruplex Aptasensor. *Chemosensors* **2020**, *8*, 100. [CrossRef]
32. Ensafi, A.A.; Hajian, R. Determination of tryptophan and histidine by adsorptive cathodic stripping voltammetry using H-point standard addition method. *Anal. Chim. Acta* **2006**, *580*, 236–243. [CrossRef]
33. Xia, Y.; Zhao, F.; Zeng, B. A molecularly imprinted copolymer based electrochemical sensor for the highly sensitive detection of L-Tryptophan. *Talanta* **2020**, *206*, 120245. [CrossRef]
34. He, Q.; Liu, J.; Feng, J.; Wu, Y.; Tian, Y.; Li, G.; Chen, D. Sensitive Voltammetric Sensor for Tryptophan Detection by Using Polyvinylpyrrolidone Functionalized Graphene/GCE. *Nanomaterials* **2020**, *10*, 125. [CrossRef] [PubMed]
35. Duan, S.; Wang, W.; Yu, C.; Liu, M.; Yu, L. Development of Electrochemical Sensor for Detection of L-Tryptophan Based on Exfoliated Graphene/PEDOT:PSS. *Nano* **2019**, *14*, 1950058. [CrossRef]
36. Tat'yana, V.V.; Efimov, O.N. Polypyrrole: A Conducting Polymer; Its Synthesis, Properties and Applications. *Russ. Chem. Rev.* **1997**, *66*, 443.
37. Gniadek, M.; Wichowska, A.; Antos-Bielska, M.; Orlowski, P.; Krzyzowska, M.; Donten, M. Synthesis and characterization of polypyrrole and its composites coatings on flexible surface and its antibacterial properties. *Synth. Met.* **2020**, *266*, 116430. [CrossRef]
38. Dinu, A.; Apetrei, C. Development of Polypyrrole Modified Screen-Printed Carbon Electrode Based Sensors for Determination of L-Tyrosine in Pharmaceutical Products. *IJMS* **2021**, *22*, 7528. [CrossRef]
39. Bieńkowski, K.; Strawski, M.; Szklarczyk, M. The determination of the thickness of electrodeposited polymeric films by AFM and electrochemical techniques. *J. Electroanal. Chem.* **2011**, *662*, 196–203. [CrossRef]
40. Apetrei, C. Novel method based on polypyrrole-modified sensors and emulsions for the evaluation of bitterness in extra virgin olive oils. *Food Res. Int.* **2012**, *48*, 673–680. [CrossRef]
41. Ghoreishi, S.M.; Behpour, M.; Ghoreishi, F.S.; Mousavi, S. Voltammetric determination of tryptophan in the presence of uric acid and dopamine using carbon paste electrode modified with multi-walled carbon nanotubes. *Arab. J. Chem.* **2017**, *10*, S1546–S1552. [CrossRef]
42. Dinu, A.; Apetrei, C. Voltammetric Determination of Phenylalanine Using Chemically Modified Screen-Printed Based Sensors. *Chemosensors* **2020**, *8*, 113. [CrossRef]
43. Zhang, J.; Wang, D.; Li, Y. Ratiometric Electrochemical Sensors Associated with Self-Cleaning Electrodes for Simultaneous Detection of Adrenaline, Serotonin, and Tryptophan. *ACS Appl. Mater. Interfaces* **2019**, *11*, 13557–13563. [CrossRef] [PubMed]

inventions

MDPI

Article

Implementation Aspects Regarding Closed-Loop Control Systems Using Evolutionary Algorithms

Viorel Minzu [1,*], Saïd Riahi [1,2] and Eugen Rusu [3]

1 Control and Electrical Engineering Department, "Dunarea de Jos" University, 800008 Galati, Romania; riahizsaid@gmail.com
2 UR-LAPER, Faculty of Sciences of Tunis, University of Tunis El Manar, Tunis 1068, Tunisia
3 Mechanical Engineering Department, "Dunarea de Jos" University, 800008 Galati, Romania; evcrusu@gmail.com
* Correspondence: viorel.minzu@ugal.ro

Abstract: When an optimal control problem requires an important computational effort, a metaheuristic algorithm (MA) can be a useful approach. An MA is conceived to solve a specific optimal control problem having a characteristic objective function. This algorithm solely yields only an optimal offline solution. The desideratum to have a closed-loop implementation can be fulfilled through a supplementary "tool", the Receding Horizon Control (RHC) structure. This paper addresses a particular case and integrates Evolutionary Algorithms into the RHC structure. The main objective is to propose a general harmonization between the Controller of the closed loop and the Evolutionary Algorithm. Some details concerning the implementation of the closed loop and Controller are described. The impact of the RHC's prediction technique upon the control sequences' encoding is also analyzed. Two general structure Controllers are proposed, one of them conceived to cope with restrictive time constraints. Practical ideas have been illustrated through a case study: the well-known optimal control of a fed-batch reactor for ethanol production. This time, our implementation achieves a closed-loop solution. The results from the programs and simulation series validate the Controllers, EAs, and the closed-loop structure. Generally speaking, the association between RHC and EA can be a realistic solution to optimal process control.

Keywords: optimal control; metaheuristic algorithms; evolutionary algorithms; simulation

Citation: Minzu, V.; Riahi, S.; Rusu, E. Implementation Aspects regarding Closed-Loop Control Systems Using Evolutionary Algorithms. *Inventions* **2021**, *6*, 53. https://doi.org/10.3390/inventions6030053

Academic Editor: Stefano Mariani

Received: 6 June 2021
Accepted: 16 July 2021
Published: 21 July 2021

1. Introduction

Optimal control of a dynamic system is a usual task in process engineering. If the process has mathematical properties sufficient to apply theoretical control laws, this task can sometimes be achieved without significant computational complexity. Other times the computational effort is important, or we do not dispose of known control techniques. In this case, metaheuristic algorithms integrated within appropriate control structures constitute a realistic solution. This kind of algorithms (see [1–3]) have been used intensively in control engineering (see [4–7]) because of their robustness and capacity to cope with big complexity problems.

A metaheuristic algorithm (MA), like simulated annealing, genetic algorithm, evolutionary algorithm (AE), ant colony systems, particle swarm optimization, etc., is not by itself prepared to achieve a closed-loop control structure. That is why it needs to be integrated into a structure like Receding Horizon Control (RHC) (see [8–10]) or Model Predictive Control ([11,12]). Both structures use prediction techniques that could involve MAs because prediction is usually optimization.

This work has rather practical relevance because it is first addressed to a practicing professional engineer who wants to implement an optimal control structure using an Evolutionary Algorithm (see [11,13–15]). More precisely, there are several requests and given data that define our work:

- A real process has to be controlled over a given control horizon;
- We dispose of a process model (usually nonlinear), usually a set of all algebraic and differential constraints imposed by the dynamic environment;
- A closed-loop control structure has to solve an Optimal Control Problem (OCP). Its Controller should optimize an objective function and meet some constraints over the control horizon. The objective function contains a terminal penalty term, which renders the prediction of the optimal control sequence a complex task;
- A basic version of the EA can find the optimal value of the objective function over the control horizon.

The RHC structure is a general and simple control structure involving a prediction technique that can fit with an evolutionary algorithm. For this reason, the solution proposed by this paper is based on the RHC structure whose Controller makes optimal predictions using an EA. The implementation aspects emphasized in our presentation mainly refer to two topics:

- The harmonization between the prediction RHC structure technique and the EA;
- The implementation of a Controller based on an EA with an acceptable computational complexity.

Section 2 briefly presents the elements that define an OCP and introduces the notations used in the next sections.

After recalling the RHC structure, Section 3 proposes the first version of the Controller. This is organized as a function and includes the slightly modified basic version of the EA. At each sampling moment, the EA is executed to determine the optimal control sequence covering the prediction horizon. The Controller takes the first value of this sequence and makes it its current control output. After that, this value is applied to the process. At each call, the prediction horizon decreases and keeps the final moment of the control horizon (see for comparison [9–11]).

In this work, we propose two versions of the Controller. The EA calls for a numerical integration function using a discretization step equal to the closed-loop sampling period. Consequently, the encoding of a control sequence uses a number of genes proportional to the prediction horizon's length. The greater computation effort of the Controller is sustained in the first sampling period when the prediction horizon is the largest. Using simulations, the implementer has to decide whether the Controller's execution time is less than the sampling period; that is, the Controller can be implemented.

The second version of the Controller, proposed in Section 4, improves the computational complexity such that the execution time would be lower than a sampling period. Regardless of the prediction horizon's length, the discretization steps' number is constant. In other words, the encoding of a control sequence always uses the same number of genes. If the increase of the discretization steps does not significantly affect the computation accuracy, this version can improve the computational complexity and assure the feasibility of the Controller.

Section 5 presents a case study concerning a particular OCP and the controller implementation using an adequate EA. The performance index is a terminal penalty. The main objective is to show how a peculiar OCP can be solved and implemented using such a controller. Notice that we do not aim to solve a particular OCP and compare our solution with other approaches but, rather, highlight some implementation aspects. The resulting RHC structure will be simulated to validate the quasi-optimal behaviour of the closed loop.

The addressed problem deals with the *optimal control* of a fed-batch reactor for *ethanol production* (OCEP). First of all, an EA is chosen to solve this problem and yield an open-loop solution. Section 5.1 describes an evolutionary algorithm (EAv0) that solves the OCEP problem. Although EAv0 does not refer to a closed-loop implementation, it will prove its capacity to solve the open-loop problem. It also allows us to assess the computational complexity. A simulation series is conducted to evaluate the performances of a typical solution and its state evolution.

The implementation of the Receding Horizon Controller for the OCEP problem is presented in Section 5.2. EAv0 is slightly modified to obtain EAv1, which is a constituent of the Controller. The control sequence is encoded by a vector of real values whose dimension is proportional to the prediction horizon's length. The closed-loop evolution (Controller's outputs and the process' states) is simulated, considering the real process, which is identical to the process model. The computation complexity is evaluated on the overall control horizon, using simplifying hypothesis.

A Controller's second version—used to solve the OCEP problem—which can improve the computation complexity, is presented in Section 5.3. Algorithm EAv0 is modified to generate EAv2, such that a control sequence is encoded by a constant number of real values, regardless of the prediction horizon's length. The evolution of the closed loop, including the new Controller, is simulated. The computation complexity of EAv2 is analyzed in contrast with EAv1 to prove that the first one is smaller, at least in the first sampling periods.

The results of the simulation series have proved that the RHC structure with the Controller that is able to integrate an EA is feasible with good quasi-optimal behaviour.

2. Defining Elements of an Optimal Control Problem

In the upcoming sections, we briefly recall the defining elements of an optimal control problem (OCP) and their usual notations, but we do so in a way that is adapted to the presentation of the next sections. Let us consider an OCP whose control horizon is finite, for example $[0,\ H \cdot T]$, where H is a positive integer and T is the sampling period. The discrete moments $t_k = k{\cdot}T$ will be specified simply by $k = 0, 1, \cdots, H$.

We suppose that a discrete process model, used to simplify the presentation, is available and given in Equation (1). An eventual continuous model can always be converted into a discrete one.

$$x(k+1) = f(k, x(k), u(k));\ k = 0,\ 1, \cdots,\ H-1, \tag{1}$$

The vectors $u(k)$, $x(k)$ have m and n elements ($m, n \in N^+$), respectively, and f is an n-dimensional vector function with all regularity properties demanded by the calculations.

Any control sequence U_0, defined as below, meets certain constraints and determines the system evolution over the control horizon:

$$U_0 \stackrel{D}{=} < u(0),\ u(1), \ldots,\ u(H-1) > \tag{2}$$

This system evolution has to be optimal under a performance index. The latter minimizes a specific objective function, which has the discrete form presented in Equation (3).

$$J(k, x(k), U(k)) = \sum_{i=k}^{H-1} L(i, x(i), u(i)) + TP(x(H)), k = 0,\ 1, \cdots,\ H-1 \tag{3}$$

$$U(k) \stackrel{D}{=} < u(k),\ u(k+1), \ldots,\ u(H-1) > \tag{4}$$

Equation (3) refers to the prediction horizon $[k,\ H]$ and the state trajectory that starts with $x(k)$. The system evolution is generated by a control sequence $U(k)$ having the structure presented in Equation (4). The resulting state trajectory can be denoted by

$$X(k) \stackrel{D}{=} < x(k),\ x(k+1), \ldots,\ x(H) >;\ k = 0,\ 1, \cdots,\ H-1. \tag{5}$$

The first part of the sum presented in Equation (3) is a Lagrange-type term that measures quality along the trajectory of the dynamic system. The scalar value $L(i, x(i), u(i))$ is the contribution of the interval $[i, i+1]$ to the quality measure J. The second part is a Mayer-type term that measures the quality of the trajectory in its final extremity. This term will be called the terminal penalty in the upcoming sections. The function $TP(x)$ penalizes the final state x.

The performance index I_0 of an optimal solution over the control horizon minimizes the objective function J as below:

$$I_0 = \min_{U_0} J(0, x_0, U_0);\ x_0 = x(0) \tag{6}$$

Obviously, J's minimization (or maximization) can be subjected to constraints regarding the control inputs and the state variables.

3. Controller Implementation Using Evolutionary Algorithms

3.1. RHC Using Metaheuristic Algorithms

If problem solving requires an important computational effort, a metaheuristic algorithm can be a useful approach. An MA is conceived to solve a specific OCP having a characteristic objective function. However, the desideratum to have a closed-loop implementation can be fulfilled through a supplementary "tool": the RHC structure.

Figure 1 recalls the principle of the RHC and suggests how it is able to implement the predictive control technique in a particular case when a metaheuristic algorithm is used.

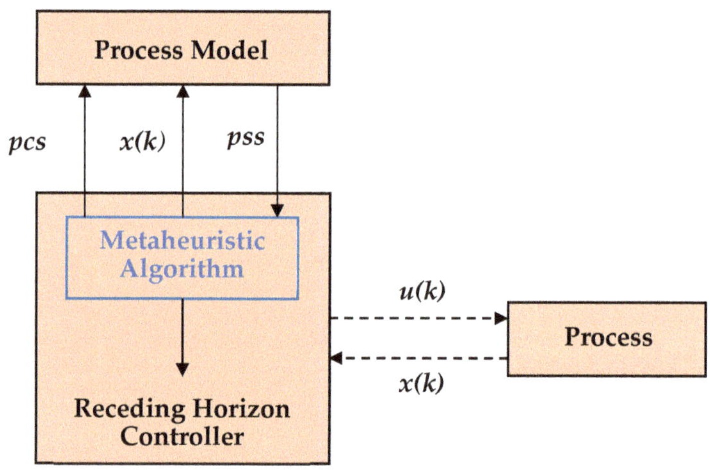

Figure 1. The principle of Receding Horizon Control using MA.

The objective of RHC is to implement closed-loop control systems under circumstances involving optimal evolution. The RHC is among the very few control strategies supporting a closed-loop implementation using an MA. The Receding Horizon Controller uses the MA to accomplish the prediction technique: it calculates the optimal control sequences (*ocs*) that optimize the performance index on the current prediction horizon $[k, H]$.

At moment k, the MA calculates the optimal solution, which is represented by a series of control inputs, as in Equation (4). We can consider the MA as a generic function that returns the following optimal solution:

$$U^*(k) = \arg\min_{U(k)} J(k, x(k), U(k)), k = 0,\ 1, \cdots,\ H-1. \tag{7}$$

The process state $x(k)$ is acquired from the dynamic environment and represents the initial state for the control horizon $[k, H]$.

We may suppose that the MA is based on a population of solutions because we shall use an EA in this work. Hence, the MA generates and evolves a population of individuals within an iterative process for each sampling period. An individual of the population is represented by a predicted control sequence (*pcs*). A numerical integration function can use the process model, the *pcs* and the current initial state $x(k)$, and return the predicted state sequence (denoted by *pss*). The MA will make an optimal prediction for the interval $[k, H]$ that becomes the prediction horizon (briefly denoted as Phz) and has *H-k* sampling periods. The objective function *J* over the Phz is optimized and subject to certain constraints through an *optimal control sequence* (7).

Finally, the Controller sends to the process the control input $u(k)$, the first element of the sequence $U^*(k)$. The Controller calls the MA that yields the vector $U^*(k)$ to the first time at $k = 0$.

A systematic procedure to implement the control structure depicted in Figure 1 is presented in [8].

3.2. Controller Using an Evolutionary Algorithm

In the upcoming sections, we shall consider the EA as an MA able to solve this kind of problem. We will refer to the EA with a large abstraction degree, such that we do not need to give, for the moment, all of the details concerning its implementation. In this way, the implementation aspects described below maintain generality. In one way or another, a particular EA calculates the optimal solution for the considered OCP meeting the initial conditions.

The considerations made up until now are valid whatever the EA's implementation would be (except Equation (7), where the *min* operator can be replaced by *max*). We shall consider a particular implementation called EAv1 (EA version 1), which adopts a natural solution encoding in the upcoming sections. The control input *u*(t) is a step function whose discretization step equals the sampling period. A candidate solution is a sequence (*pcs*) that covers the Phz and has the following structure:

$$pcs =< u(k|k),\ \cdots, u(k+i|k), \cdots\ , u(H-1|k) >, \tag{8}$$

where $u(k+i|k), i = 0, \ldots, H-k-1$ is the predicted value for the control input $u(k+i)$ based on knowledge up until moment *k*. The control input $u(k+i|k)$ is kept constant within the sampling period $[k+i,\ k+i+1)$, such that a *pcs* represents a step function.

Note that:

$$u(k+i|k)\ \neq u(k+i).$$

The value $u(k+i|k)$ is a future control input predicted at the present moment, whereas the value $u(k+i)$ is the future real control input that is unknown at the present moment. We can assert the same thing for the state variables. Using Equations (1) and (8), the process model returns the predicted state sequence (*pss*) as below:

$$pss =< x(k+1|k), \ldots, x(H\ |k) >, \tag{9}$$

where $x(k+i|k),\ i = 0, \ldots, H-k$ is the predicted value of the state $x(k+i)$ based on knowledge up until moment *k*.

The EAv1 has now the necessary data to compute the objective function *J* over the prediction horizon. At the end of the iterative process, the EAv1 returns to the optimal control sequence (*ocs*):

$$ocs =< u^*(k|k), \ldots\ , u^*(H-1|k) > . \tag{10}$$

After that, the Controller will send the first element of *ocs* sequence, namely $u(k) = u^*(k|k)$, to the process as a real control input for the current sampling period. The Controller then shifts the beginning of the Phz by one sample and restarts the optimization for the next interval $[k+1,\ H]$.

Table 1 presents an outline of the Controller's actions for every sampling period. The execution of the EAv1 is presented in Line #2 as a function call, whose input parameters are the current sampling period and the process state vector. It returns the *ocs* over the interval $[k, H]$ according to Equation (10). Line #3 sets the output of the Controller $u(k)$ at the value $u^*(k|k)$. In Line #2, the function EAv1$(k, x(k))$ can be replaced by a new version, EAv2$(k, x(k))$, developed in the next section.

Table 1. Outline of Receding Horizon Controller using an EAv1.

1	Get the current value of the state vector, $x(k)$.
2	$ocs \leftarrow$ EAv1$(k, x(k))$
3	$u(k) \leftarrow u^*(k|k)$
4	Send $u(k)$ towards the dynamic system
5	Shift the prediction horizon and wait for the next sampling period

Remarks:

1. This work addresses OCPs whose objective function (3) includes a terminal penalty. The Phz has to be $[k, H]$ even if the integral term is missing. The Phz "recedes" at each sampling period but keeps the final extremity. Hence, its length decreases by one unit at each sampling period;
2. For the current sampling period, $[k, k+1)$, the Phz is $[k, H]$, and its length is $H - k$. Consequently, the computational complexity of the EAv1 is variable and depends on this value: $H - k$. Let us note that the discretization step used by the EAv1 for encoding a *pcs* is equal to the sampling period T. Therefore, the length of the *ocs* is variable along the control horizon and decreases at each sampling period;
3. The most restrictive time constraint for the algorithm presented in Table 1 is that the execution time has to be smaller than the sampling period. The length of the Phz could be very large at the beginning of the control horizon, especially in the first sampling period. If the time constraint is met initially, it will also be fulfilled in the next sampling period because the length of the Phz decreases;
4. The choice of T is related to control engineering aspects, such as continuous signal discretization and time constants of the considered dynamic system. Therefore, the increase of T is not an option for solving constraint #3. Before implementation, the control system designer must verify the fulfilment of this constraint using simulations. Generally speaking, RHC using an MA is suitable for processes having relatively large time constants.

4. Controller Implementation with Improved Computational Complexity

The Controller improvement can be done using a new version of the EA. The latter keeps its general structure and the most parameter values, but it will consider fewer gene encoding predicted control sequences.

4.1. EA with Variable Discretization Step

This section proposes a version of the EA called EAv2 (version 2), which can reduce the Controller's computational complexity compared to the first version. The aforementioned Remark #2 underlines that EAv1 uses a constant discretization step equal to the sampling period T.

In the proposed new version, inside the EAv2, the number of discretization steps is constant and independent of the length of the Phz. Consequently, the discretization step is variable and, as we shall see, greater than or equal to T.

With h, let us denote the number of the discretization steps. As we already saw, for the current sampling period $[k, k+1)$, the Phz is $[k, H]$. Expressed in time units, the Phz is

$$[t_k,\ t_H] = [k \cdot T,\ H \cdot T]$$

This interval will be discretized using h steps. Figure 2 shows the new discretization of the Phz in contrast with the old one involved in the sampling period.

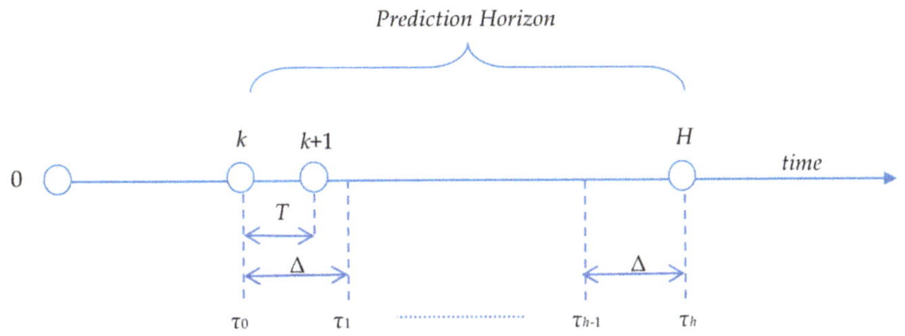

Figure 2. The new prediction horizon's discretization versus the discretization involved in the sampling period.

Hence, the EAv2 will use *pcs* having the form

$$pcs =< u(\tau_0|k),\ u(\tau_1|k),\ \cdots, u(\tau_j|k), \cdots\ ,\ u(\tau_{h-1}|k) > . \tag{11}$$

If the discretization step is denoted by Δ, the discretization moments τ_j are defined below:

$$\tau_j = t_k + j \cdot \Delta;\ j = 0, \cdots, h-1; \tau_0 = t_k;\ \tau_h = t_H$$

We recall here that the predicted value for the control input $u(\tau_j)$ based on knowledge up until the moment t_k is the value

$$u(\tau_j|k),\ j = 0, \ldots, h-1.$$

Because our goal is to reduce the computational complexity of EAv2 in contrast with EAv1, the following constraint has to be met:

$$h \leq H - k. \tag{12}$$

If the constraint introduced by Equation (12) is met, a smaller number of predicted values will cover the prediction horizon. A possibility, which also simplifies the implementation, is to set the number of discretization steps using the rule below:

$$h = \begin{cases} h_0, & \text{for} \quad k = 0,\ 1, \cdots, H - h_0 \\ H - k & \text{for} \quad k = H - h_0 + 1, \cdots, H - 1 \end{cases} \tag{13}$$

The value of the constant integer h_0 is chosen such that

$$h_0 < H.$$

4.2. Computational Complexity

This section makes a comparison between the computational complexity of the two EAs. Usually, an effective way to compare two MAs is to evaluate the calls number of the objective function. In our work, all the EAs have the structure corresponding to Lines #9—#19 of the pseudo-code described in Table 2. Hence, the structure of the two algorithms, EAv1 and EAv2, will basically be the same and only differ by few initializing instructions. One can count the call number of the objective function:

- μ calls within Line #11: Calculation of the performance indexes for the initial population having μ individuals;
- λ calls within Line #14: offspring generation (λ individuals);
- $p \cdot \lambda$ *calls* within Line #15: p is the offspring's probability of undergoing mutation.

Table 2. Outline of the EAv2 with improved computational complexity.

```
start EA(k, x(k))
/* Initialize the parameters of EA*/
if k ≤ H − h0
        h ← h0;
        Δ ← T · (H − k)/h0;
    else
        h ← H − k;
        Δ ← T;
end
Generation of the initial population
Calculation of the performance indexes
g ← 1 /* g: current generation number */
while (g ≤ NGen) /* optimization iterative process */
        parent selection;
        offspring generation; /*calculate the performance indexes*/
        self-adaptive mutation; /*update the performance indexes*/
        replacement;
        population sorting
        g ← g + 1
end /*while*/
return ocs (the best of the population)
```

If the iterative optimization process evolves on *NGen* generations, then the average call number for *J* is

$$\text{Ncalls} = \mu + NGen \cdot \lambda \cdot (1 + p). \tag{14}$$

As such, the call number for *J* will actually be the same, as we shall confirm using simulations in Section 6.

The most time-consuming part of the objective function execution is the process model integration determining the state evolution. The question that one can raise is whether the integration time is significantly different for the new version. The integration time interval is $[k, H]$ in both cases. As such, the process model's integration has the same initial state and time interval and similar input functions *u*(t), which are step functions. The different number of genes involves different step numbers for the two input functions. This fact has a small influence on the integration time. Simulation tests show that the integration of input functions takes a similar amount of time while in these conditions.

In conclusion, the computation effort to calculate the objective function is similar for the two versions of the EA.

On the other side, *pcs* encoding is decisive for the algorithm's complexity. EAv1 encodes *pcs* using $H - k$ real values (genes), while EAv2 only uses h_0 values. Since it holds $h_0 < H$, the EAv2 obviously has a smaller computational complexity for all genetic operators (selection, crossover, mutation, replacement) and initializations.

Finally, one can say that EAv2 is better than EAv1 from the point of view of computational complexity. The number of discretization steps is h_0, while $k \leq H - h_0$ and decreases from $h_0 - 1$ toward 1, for $k > H - h_0$.

NB: If Equation (13) is fulfilled, the implementation issue stated by Remark #3 can be avoided. On the other hand, this value must be a trade-off between the computational complexity and the *ocs* accuracy.

4.3. General Implementation of the EAv2

Equation (13) allows the calculation of the discretization step Δ. Because h has a constant value for the initial segment, the Δ value will be variable in return. From (11) and (13), it holds that:

$$\Delta = \begin{cases} T \cdot \frac{H-k}{h_0} > T & \text{for } k = 0,\ 1, \cdots, H - h_0 - 1 \\ T & \text{for } k = H - h_0, \cdots, H - 1 \end{cases}. \tag{15}$$

According to Equation (11), a *pss* will have the form

$$pss = < x(\tau_1|k), \ldots, x(\tau_h\ |k) >$$

The main output of the EAv2 is the *ocs* for the current Phz, which has the following structure:

$$ocs = < u^*(\tau_0|k),\ \cdots, u^*(\tau_j|k), \cdots\ ,\ u^*(\tau_{h-1}|k) > \tag{16}$$

Table 2 schematically describes the EAv2 included in a controller with improved computational complexity. EAv2 has specific parts that cannot be detailed here. Therefore, the following sketch describes them with a large degree of extraction. The outline only presents the details allowing the management of the length of the *pcs*.

The iterative optimization process evolves on *NGen* generations counted by the variable g. Lines #13–#17 call for functions implementing well-known genetic operators. We recall here that the *offspring generation* also calculates the performance indexes of the new *pcs* (solutions). The *self-adaptive mutation* also updates the performance indexes of the solutions that are subjected to mutation.

Remarks:

5. The encoding of the *pcs* and *ocs* according to (11) and (16) can obviously diminish the computational complexity when generating *ocs*. However, on the other hand, the discretization with fewer intermediary points affects the *ocs* accuracy, that is, its capacity to represent the optimal solution to our problem. The question is, to what extent this accuracy is affected? Using simulations, the control system designer must verify that the *ocs* representation is not degraded and the quasi-optimality is still kept before implementation.
6. Equation (15) shows that, for each sampling period, it holds:

$$\Delta \geq T. \tag{17}$$

Considering the inequality of Equation (17), the first element of the *ocs* covers the sampling period:

$$[\tau_0,\ \tau_1] = [t_k, t_k + \Delta]\ \supseteq\ [t_k, t_k + T].$$

Hence, the first *ocs* value, i.e., $u^*(\tau_0|k)$, will be sent to the process as the optimal control input for the current sampling period (as in Line #4 from Table 1).

5. Case Study

This section will present an example of implementing a metaheuristic-based controller that integrates an RHC structure.

We have considered a well-known problem that has been described in many articles, among which we recall [16,17]. This problem is known as the optimal control of a fed-batch reactor for ethanol production (OCEP) and concerns a nonlinear dynamic system. Different sub-cases are presented in these papers, which differ in control horizon and control bounds. We have considered the OCEP problem stated in Appendix A, which presents all of the basic elements. Because our main objective is to construct the Controller using an AE, we have considered a fixed final time in contrast with other versions of the problem. Keeping notations, it holds:

$$n = 4;\ m = 1;\ t_f = H \cdot T = 54h$$

The process model has n state variables x_1, x_2, x_3, x_4 representing the cell mass, substrate, product concentration (g/L), and volume (L). The control variable is the feed rate (L/h) that meets the following constraint:

$$0 \leq u(t) \leq 12$$

The state variable x_4 has to meet the following constraint:

$$x_4(t_f) \leq 200 \tag{18}$$

The objective function given below has to be maximized:

$$J\left(x(t_f)\right) = x_3(t_f) \cdot x_4(t_f)$$

Although this one is a terminal penalty and the integral component is missing, the implementation maintains its difficulty: the process evolution of the entire control horizon must be determined when a *pcs* is adopted.

5.1. *Open-Loop Solution Using an Evolutionary Algorithm*

In this section, we present preliminary work before constructing the Controller for the closed-loop control structure. We have implemented and conducted an evolutionary algorithm (see [1–3]) that solves the OCEP problem. It will be denoted by EAv0 (EA version 0) in the upcoming sections. The implementation of the EAv0 and the simulations were conducted using the MATLAB language and system. We have employed special functions devoted to integrating the differential equations for the simulation of the process evolution.

Let us recall that EAv0 determines the following quasi-optimal control input because our problem looks for the maximum value of J:

$$U^*(0) = \arg\max_{U(0)} J(0, x_0, U(0))$$

This means that the control horizon is [0, H], and the initial state is x_0.

This algorithm is the basis for the two versions of EA used inside the Controller. EAv0 is modified and adapted to become part of the optimal Controller.

As mentioned previously, EAv0 uses direct encoding and has the usual characteristics listed below:

- For each generation, the population has μ individuals;
- The children population has λ individuals ($\lambda < \mu$);
- The population evolves over NGen generations;
- The selection strategy uses stochastic universal sampling and the rank of individuals. The latter is scaled linearly using selection pressure (s); see [3] (pp. 205–221);
- A BLX-α crossover operator;
- Inside the mutation operator, the mutation step is subjected to a global variance adaptation. This one is conducted according to the "1/5 success rule"; see [1] (pp. 245–274);
- The replacement strategy involves that the children replace the λ worst individuals of the generation.

The most important parameters are given in Table 3.

Table 3. The main parameters of EAv0.

	λ	μ	NGen	H	s	α
EAv0	20	30	70	20	1.8	0.4

This algorithm yields a solution that is actually a quasi-optimal sequence of control inputs. It can be only used within an open-loop control structure. For the reasons mentioned before, this sequence is useless for a closed-loop control structure. Nevertheless, this phase is mandatory because it can validate the aptitude of the EA to solve the problem and evaluate the computational complexity.

The resulting simulation program has a stochastic character; therefore, we cannot conclude after a single execution. We must repeat the execution more times to confer consistency to statistical parameters. That is why we have conducted a simulation series that has repeated times the execution of EAv0 30 times. The results, namely a performance index statistic, the quasi-optimal control input $u(t)$, and state evolution, are presented in Table 4 and Figure 3.

Table 4. Results of the open-loop simulation series using EAv0.

*J*min	*J*avg	*J*max	Sdev	*J*typical
18,973.0	19,906.4	20,395.2	312.5	19,885.7

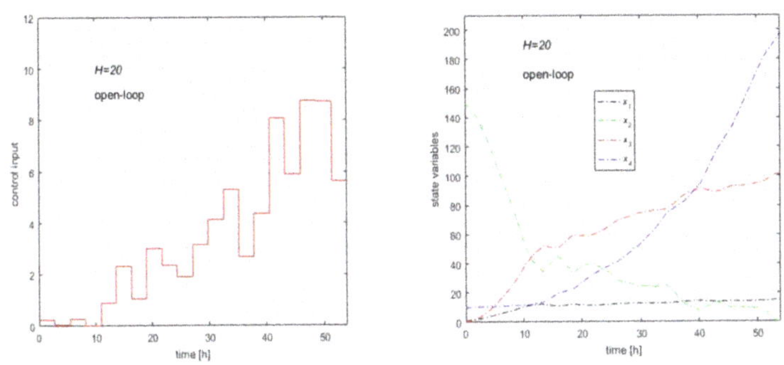

(a) **(b)**

Figure 3. The typical solution given by the EA. (**a**) The quasi-optimal solution. (**b**) State evolution in open loop.

Table 4 shows some statistical parameters characterizing the 30 executions of EAv0. The minimum, average, and maximum values and standard deviation are indicated for the performance index. We consider typical execution as the particular simulation among the 30 that produce the values that are the closest to the average performance index. In our simulation series, there is an execution that yields *J*typical = 19885.7. The quasi-optimal control input for the typical execution is depicted in Figure 3a, and its state evolution in Figure 3b.

Further details concerning EAv0 are given in Appendix B.

5.2. Closed-Loop Solution Using an EA with Constant Discretization Step

This section aims to achieve the closed-loop control of the dynamic system involved in the OCEP problem using the Controller based on EAv1. We have implemented the Controller and a program that simulates how the closed loop works. The process model has been used to predict the *ocs*. We have also considered the same model to simulate the real process.

The Controller has the same structure as the one presented in Table 1, except that it includes EAv1 (does not call a function). It is organized as a function called for each sampling period (see Appendix B).

The Controller has the following characteristics:

- Input parameters: k and $x(k)$;
- Output parameter: $u(k)$ (the first element of the best solution, i.e., $u^*(k|k)$);
- Control horizon: $[k, H]$, with $k = 0, 1, \cdots, H-1$; $H = 20$;
- Number of genes: $H - k$ (see (8));
- Discretization step: T;

- Other parameters such as those in Table 3.

EAv1—which is included not called by the Controller—has the same structure and characteristics as EAv0 (given in Section 5.1).

The program that simulates the closed-loop optimal control has the outline presented in Table 5. Essentially, the closed loop is achieved by a loop that simulates the Controller's call for each sampling period and the feedback from the real process.

Table 5. Outline of the closed-loop optimal control using EAv1.

1	Initializations /* for OCEP problem, controller and simulation*/
2	**for** $k = 0, 1, \cdots, H-1$
3	$u(k) \leftarrow Controller\_EA(k)$
4	Compute the state vector $x(k+1)$ of the process model using $u(k)$;
5	$x_0 \leftarrow x(k+1)$
6	**end**
7	**display** J^* and $u^*(k)$; $k = 0, 1, \cdots, H-1$

The Controller's call is done in Line #3. It uses as an input parameter, which is the initial state that is stored in the global variable x_0. The numerical integration of the process model is accomplished in Line #4. This algorithm is implemented as a MATLAB script.

We conducted a simulation series that executed the script "Closed-loop.m" 30 times. In our simulation series, there was an execution that yielded Jtypical = 20,153.5. The performance index statistic, the quasi-optimal control input u(t), and the state evolution are presented for typical execution in Table 6, Figure 4a,b, respectively.

Table 6. Results of the closed-loop simulation series using the Controller with EAv1.

Jmin	Javg	Jmax	Sdev	Jtypical
19,127.7	20,136.7	21,014.9	449.5	20,153.5

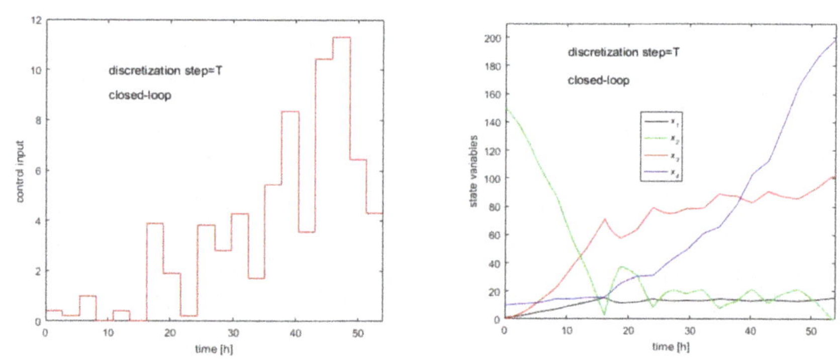

(a) (b)

Figure 4. The typical solution given by the Controller using EAv1. (**a**) The quasi-optimal solution given by the Controller using EA. (**b**) State evolution in closed loop.

5.3. Closed-Loop Solution Using an EA with Improved Computational Complexity

This section aims to implement the closed-loop control of the dynamic system under consideration using the Controller based on EAv2. We have implemented the Controller and a program that simulates the closed-loop working (see "Closed-loop2.m", inside the folder "EA_Etanol_loop2"). The latter is basically the same as that described in Table 5, except for a few initializations.

The new Controller has the structure that was presented in Table 1, except that it includes EAv2. It is also organized as a function called for each sampling period (see the script "Controller_EA2.m"). It has the following characteristics:

- Input parameters: k and $x(k)$;
- Output parameter: $u(k)$ (the first element of the best solution, i.e., $u^*(k|k)$);
- Control horizon: $[k, H]$, with $k = 0, 1, \cdots, H-1$; $H = 20$;
- Number of genes: h_0, for the first $H - h_0$ sampling periods, and $h_0 - 1, \cdots, 2, 1$ for the last sampling periods (see (13)); $h_0 = 10$;
- Discretization step: $\Delta \geq T$ variable (see (15));
- Other parameters such as those in Table 3.

The EAv2 algorithm —which is included and not called by the Controller—has the same structure and characteristics as EAv0 (given in Section 5.1)

Because OCEP maximizes the objective function, the best solution is as below:

$$U^*(k) = \arg\max_{U(k)} J(k, x(k), U(k)), k = 0, 1, \cdots, H-1.$$

We conducted a simulation series that executes the script "Closed-loop2.m" 30 times. In our simulation series, there was an execution that yielded Jtypical = 20319.0. A statistic of the performance index achieved by the closed loop is presented in Table 7. The quasi-optimal control input calculated by the Controller for typical execution is depicted in Figure 5a, and its state evolution is depicted in Figure 5b.

Table 7. Results of the closed-loop simulation series using the Controller with EAv2.

Jmin	Javg	Jmax	Sdev	Jtypical
19,497.0	20,323.0	20,982.0	353.0	20,319.0

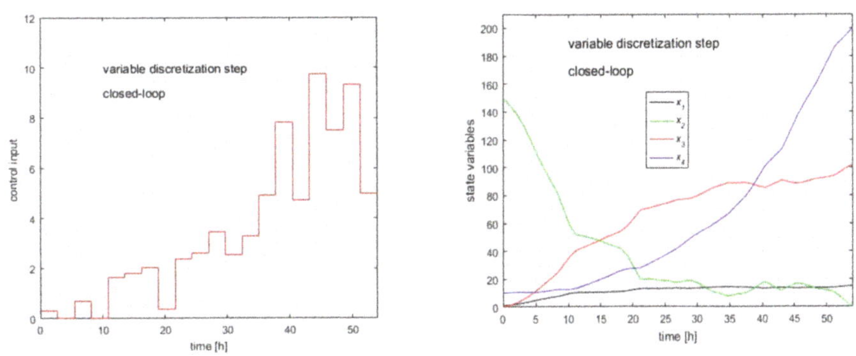

(a) **(b)**

Figure 5. The typical solution given by the Controller using an evolutionary algorithm with improved computational complexity. (**a**) The quasi-optimal solution given by the Controller using EA with a variable discretization step. (**b**) State evolution in closed loop.

6. Discussion

We can now verify the statement made in Section 4.2, which was that the Controller based on EAv2 has a smaller computational complexity than the one based on EAv1.

Table 8 shows the call number of the objective function recorded at the time of the closed-loop simulations. This number refers to the call of only one Controller, determining the optimal control input for the current sampling period. Its value is independent of the prediction horizon's length. The first three columns represent the minimum, average, and maximum recorded values. The last column presents the theoretical value calculated with

Equation (14) and Table 3. The adopted offspring's probability of undergoing mutation is $p = 0.9$.

Table 8. The call number of the objective function.

	Nmin	**Navg**	**Nmax**	**Ncalls**
EAv1	2674	2691	2719	2690
EAv2	2666	2690.8	2711	

Hence the call number is the same for the two Controller versions.

For the same prediction horizon $[k, H]$, the process model's integration works over the same time interval and has similar input functions $u(\mathrm{t})$, which are step functions. The step number of $u(\mathrm{t})$ is not so important for the integration time. Hence, under these circumstances, the integration durations are similar.

Considering the two aspects mentioned above, we can conclude that the computational complexity regarding the objective function evaluation is similar for the two Controllers.

The aspect that makes the difference between the two controllers is related to encoding a *pcs*, namely the gene number. The Controller based on EAv1 uses a variable number of genes: $H,\ H-1,\cdots,\ 2,\ 1$. Constraint #3 can be prohibitive, especially in the first sampling period.

The Controller based on EAv2 only has h_0 genes and $h_0 < H$. Hence, EAv2 obviously has a smaller computational complexity for all genetic operators (selection, crossover, mutation, replacement) and initializations. The confirmation of this statement is given by Table 9, which gives the execution time for the two controllers. These durations have been recorded at the time of the closed-loop simulations for each sampling period. The columns labelled by ***k*** give the sampling period rank within the control horizon. The columns labelled by **tv1** and **tv2** give the execution times (in seconds) for the Controller based on EAv1 and EAv2, respectively.

Table 9. Execution time for the two controllers.

k	**tv1[s]**	**tv2[s]**	**k**	**tv1[s]**	**tv2[s]**	**k**	**tv1[s]**	**tv2[s]**	**k**	**tv1[s]**	**tv2[s]**
1	84.4	60.7	6	62.9	36.0	11	53.9	35.2	16	43.1	18.9
2	78.8	51.3	7	62.2	41.3	12	51.8	41.1	17	37.1	20.4
3	91.3	49.0	8	62.0	50.4	13	46.1	25.8	18	26.6	11.9
4	84.6	49.8	9	62.4	38.8	14	44.8	22.7	19	23.1	7.9
5	70.7	55.7	10	59.0	33.8	15	44.8	19.0	20	13.0	7.0

The durations indicated in the two columns show that the Controller based on EAv2 has undoubtedly smaller computational complexity than EAv1.

In our case, because the durations tv1 are less than T, even the first Controller can be used in a closed-loop control structure. Generally speaking, constraint #3 can be prohibitive for other OCPs, and the Controller based on EAv2 could be the solution. The price to pay for this Controller's computational complexity is underlined by Remark #5. The simulations should prove that the optimal accuracy of the control sequence is not seriously affected. In our case, the values of the performance indexes shown in Table 7 prove that the best solution accuracy is more than acceptable.

The third line of Table 10 proves that EAv2 works very well in a closed loop with only ten genes.

Table 10. Performance indexes recorded within the simulation series.

	*J*min	*J*avg	*J*max	Sdev	*J*typical
Open-loop EAv0	18,973.0	19,906.4	20,395.2	312.5	19,885.7
Closed-loop EAv1	19,127.7	20,136.7	21,014.9	449.5	20,153.5
Closed-loop EAv2	19,497.0	20,323.0	20,982.0	353.0	20,319.0

The curves depicted in Figures 3–5 have similar aspects. Moreover, Figures 4 and 5 show that the control action is present and the closed-loops work. To facilitate the comparison among the performance indexes for the three situations, Table 9 unites Tables 4–6. We recall that each simulation series has 30 program executions. Some explanations concerning the differences among the three lines are necessary.

One can remark that the values of J from in the first line are less than those accomplished by the two controllers. This explanation is related to the extent of the search operation. EAv0 searches for the best solution using only 2690 calls (Ncalls) of the objective function, while the controllers use 20 times more calls (2690 calls for each sampling period). In other words, the execution of EAv0 is equivalent to the Controller's call for the first sampling period.

EAv1 finds the best performance index in the closed loop J = 21,014.9, which is greater than the corresponding value of EAv2, J = 20,982. This time, the explanation is related to the bigger number of genes that involve a finer representation of the *pcs*. From this point of view, EAv1 is superior to EAv2. More accurate representation leads to better optimization.

At the same time, the bigger number of genes involves a greater explorative character and a greater representation power for u(t). That is why EAv1 has a standard deviation greater than the corresponding value of EAv2. In comparison with EAv0, EAv1 has 20 times more chances to strengthen its explorative character. This fact explains the smaller value of the standard deviation in the open loop.

Our closed-loop simulations had two entities: the Controller based on EA and the controlled process. The Controller already uses a process model to implement the RHC structure (to make predictions). In this work, the real process is also simulated using the process model. That is a simplification because the real process is almost always affected by unmodeled dynamics and noises. One may ask: at what extent could the real process degrade the closed-loop functioning? The answer can always be found through simulation using a real process model. For example, a real process model could be obtained from the process model to which a simulated "noise" is added. We have simulated closed-loop functioning using this technique, but quantitative results are not included in this paper to avoid a presentation that is too complex.

We can notice a degradation of quasi-optimal behaviour according to the amplitude of the noise. The increase of the noise implies the decrease of the performance index. The closed loop does or does not work well depending on whether the performance index value is acceptable.

Generally speaking, it is relatively easy to deal with the constraints regarding the control input $u(k)$, $k = 0, \cdots, H-1$. This can be done when generating new pcs. Finally, each component belongs to a certain domain, $u(k) \in U_k$, and the construction of a new pcs could meet this constraint. What is more difficult is the handling of the constraints regarding the state variables $x(k)$, $k = 0, \cdots, H-1$. A valid *pcs* will yield the resulting state trajectory $X(k)$ defined by Equation (5). For a specific OCP, state path constraints involving a trajectory that is a *pss* could be imposed. The most general constraint could have a form that involves an admissible domain for each k: $x(k) \in D_k$, $k = 0, \cdots, H$. In our case study, Constraint (18) limits the produced ethanol's volume in the final state to the reactor's volume.

The pss is calculated through the integration procedure. A good strategy is to verify if the component of the pss meet the path constraints just after the calculation of the pss within the objective function. If at least one constraint is not fulfilled, the cause associated with this effect, namely the pcs, must be moved away. A correct approach is to assign an infinite value (for minimization) to its performance index. In this way, the MA will replace the considered pcs, and the stochastic character of the population evolution will not be corrupted.

At successive time moments, the *pcs* values are randomly set. Thus, the difference between two successive values could be quite big in one sense or another. This fact does not fit with the certain and desired smoothness of the control profile. Actually, this is the price to pay for the stochastic nature of the EA. The latter can integrate the so-called *step control* technique, which implements the following differential constraint:

$$\left|\frac{du(t)}{dt}\right| \leq \Delta s_{\max}$$

The value $\Delta s_{\max}$ is the maximal accepted slope. If the step control technique is implemented, it will involve a certain smoothness at the level of the state trajectory. Starting from this remark and assimilating gradient-based optimization with evolutionary optimization, in principle, we could facilitate the state evolution toward the optimal *pcs*. Within a future work, this idea could help to make the control action of the proposed receding horizon controller more effective.

Besides the EA, other metaheuristic algorithms have been used by the authors to implement receding horizon controllers. Both population-based metaheuristic methods (firefly algorithm, harmony search, Gaussian adaptation, memetic algorithm, etc.) and Swarm algorithms (Ant Colony Optimization, the Artificial Bee Colony Algorithm, the Bees Algorithm, Cuckoo search, Lévy flights for global optimization, particle swarm optimization, etc.) have been used to solve OCPs. The communication skills among agents are not making swarm algorithms always more effective than the population-based algorithms. Our first perception is that the effectiveness of the Controller mainly depends on the dynamic process nature. A systematic comparison can be made within a future investigation, and more implementation aspects could be revealed.

7. Conclusions

This paper presented some implementation aspects regarding the optimal control of a process using a Controller based on an evolutionary algorithm. This solution is addressed when other analytical or numerical solutions are not available and when using a metaheuristic algorithm yields a flexible and robust solution.

Using a metaheuristic algorithm involves a control structure that can use such an algorithm. The RHC structure is adequate from this point of view because the prediction technique can fit with an evolutionary algorithm.

We have presented some implementation aspects concerning the harmonization between the prediction technique and the evolutionary algorithm. Metaheuristic algorithms are known to have a large computational complexity. This paper has addressed a method to thwart this drawback and fulfil the time constraint for an evolutionary algorithm.

For the case study presented in this work, we have illustrated the proposed algorithm using simulation programs. The material attached to this work includes the simulation programs that show how the RHC structure, Controller, and prediction with an EA can be implemented with good results.

Supplementary Materials: The following are available online at https://www.mdpi.com/article/10.3390/inventions6030053/s1, the archive file "CLoop_EA.zip" contains all folders and files described in Appendix B.

Author Contributions: Conceptualization, V.M.; methodology, V.M.; software, V.M. and S.R.; validation, V.M. and E.R.; formal analysis, V.M.; resources, S.R.; data curation, S.R.; writing—original draft preparation, V.M.; writing—review and editing, V.M. and E.R.; visualization, S.R.; supervision, E.R.; project administration, E.R. All authors have read and agreed to the published version of the manuscript.

Funding: This work was conducted in the framework of the research project DREAM (Dynamics of the REsources and technological Advance in harvesting Marine renewable energy), supported by the Romanian Executive Agency for Higher Education, Research, Development and Innovation Funding—UEFISCDI, grant number PN-III-P4-ID-PCE-2020-0008.

Institutional Review Board Statement: Not applicable.

Informed Consent Statement: Not applicable.

Data Availability Statement: The data that support the findings of this study are available in the public domain.

Acknowledgments: The results of this work have also been presented to the 9th edition of the Scientific Conferences organized by the Doctoral Schools of the "Dunărea de Jos", University of Galati http://www.cssd-udjg.ugal.ro/ (accessed on 1 July 2021), which was held on 10–11 June 2021, in Galati, Romania.

Conflicts of Interest: The authors declare no conflict of interest.

Appendix A

The Elements of the Optimal Control Problem

Process Model:

$$\begin{aligned}
\frac{dx_1}{dt} &= g_1(t) \cdot x_1(t) - u(t) \cdot \frac{x_1(t)}{x_4(t)} \\
\frac{dx_2}{dt} &= -10 \cdot g_1(t) \cdot x_1(t) + u(t) \cdot \frac{150-x_2(t)}{x_4(t)} \\
\frac{dx_3}{dt} &= g_2(t) \cdot x_1(t) - u(t) \cdot \frac{x_3(t)}{x_4(t)} \\
\frac{dx_4}{dt} &= u(t) \\
g_1(t) &= \left(\frac{0.408}{1+x_3(t)/16}\right)\left(\frac{x_2(t)}{0.22+x_2(t)}\right) \\
g_2(t) &= \left(\frac{1}{1+x_3(t)/71.5}\right)\left(\frac{x_2(t)}{0.44+x_2(t)}\right)
\end{aligned}$$

Control horizon:

$$0 \le t \le 54;\ t_0 = 0;\ t_f = 54\text{h}$$

Initial conditions:

$$x_{00} = [1.\ 150.\ 0.\ 10.]^T$$

Bound constraints:

$$\begin{aligned}
0 &\le u(t) \le 12 \\
0 &\le x_4(t_f) \le 200
\end{aligned}$$

Performance index:

$$\max_{u(t)} J\Big(x(t_f)\Big), \text{ with } J\Big(x(t_f)\Big) = x_3(t_f) \cdot x_4(t_f)$$

Appendix B

Appendix B.1. Implementation Details for the Algorithm EAv0

The OCEP problem described in Appendix A has been solved using the evolutionary algorithm EAv0. The optimal solution can only be used in an open loop. Every chromosome for the solutions population encodes a control profile that is a series of values $u_0, u_1, \cdots, u_{H-1}$. We chose $H = 20$ and $T = 2.7$ h (the control horizon is 54 h). The EAv0 was implemented by the script "EA_EtanolP_single.m". This program was executed 30 times using the script "cycle30single.m" in order to calculate some statistical parameters, which are presented in Table 4. The computation of the statistical parameters and the drawing of Figure 3 was conducted using the script "MEDIERE30single20.m". All of the scripts are included in the folder "EA_Etanol_single", which is attached to this work.

Appendix B.2. Implementation Details for the Controller Using EAv1

The OCEP problem described in Appendix A was solved using the Controller based on EAv1. It is about a closed-loop solution. The function "Controller_EA.m" implements a Controller that is called for each sampling period. Every chromosome encodes the control profile, which is a series of values $u_k, u_{k+1}, \cdots, u_{H-1}$. It also holds $H = 20$ and $T = 2.7$ h. The EAv1 is not a distinct function, but it is a part of the Controller. The script "Closed_loop.m" simulates the control action over the entire horizon control.

In order to calculate the statistical parameters presented in Table 6, the script "Closed_loop.m" was executed 30 times using the script "ciclu30_loop.m", which saves the workspace. After that, the computation of the statistical parameters and the drawing of Figure 4 could be conducted using the script "MEDIERE30loop.m". All of the scripts are included in the folder "EA_Etanol_loop", which is attached to this work.

Appendix B.3. Implementation Details for the Controller Using EAv2

The OCEP problem described in Appendix A was also solved using the Controller based on EAv2. The function "Controller_EA2.m" implemented the Controller having a better computational complexity. This one called for each sampling period. The number of genes was $h_0 = 10$. Every chromosome of the solution population encoded the control profile, which was a series of values $u_{\tau_0}, u_{\tau_1}, \cdots, u_{h_0-1}$. The EAv2 is not a distinct function, but it is a part of the Controller. The script "Closed_loop2.m" simulates the control action over the entire horizon control.

In order to calculate the statistical parameters presented in Table 7, the script "Closed_loop2.m" was executed 30 times using the script "ciclu30_loop2.m", which saves the workspace. After that, the computation of the statistical parameters and the drawing of Figure 5 could be carried out using the script "MEDIERE30loop2.m". All of the scripts are included in the folder "EA_Etanol_loop2", which is attached to this work.

The supplementary material "CLoop_EA.zip" includes all of the folders and files mentioned above.

References

1. Kruse, R.; Borgelt, C.; Braune, C.; Mostaghim, S.; Steinbrecher, M. *Computational Intelligence—A Methodological Introduction*, 2nd ed.; Springer: Berlin/Heidelberg, Germany, 2016.
2. Patrick, S. (Ed.) *Metaheuristics*; Springer: Berlin/Heidelberg, Germany, 2014; ISBN 978-3-319-45401-6.
3. Talbi, E.G. *Metaheuristics—From Design to Implementation*; Wiley: Hoboken, NJ, USA, 2009; ISBN 978-0-470-27858-1.
4. Faber, R.; Jockenhövelb, T.; Tsatsaronis, G. Dynamic optimization with simulated annealing. *Comput. Chem. Eng.* **2005**, *29*, 273–290. [CrossRef]
5. Onwubolu, G.; Babu, B.V. *New Optimization Techniques in Engineering*; Springer: Berlin/Heidelberg, Germany, 2004.
6. Valadi, J.; Siarry, P. (Eds.) *Applications of Metaheuristics in Process Engineering*; Springer: Berlin/Heidelberg, Germany, 2014; ISBN 978-3-319-06507-6.
7. Minzu, V.; Riahi, S.; Rusu, E. Optimal control of an ultraviolet water disinfection system. *Appl. Sci.* **2021**, *11*, 2638. [CrossRef]
8. Minzu, V.; Serbencu, A. Systematic Procedure for Optimal Controller Implementation using Metaheuristic Algorithms. *Intell. Autom. Soft Comput.* **2020**, *26*, 663–677. [CrossRef]
9. Mayne, Q.D.; Michalska, H. Receding horizon control of nonlinear systems. *IEEE Trans. Autom. Control.* **1990**, *35*, 814–824. [CrossRef]
10. Attia, S.A.; Alamir, M.; Wit, C.C.D. Voltage Collapse Avoidance in Power Systems: A Receding Horizon Approach. *Intell. Autom. Soft Comput.* **2006**, *12*, 9–22. [CrossRef]
11. Chiang, P.-K.; Willems, P. Combine evolutionary optimization with model predictive control in real-time flood control of a river system. *Water Resour. Manag.* **2015**, *29*, 2527–2542. [CrossRef]
12. Yang, Y.; Lin, X.; Miao, Z.; Yuan, X.; Wang, Y. Predictive Control Strategy Based on Extreme Learning Machine for Path-Tracking of Autonomous Mobile Robot. *Intell. Autom. Soft Comput.* **2014**, *21*, 1–19. [CrossRef]
13. Hu, X.B.; Chen, W.H. Genetic algorithm based on receding horizon control for real-time implementations in dynamic environments. *IFAC Proc. Vol.* **2005**, *38*, 156–161. [CrossRef]

14. Hu, X.B.; Chen, W.H. Genetic algorithm based on receding horizon control for arrival sequencing and scheduling. *Eng. Appl. Artif. Intell.* **2005**, *18*, 633–642. [CrossRef]
15. Minzu, V. Optimal Control Implementation with Terminal Penalty Using Metaheuristic Algorithms. *Automation* **2020**, *1*, 4. [CrossRef]
16. Balsa-Canto, E.; Banga, J.R.; Aloso, A.; Vassiliadis, V. Dynamic optimization of chemical and biochemical processes using restricted second-order information. *Comput. Chem. Eng.* **2001**, *25*, 539–546. [CrossRef]
17. Banga, J.R.; Balsa-Canto, E.; Molesc, C.G.; Alonso, A. Dynamic optimization of bioprocesses: Efficient and robust numerical strategies. *J. Biotechnol.* **2005**, *117*, 407–419. [CrossRef] [PubMed]

inventions

Article

Optimization of Betalain Pigments Extraction Using Beetroot By-Products as a Valuable Source

Silvia Lazăr (Mistrianu) [1], Oana Emilia Constantin [1], Nicoleta Stănciuc [1], Iuliana Aprodu [1], Constantin Croitoru [2] and Gabriela Râpeanu [1,*]

[1] Faculty of Food Science and Engineering, Dunarea de Jos University of Galati, 111 Domnească Street, 800201 Galati, Romania; silvia.lazar@ugal.ro (S.L.); econstantin@ugal.ro (O.E.C.); nstanciuc@ugal.ro (N.S.); iaprodu@ugal.ro (I.A.)

[2] Academy of Agricultural and Forestry Sciences, 61 Marasti Blvd, 011464 Bucharest, Romania; c.croitoru@sodinal.com

* Correspondence: gabriela.rapeanu@ugal.ro; Tel.: +40-336-130-177

Abstract: (1) Background: This study is designed to extract the bioactive compounds from beetroot peel for future use in the food industry. (2) Methods: Spectrophotometry techniques analyzed the effect of conventional solvent extraction on betalains and polyphenolic compounds from beetroot peels. Several treatments by varying for factors (ethanol and citric acid concentration, temperature, and time) were applied to the beetroot peel samples. A Central Composite Design (CCD) has been used to investigate the effect of the extraction parameters on the extraction steps and optimize the betalains and total polyphenols extraction from beetroot. A quadratic model was suggested for all the parameters analyzed and used. (3) Results: The maximum and minimum variables investigated in the experimental plan in the coded form are citric acid concentration (0.10–1.5%), ethanol concentration (10–50%), operating temperature (20–60 °C), and extraction time (15–50 min). The experimental design revealed variation in betalain content ranging from 0.29 to 1.44 mg/g DW, and the yield of polyphenolic varied from 1.64 to 2.74 mg/g DW. The optimized conditions for the maximum recovery of betalains and phenols were citric acid concentration 1.5%, ethanol concentration 50%, temperature 52.52 °C, and extraction time 49.9 min. (4) Conclusions: Overall, it can be noted that the extraction process can be improved by adjusting operating variables in order to maximize the model responses.

Keywords: beetroot peel; betalains; polyphenols; CCD-RSM

Citation: Lazăr, S.; Constantin, O.E.; Stănciuc, N.; Aprodu, I.; Croitoru, C.; Râpeanu, G. Optimization of Betalain Pigments Extraction Using Beetroot By-Products as a Valuable Source. *Inventions* **2021**, *6*, 50. https://doi.org/10.3390/inventions6030050

Academic Editor: Monique Lacroix

Received: 24 June 2021
Accepted: 10 July 2021
Published: 14 July 2021

1. Introduction

Beetroot (*Beta vulgaris* L.) is a plant belonging to Chenopodiaceae's family that includes approximately 105 genera with 1400 species [1], which is widely grown in Europe, America, and Asia [2]. Similarly, red beetroot is perhaps the most generally planted root vegetable in Southeast Romania. The edible portion of the beets used in the food sector is the tuberous root. The leaves are also used but are generally considered feedstuff material. Beets are available mainly in the cold season but are also adapted to high temperatures. The optimum growth temperature varies between 15 and 19 °C, with lower temperatures influencing the red pigment increase. The beetroot was cultivated especially for medicinal purposes but was used predominantly for food and drink from the beginning of the third century [3]. The best-known existing species are sugar beet (*Beta vulgaris saccharifera*), fodder beet (*Beta vulgaris crassa*), leaf beet (*Beta vulgaris cicla*), and garden beet (*Beta vulgaris rubra*).

In general, beetroot is a vegetable rich in carotenoids, flavonoids, vitamins (niacin, biotin, pyridoxine), and minerals (such as potassium, sodium, phosphorus, calcium, magnesium, copper, iron, zinc, etc.) [4]. Beetroot is also one of the richest sources of betalains, which are plant pigments derived from the betalamic acid that are soluble in water and responsible for an intense red (betacyanins) or yellow color (betaxanthins) [5]. The intensity of beet color depends on the ratio between betacyanins and betaxanthins. More than 80% of all beet pigments are betacyanin compounds [6]. Betalains from Beta vulgaris rubra are natural food coloring, being antiradical and antioxidant agents capable of protecting in vivo from disorders caused by oxidative stress. The functions of betalains refer to reduction of homocysteine concentration, which regulates vascular homeostasis, maintaining platelet function, thrombotic activity, vascular tone, and delicate stability by the release of vasodilators and vasoconstrictors. Some effects attributed to these compounds include antioxidants [7,8], antiproliferative [9], cardioprotective [10], anti-inflammatory [11], and antimicrobial effects [12].

Betalains found in red beetroot are utilized to color various food items, for example, frozen yogurt, wine, jams, and yogurt [13]. Extraction is an essential step in the process of separating biologically active compounds. There are many extraction methods, but the most used method is conventional solvent extraction. It is widely used due to its efficiency, short extraction time, and low economic costs despite the disadvantages, such as using large amounts of solvent [14].

Betalains are usually extracted from plant matrices by conventional solvent extraction (CSE) methods and Soxhlet extraction [15,16]. In this study, conventional solvent extraction was used by varying four factors (ethanol concentration, citric acid, temperature, time, agitation rpm). Furthermore, for more efficient solid/liquid extraction under stirring of biologically active compounds from beetroot peel, a central composite design (CCD) was used in order to optimize the extraction procedure to maximize both phenolic compounds extraction and betalains.

2. Materials and Methods

2.1. Reagents and Chemicals

For the phytochemical characterization of biologically active compounds from beetroot, peels were used: ethanol of HPLC purity, ethanol, citric acid, Folin–Ciocâlteu reagent, and gallic acid solution were obtained from Sigma Aldrich (St. Louis, MO, USA).

2.2. Beetroot Peels Powder

A local producer in Galati County provided beetroot. After being thoroughly washed, the shells were removed with a knife. The resulting beetroot peel was washed with ultrapure water, wiped, and frozen for drying. Drying was performed by the lyophilization technique using CHRIST Alpha 1–4 LD. Plus equipment, Germany, at −42 °C, under a pressure of 0.10 mBar, for 48 h. The resulting dry peels had a relative humidity of 1%. They were grounded and stored in a glass container with an airtight lid at room temperature and in the dark.

2.3. Conventional Solvent Extraction

The extraction was performed using 1 g of beet peel with 10 mL of ethanol in different concentrations, respectively 9–60%. The ratio of plant material to the solvent was 1 to 10. Each extraction was acidified with a citric acid solution using different concentrations of 0.03–2%, in a ratio of 1 to 9. The extractions took place at 3–87 °C for 3 to 74 min on an orbital shaker (SI-300R Medline Scientific, Chalgrove, UK) at 100 rpm. Subsequently, the samples were centrifuged using Hettich Universal 320R equipment, Germany, for 10 min at 14,000 rpm, at 4 °C, and the supernatant was phytochemically analyzed.

2.4. Determination of Total Betalains (TB) Content

The total betalains (TB) content of the beet peels extract was determined using a spectrophotometric method based on absorbance measurement at two different wavelengths, respectively 480 nm for betaxanthins and 537 nm for betacyanins.

The total betalains (TB) content was expressed in mg/g DW and was calculated as the sum of betacyanins and betaxanthins according to Formula (1) as described by Wruss et al. [2], with slight modifications.

$$BC[\text{mg/g}] = \frac{A \cdot DF \cdot M_w \cdot V_d}{\varepsilon \cdot L \cdot W_d} \quad (1)$$

where A is the maximum absorption value of 537 nm for betacyanins and 480 nm for betaxanthins; DF is the dilution factor; V_d is the volume of dry pulp solution (mL); W_d is the weight of the dried pulp (g); L is the length of the tank path (1 cm); M_W is the molecular weight of betalains (550 g/mol); ε is the the extinction coefficient of betalain; ε of 60,000 L/mol cm in H_2O were applied to quantify betacyanins, and the quantitative equivalents of major betaxanthins (Bx) were determined by applying the average molar extinction coefficient (ε) 48,000 L/(mol·cm) in H_2O.

2.5. Determination of Total Polyphenol Content (TPC)

The total polyphenol content (TPC) of the beet peels extract was determined using the Folin–Ciocalteu (FC) method. The description of this method is based on the redox reaction between phenolic compounds and a mixture of tungsten and molybdenum in an alkaline environment. This reaction leads to the creation of a blue complex that was quantified at 765 nm. The results were expressed as mg gallic acid equivalent (GAE)/g DW.

2.6. Statistical Analysis

To analyze the experimental model, we used statistical software Design Expert (v. 13) from Design-Expert® (Stat-Ease, Inc., minneapolis, MN, USA).

2.7. Experimental Design

Central Composite Design (CCD) has been used to optimize the extraction of betalains and total polyphenols from beetroot and establish the experimental antioxidant activity. An experimental factorial model with a central composition with five factors, three central points, and the design of 19 experimental variants was used. The maximum and minimum ranges of the variables investigated in the experimental plan in the current form and the coded form are presented in Table 1. In addition, CCD builds a quadratic model for response variables.

Table 1. Range factors studied and the corresponding encoded values.

Code	Independent Variables	Units	Minimum	Maximum	Coded Low	Coded High
A	Citric acid	%	0.03	1.98	−1 ↔ 0.10	+1 ↔ 1.50
B	Ethanol	%	9.77	60.23	−1 ↔ 20.00	+1 ↔ 50.00
C	Temperature	°C	2.96	87.04	−1 ↔ 20.00	+1 ↔ 60.00
D	Time	min	3.07	61.93	−1 ↔ 15.00	+1 ↔ 50.00

The software used for testing the experimental conditions can be described by a second-order polynomial model (2):

$$R = b_0 + \sum_{i=1}^{n} b_i \cdot x_i + \sum_{i=1}^{n} b_{ii} \cdot x_i^2 + \sum_{i \neq j=1}^{n} b_{ij} \cdot x_i \cdot x_j \quad (2)$$

where R—predicted response, b_0—intercept, b_i, b_{ii}, and b_{ij}—regression coefficients, x_i and x_j—independent variables analyzed, n—number of factors.

3. Results

In order to determine the optimal parameters for optimization of the extraction process, a Central Composite Design (CCD) and surface response modeling were used. In addition, the BT and TPC contents were measured in response. Table 2 shows the complete CCD matrix used to optimize the main variables studied and the corresponding values.

Table 2. The CCD matrix with the actual values of the main variables studied.

Run	Factor 1 A: Citric Acid %	Factor 2 B: Ethanol %	Factor 3 C: Temperature °C	Factor 4 D: Time min	Response 1 Betalains mg/g DW	Response 2 TPC mg GAE/g DW
1	0.80	35.00	40.00	3.07	0.80	2.02
2	0.80	35.00	40.00	32.50	0.98	2.30
3	1.98	35.00	40.00	32.50	1.03	2.38
4	0.10	50.00	20.00	50.00	0.88	2.12
5	0.10	50.00	60.00	50.00	0.70	1.92
6	1.50	20.00	20.00	50.00	0.85	2.00
7	1.50	20.00	60.00	50.00	0.62	1.80
8	0.80	35.00	6.36	32.50	0.65	1.89
9	0.80	35.00	40.00	32.50	1.02	2.28
10	1.50	50.00	60.00	15.00	0.76	1.96
11	0.10	20.00	20.00	15.00	0.78	2.05
12	0.80	35.00	73.64	32.50	0.29	1.64
13	0.03	35.00	40.00	32.50	0.67	1.99
14	0.80	60.23	40.00	32.50	1.44	2.74
15	0.80	35.00	40.00	61.93	1.14	1.87
16	1.50	50.00	20.00	15.00	0.99	2.22
17	0.80	35.00	40.00	32.50	1.10	2.26
18	0.80	9.77	40.00	32.50	0.65	1.81
19	0.10	20.00	60.00	15.00	0.36	1.77

3.1. *The Influence of Extraction Parameters on BT*

This study aimed to identify the proper optimal pattern of factors for the betalains extraction from beetroot. The experimental design revealed a variation in betalain content ranging from 0.29 to 1.44 mg/g DW (Table 2).

The regression equations obtained after the analysis of the ANOVA variance described the content in BT of the beet extract obtained, depending on the factors of the extraction environment (Table 3).

The regression model acquired for BT revealed a determination coefficient of $R^2 = 0.96$, suggesting that only 0.04 of the variation of BT cannot be described by the current model. The Lack of Fit F-value of 1.12 implies that the Lack of Fit is not significant relative to the pure error. *p*-values less than 0.0500 indicate that model terms are significant, and in this case, A, B, C, D, AB, AD, A^2, and C^2 are significant model terms.

The minor model terms were excluded, and a model reduction was achieved. As a result, the model equation indicating the relationship between the betalains content (R1) and variables in coded units is represented in Equation (3). Moreover, the Pred R^2 (0.8796) values were in reasonable agreement with Adj R^2 (0.9452).

$$\text{R1 (BT)} = +1.01 + 0.08A + 0.23B - 0.12C + 0.1011D + 0.08AB + 0.14AD - 0.06A^2 - 0.19C^2 \quad (3)$$

The b coefficients from the regression equation indicated that the temperature had a significant negative effect on betalain extractions. In addition, the interactions between temperature and temperature (C^2) have significant negative effects, and quadratic citric acid concentration (A^2) had a more negligible contribution. Furthermore, ethanol concentration (B) and extraction time (D) appreciably positively influenced the BT. The interaction between temperature and extraction time (AB) moderately affected the extraction, while citric acid concentration and ethanol concentration (AD) have a better effect.

Table 3. ANOVA for the reduced quadratic model for BT and TPA extractions.

	Total Betalains (TB) Content					Total Polyphenol Content (TPC)				
Source	**Sum of Squares**	**df**	**Mean Square**	**F-Value**	***p*-Value**	**Sum of Squares**	**df**	**Mean Square**	**F-Value**	***p*-Value**
Model	1.30	8	0.1625	39.84	<0.0001 [a]	1.22	9	0.1356	80.87	<0.0001
A-Citric acid	0.0898	1	0.0898	22.02	0.0009	0.0765	1	0.0765	45.61	<0.0001
B-Ethanol	0.3120	1	0.3120	76.50	<0.0001	0.4336	1	0.4336	258.56	<0.0001
C-Temperature	0.2026	1	0.2026	49.67	<0.0001	0.1340	1	0.1340	79.93	<0.0001
D-Time	0.0578	1	0.0578	14.17	0.0037	0.0125	1	0.0125	7.44	0.0233
AB	0.0213	1	0.0213	5.23	0.0453	0.1383	1	0.1383	82.49	<0.0001
AD	0.0687	1	0.0687	16.85	0.0021	0.0345	1	0.0345	20.58	0.0014
A^2	0.0567	1	0.0567	13.90	0.0039	0.0097	1	0.0097	5.77	0.0397
C^2	0.5475	1	0.5475	134.20	<0.0001	0.4186	1	0.4186	249.63	<0.0001
D^2	-	-	-	-	-	0.1666	1	0.1666	99.38	<0.0001
Residual	0.0408	10	0.0041			0.0151	9	0.0017		
Lack of Fit	0.0333	8	0.0042	1.12	0.5545 [b]	0.0145	7	0.0021	7.58	0.1215
Pure Error	0.0075	2	0.0037			0.0005	2	0.0003		
Cor Total	1.34	18				1.24	18			

SS—Sum of Squares, MS—Mean Square; [a] Significant; [b] Not significant.

Figure 1a represents the correlation between ethanol and citric acid concentration on the extraction yield. The betalain content increased as the ethanol concentration increase to 40% and from 1.30% for the citric acid concentration. Betalains extraction was influenced by the citric acid concentration and extraction time correlative effect (Figure 1b). Extraction was inhibited both at low concentrations of citric acid (0.10%) and at high concentrations exceeding 1.50%. Similarly, the extraction time negatively affects the process at values greater than 50 min and less than 15 min. It can be concluded that by increasing the extraction time and decreasing the concentration of citric acid in the extraction medium, the extraction yield decreases significantly.

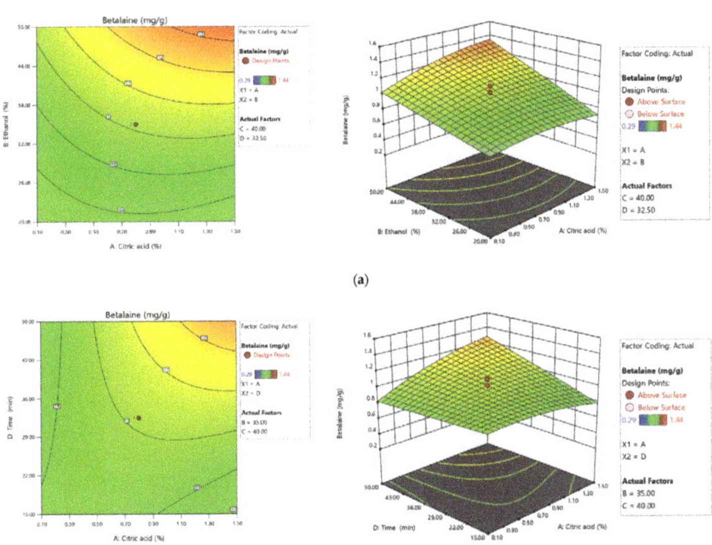

(a)

(b)

Figure 1. Second-order contour plots (left) and 3D surface plots (right) screening the effect of ethanol and citric acid concentra-tion (**a**) and extraction time and citric acid concentration (**b**) on the extraction yield of total betalain content (BT).

In analyzing the deviation from the reference point, a slope with a large or curved inclination for a specific factor shows that the response is sensitive in this case, while a relatively flat line demonstrates insensitivity to changes in this factor. The main factor affecting BT extraction is the ethanol (Figure 2a, curve B) followed by temperature (Figure 2a, curve C) and extraction time (Figure 2a, curve D). Meanwhile, the findings showed that citric acid is less sensitive (Figure 2a, curve A).

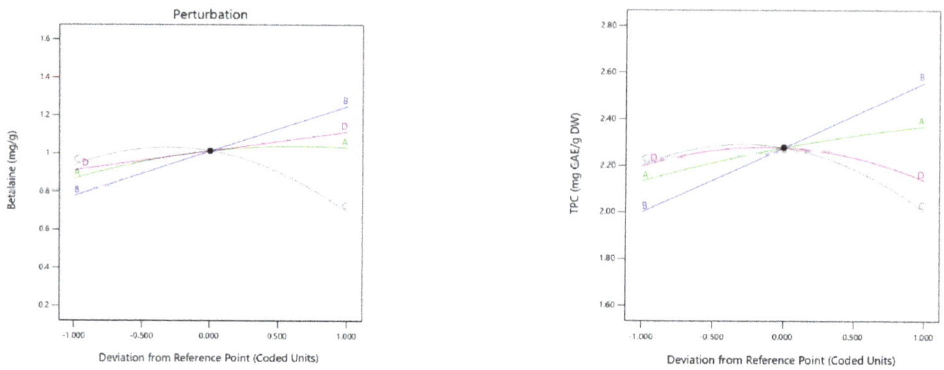

(**a**) (**b**)

Figure 2. Graph of perturbations describing the effect of each independent variable (A, B, C, and D) on BT (**a**) and TPC (**b**) extraction.

3.2. *The Influence of Extraction Parameters on TPC*

Due to the effect of different variables, the yield of polyphenolic varied from 1.64 to 2.74 mg/g DW (Table 2).

The Model F-value of 80.87 implies that the model is significant. The determinant coefficient of R^2 = 0.98 suggests that only 0.02 of the variation of TPC cannot be described by the current model. Subsequently, from Table 3, it was observed that the Lack of Fit F-value of 7.58 implies the Lack of Fit lack was non-significant, which showed that the model is also significant. *p*-values less than 0.0500 indicate that model terms are significant, and in this case, they are significant model terms. In this case, A, B, C, D, AD, BD, A^2, C^2, and D^2 are significant model terms.

The insignificant model terms were ignored, and a model reduction was attained. The model equation indicating the relationship between the total phenolic content (R^2) and variables in coded units was represented in Equation (4). Moreover, the Pred R^2 values (0.8875) were in reasonable agreement with Adj R^2 (0.9756).

$$R^2\ (TPC) = +2.28 + 0.11A + 0.27B - 0.99C - 0.30D + 0.20AD + 0.10BD - 0.26A^2 - 0.17C^2 - 0.11D^2 \quad (4)$$

The b coefficients from the regression equation indicated that the temperature and time had a minor negative effect on TPC extractions. On the other hand, the extraction time positively affects the TPC recovery when associated with ethanol and citric acid concentration. In addition, significant negative effects have the interactions between temperature and temperature (C^2) and time and time (D^2). Furthermore, ethanol concentration (B) and citric acid concentration (A) had an appreciably positive effect on polyphenols extractions.

Second-order contour plots were designed to predict the relationship between the independent and dependent variables (Figure 3) and to illustrate the synergistic effects of the independent factors on the TPC. The three-dimensional response area describes the correlative effect of the selected factors on TPC. The coordinates of the central point in the contour graph correspond to the optimal concentration of the four components of the extraction process to obtain a maximum phenolic extraction efficiency.

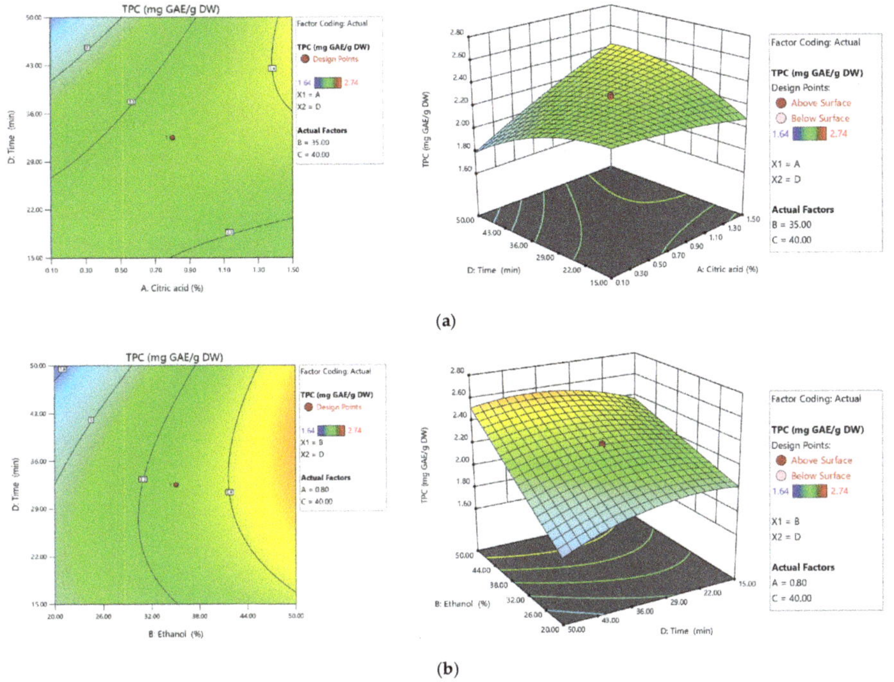

Figure 3. Second-order contour plots (**left**) and 3D surface plots (**right**) screening the effect of extraction time and citric acid concentration (**a**) and extraction time and ethanol (**b**) on the extraction yield of total phenolic content (TPC).

The effects of varying extraction time and citric acid concentration on the extraction of phenolic compounds are shown in Figure 3a. The concentration of polyphenols is lower as the extraction time and the concentration of citric acid increase simultaneously. Figure 3b confirmed that the extraction was not influenced by time variation but was influenced by ethanol concentration.

According to the perturbations graph describing the effect of each independent variable, the TPC has strongly influenced ethanol concentration and, to a lesser extent, temperature and time have influenced the concentration as well (Figure 2b, curve B, C, and D). Thus, a maximum of 2.74 mg GAE/g DW was found at 40 °C at the maximum ethanol concentration (60.23%) in the present study.

3.3. Optimization and Validation of Extraction Parameters

The model suggested the optimal conditions based on the maximization of the responses desirability to validate the model equation. A value of desirability near 1 (0.956) suggested that all the selected conditions were in a proper combination (Figure 4, Table 4).

The optimal conditions for the maximum recovery of betalains and phenols were citric acid concentration 1.5%, ethanol concentration 50%, temperature 52.52 °C, and extraction time of 49.9 min.

The model estimated the maximum concentration of betalains and total polyphenols of 1.15 mg/g DW and 2.44 mg GAE/g DW, respectively, and the experimental results indicated immediate responses to those predicted by the model, namely 1.20 mg/g DW and 2.39 mg GAE/g DW (Table 4).

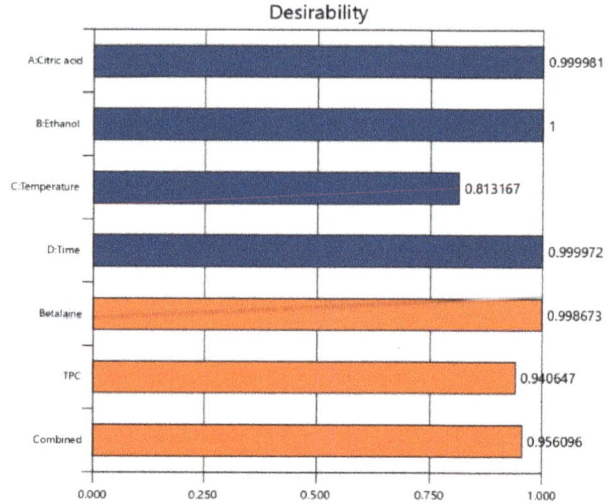

Figure 4. Desirability plot of process parameters and responses.

Table 4. Validation of the mathematical model.

Dependent Variable	Predicted Value	95% Confidence Intervals	Experimental Value
BT (mg/g DW)	1.15	1.06–1.25	1.20
TPC (mg GAE/g DW)	2.44	2.39–2.49	2.39

4. Discussion

The phytochemicals extraction from plant materials relies upon the extraction strategy being influenced by the extraction parameters used. Moreover, the various polarities of compounds extracted by applying the experimental model could have an unpredictable impact on the extraction conditions. Hence, extractions were performed with extractants having various polarities controlled by adjustments in the water and ethanol. Adding ethanol to water may increase the betalain extraction yields [17], and the extracts can be easily used in biological systems. Moreover, the extraction optimization led to adding citric acid in the extraction mixture to acidify the medium, mainly because the betalains became stable at pH 3–7 [18].

The results from the CCD indicated that the temperature had a minor negative effect on TPC extractions. The results agree with [19] that revealed that the temperature does not bring any significant transformation of betalains content. The CCD matrix revealed that the low temperature of 6.3 °C is insufficient for the total extraction of betalains, and also, the higher temperatures of 73.54 °C have a negative influence on betalains that can lead to their degradation [20]. The results agree with [21], which achieved the same conclusions at temperatures of 50 °C. In the same pattern the authors of [22] found that the maximum betalain retention was 25 to 30 °C, and by increasing the temperature of extraction, a decreased in betalain retention was accomplished.

However, a study conducted by [23] to determine the optimization of betalain pigments extraction from xoconostle using RMS methodology found contrasting findings. The authors obtained a maximum concentration of 0.92 mg/g of fruit at a lower temperature of extraction of 15 °C for 10 min using as solvent methanol and a ratio of 20:80 (methanol:water).

The results from the optimization experiment indicate that the factors affecting BT extraction are the ethanol concentration and time. Since betalains are sensitive to high temperature and long processing times, we have searched for a possible green method to keep these compounds in the product. Several authors obtained similar findings [24,25].

Similar quantities of betalains of 0.8 and 1.3 mg/g juice were obtained by analyzing different beetroot varieties from Upper Austria [2]. The authors of [26] indicated values of 0.4–20 mg/g betalains in red beetroot. In addition, in a study conducted by [27] for different beetroot species, a content of 0.65–0.80 mg/g FM betalain was obtained. Moreover, in a comparative study that analyzed the different parts and cultivars of beetroot for betalain profiling, a variation in the content of betalains was observed [28]. The betalains distribution obtained was from 5.33 to 31.04 mg/g DW for peel, from 0.35 to 8.65 mg/g DW for the flesh, and 0.85 to 11.10 mg/g DW for the petiole.

The authors of [29] reported lower values of TPC of 1.77 mg GAE/g DW than this study. Low values were also obtained by [30] that the total polyphenols range from 0.82 to 1.28 mg/g DW for two beetroot varieties. A similar study that analyzed four beetroot varieties revealed after extraction a phenolic content of 0.365–8.87 mg/g DW using the following parameters: solvent, 80% ethanol, and 12 h as the extraction time [31]. Another study [32] achieved a total phenolic amount of between 3.52 and 4.89 mg GAE/g DW for the red-colored genotypes.

5. Conclusions

CCD-RSM was employed to optimize the solvent extraction process variables to produce beetroot peel extracts with a high yield of bioactive compounds (citric acid concentration—1.5%, ethanol concentration—50%, temperature—52.52 °C, and extraction time—49.9 min). In the experiment, the interactions of time, temperature, and acid and solvent concentrations led to the improvement of compounds extraction in terms of the highest concentration of betalains (1.20 mg/g DW) and phenolic compounds (2.39 mg GAE/g DW). This study's novelty is optimizing the extraction protocol using conventional methods to improve the extraction of bioactive compounds from beetroot peels. Therefore, the method developed can be successfully utilized as a promising method for the efficient extraction of betalains and polyphenols from the beetroot peel for its economic utilization. Due to the significant concentration of valuable bioactive constituents from beetroot peel, multipurpose applications of these compounds could be found in different food industries and the pharmaceutical and nutraceutical industries.

Author Contributions: Conceptualization, G.R. and N.S.; methodology, G.R. and O.E.C.; validation, O.E.C., and S.L.; formal analysis, S.L.; investigation, S.L.; resources, N.S. and C.C.; data curation, I.A.; writing—original draft preparation, O.E.C. and G.R.; writing—review and editing, G.R. and I.A.; visualization, N.S.; supervision, C.C. All authors have read and agreed to the published version of the manuscript.

Funding: Doctoral Schools of Dunarea de Jos University of Galati (SCDS-UDJG) http://www.cssd-udjg.ugal.ro/ (accessed on 11 April 2021).

Institutional Review Board Statement: Not applicable.

Informed Consent Statement: Not applicable.

Data Availability Statement: The data that support the findings of this study are available from the corresponding author (G.R.) upon reasonable request.

Acknowledgments: The results of this work have been presented to the 9th edition of the Scientific Conference organized by the Doctoral Schools of "Dunărea de Jos" University of Galati (SCDS-UDJG) http://www.cssd-udjg.ugal.ro/ (accessed on 11 April 2021) that will be held on 10th and 11th of June 2021, in Galati, Romania.

Conflicts of Interest: The authors declare no conflict of interest.

References

1. Chawla, H.; Parle, M.; Sharma, K.; Yadav, M. Beetroot: A Health Promoting Functional Food. *Inventi Rapid Nutraceuticals* **2015**, *2016*, 0976–3872.
2. Wruss, J.; Waldenberger, G.; Huemer, S.; Uygun, P.; Lanzerstorfer, P.; Muller, U.; Hoglinger, O.; Weghuber, J. Compositional characteristics of commercial beetroot products and beetroot juice prepared from seven beetroot varieties grown in Upper Austria. *J. Food Compos. Anal.* **2015**, *42*, 46–55. [CrossRef]
3. Neha, P.; Jain, S.K.; Jain, N.K.; Jain, H.K.; Mittal, H.K. Chemical and Functional Properties of Beetroot (*Beta vulgaris* L.) for Product Development: A Review. *Int. J. Chem. Stud.* **2018**, *6*, 3190–3194.
4. Wootton-Beard, P.C.; Ryan, L. A Beetroot Juice Shot Is a Significant and Convenient Source of Bioaccessible Antioxidants. *J. Funct. Foods* **2011**, *3*, 329–334. [CrossRef]
5. Vorobiev, E.; Chemat, F. Principles of Physically Assisted Extractions and Applications in the Food, Beverage and Nutraceutical Industries. In *Separation, Extraction and Concentration Processes in the Food, Beverage and Nutraceutical Industries*; Syed, S.H., Rizvi, A., Eds.; Woodhead Publishing: Sawston, UK, 2013; pp. 71–108.
6. Liu, X.; Gao, Y.; Xu, H.; Wang, Q.; Yang, B. Impact of High-Pressure Carbon Dioxide Combined with Thermal Treatment on Degradation of Red Beet (*Beta vulgaris* L.) Pigments. *J. Agric. Food Chem.* **2008**, *56*, 6480–6487. [CrossRef] [PubMed]
7. Albano, C.; Negro, C.; Tommasi, N.; Gerardi, C.; Mita, G.; Miceli, A.; De Bellis, L.; Blando, F. Betalains, Phenols and Antioxidant Capacity in Cactus Pear [Opuntia *ficus-indica* (L.) Mill.] Fruits from Apulia (South Italy) Genotypes. *Antioxidants* **2015**, *4*, 269–280. [CrossRef]
8. Ravichandran, K.; Saw, N.M.M.T.; Mohdaly, A.A.A.; Gabr, A.M.M.; Kastell, A.; Riedel, H.; Cai, Z.; Knorr, D.; Smetanska, I. Impact of Processing of Red Beet on Betalain Content and Antioxidant Activity. *Food Res. Int.* **2013**, *50*, 670–675. [CrossRef]
9. Kumar, N.; Pruthi, V. Potential Applications of Ferulic Acid from Natural Sources. *Biotechnol. Rep.* **2014**, *4*, 86–93. [CrossRef]
10. Hobbs, D.A.; George, T.W.; Lovegrove, J.A. The Effects of Dietary Nitrate on Blood Pressure and Endothelial Function: A Review of Human Intervention Studies. *Nutr. Res. Rev.* **2013**, *26*, 210–222. [CrossRef]
11. Vidal, P.J.; López-Nicolás, J.M.; Gandía-Herrero, F.; García-Carmona, F. Inactivation of Lipoxygenase and Cyclooxygenase by Natural Betalains and Semi-Synthetic Analogues. *Food Chem.* **2014**, *154*, 246–254. [CrossRef] [PubMed]
12. Faridah, A.; Holinesti, R.; Syukri, D. Betalains from Red Pitaya Peel (*Hylocereus polyrhizus*): Extraction, Spectrophotometric and HPLC-DAD Identification, Bioactivity and Toxicity Screening. *Pak. J. Nutr.* **2015**, *14*, 976–982. [CrossRef]
13. Kusznierewicz, B.; Mróz, M.; Koss-Mikołajczyk, I.; Namieśnik, J. Comparative Evaluation of Different Methods for Determining Phytochemicals and Antioxidant Activity in Products Containing Betalains—Verification of Beetroot Samples. *Food Chem.* **2021**, *362*, 130132. [CrossRef]
14. Azmir, J.; Zaidul, I.S.M.; Rahman, M.M.; Sharif, K.M.; Mohamed, A.; Sahena, F.; Jahurul, M.H.A.; Ghafoor, K.; Norulaini, N.A.N.; Omar, A.K.M. Techniques for Extraction of Bioactive Compounds from Plant Materials: A Review. *J. Food Eng.* **2013**, *117*, 426–436. [CrossRef]
15. Ramli, N.S.; Ismail, P.; Rahmat, A. Influence of Conventional and Ultrasonic-Assisted Extraction on Phenolic Contents, Betacyanin Contents, and Antioxidant Capacity of Red Dragon Fruit (*Hylocereus polyrhizus*). *Sci. World J.* **2014**, *2014*, 964731. [CrossRef] [PubMed]
16. Hilou, A.; Millogo-Rasolodimby, J.; Nacoulma, O.G. Betacyanins are the most relevant antioxidant molecules of Amaranthus spinosus and Boerhavia erecta. *J. Med. Plants Res.* **2013**, *7*, 645–652.
17. Fu, Y.; Shi, J.; Xie, S.-Y.; Zhang, T.-Y.; Soladoye, O.P.; Aluko, R.E. Red Beetroot Betalains: Perspectives on Extraction, Processing, and Potential Health Benefits. *J. Agric. Food Chem.* **2020**, *68*, 11595–11611. [CrossRef] [PubMed]
18. Castro-Enríquez, D.D.; Montaño-Leyva, B.; Del Toro-Sánchez, C.L.; Juaréz-Onofre, J.E.; Carvajal-Millan, E.; Burruel-Ibarra, S.E.; Tapia-Hernández, J.A.; Barreras-Urbina, C.G.; Rodríguez-Félix, F. Stabilization of Betalains by Encapsulation—A Review. *J. Food Sci. Technol.* **2020**, *57*, 1587–1600. [CrossRef]
19. Zin, M.M.; Borda, F.; Márki, E.; Bánvölgyi, S. Betalains, Total Polyphenols, and Antioxidant Contents in Red Beetroot Peel (Cylindra Type). *Prog. Agric. Eng. Sci.* **2021**, *16*, 27–36.
20. Chhikara, N.; Kushwaha, K.; Sharma, P.; Gat, Y.; Panghal, A. Bioactive Compounds of Beetroot and Utilization in Food Processing Industry: A Critical Review. *Food Chem.* **2019**, *272*, 192–200. [CrossRef]
21. Das, M.; Saeid, A.; Hossain, M.F.; Jiang, G.-H.; Eun, J.B.; Ahmed, M. Influence of Extraction Parameters and Stability of Betacyanins Extracted from Red Amaranth during Storage. *J. Food Sci. Technol.* **2019**, *56*, 643–653. [CrossRef]
22. Tang, C.S.; Norziah, M.N. Stability of Betacyanin Pigments from Red Purple Pitaya Fruit (*Hylocereus polyrhizus*): Influence of pH, Temperature, Metal Ions and Ascorbic Acid. *Indones. J. Chem.* **2007**, *7*, 327–331. [CrossRef]
23. Sanchez-Gonzalez, N.; Jaime-Fonseca, M.R.; San Martin-Martinez, E.; Zepeda, L.G. Extraction, Stability, and Separation of Betalains from Opuntia Joconostle Cv. Using Response Surface Methodology. *J. Agric. Food Chem.* **2013**, *61*, 11995–12004. [CrossRef] [PubMed]
24. Latorre, M.E.; Narvaiz, P.; Rojas, A.M.; Gerschenson, L.N. Effects of Gamma Irradiation on Bio-Chemical and Physico-Chemical Parameters of Fresh-Cut Red Beet (*Beta vulgaris* L. var. *conditiva*) Root. *J. Food Eng.* **2010**, *98*, 178–191. [CrossRef]
25. Thirugnanasambandham, K.; Sivakumar, V. Influence of Process Conditions on the Physicochemical Properties of Pomegranate Juice in Spray Drying Process: Modelling and Optimization. *J. Saudi Soc. Agric. Sci.* **2017**, *16*, 358–366. [CrossRef]

26. Krsnik-Rasol, M.; Pavoković, D. Complex Biochemistry and Biotechnological Production of Betalains. *Food Technol. Biotechnol.* **2011**, *49*, 145–155.
27. Lee, E.J.; An, D.; Nguyen, C.T.T.; Patil, B.S.; Kim, J.; Yoo, K.S. Betalain and Betaine Composition of Greenhouse- or Field-Produced Beetroot (*Beta vulgaris* L.) and Inhibition of HepG2 Cell Proliferation. *J. Agric. Food Chem.* **2014**, *62*, 1324–1331. [CrossRef]
28. Slatnar, A.; Stampar, F.; Veberic, R.; Jakopic, J. HPLC-MSnIdentification of Betalain Profile of Different Beetroot (*Beta vulgaris* L. Ssp. *vulgaris*) Parts and Cultivars. *J. Food Sci.* **2015**, *80*, C1952–C1958. [CrossRef]
29. Ninfali, P.; Angelino, D. Nutritional and Functional Potential of Beta *vulgaris cicla* and *rubra*. *Fitoterapia* **2013**, *89*, 188–199. [CrossRef]
30. Kavalcová, P.; Bystrická, J.; Tomáš, J.; Karovičová, J.; Kovarovič, J.; Lenková, M. The Content of Total Polyphenols and Antioxidant Activity in Red Beetroot. *Potravin. Slovak J. Food Sci.* **2015**, *9*, 77–83. [CrossRef]
31. Kovarovič, J.; Bystrická, J.; Tomáš, J.; Lenková, M. The Influence of Variety on the Content of Bioactive Compounds in Beetroot (*Beta vulgaris* L.). *Potravin. Slovak J. Food Sci.* **2017**, *11*, 106–112. [CrossRef]
32. Yasaminshirazi, K.; Hartung, J.; Fleck, M.; Graeff-Hoenninger, S. Bioactive Compounds and Total Sugar Contents of Different Open-Pollinated Beetroot Genotypes Grown Organically. *Molecules* **2020**, *25*, 4884. [CrossRef] [PubMed]

inventions

Article

Impact of Wall Materials on Physico-Chemical Properties and Stability of Eggplant Peels Anthocyanin Hydrogels

Nina-Nicoleta Condurache (Lazăr) [1], Mihaela Turturică [1], Elena Enachi [1], Vasilica Barbu [1], Gabriela-Elena Bahrim [1], Nicoleta Stănciuc [1], Constantin Croitoru [2] and Gabriela Râpeanu [1,*]

1 Faculty of Food Science and Engineering, Dunărea de Jos University of Galati, Domnească Street 111, 800201 Galati, Romania; nina.condurache@ugal.ro (N.-N.C.); mihaela.turturica@ugal.ro (M.T.); elena.ionita@ugal.ro (E.E.); vasilica.barbu@ugal.ro (V.B.); gabriela.bahrim@ugal.ro (G.-E.B.); nicoleta.stanciuc@ugal.ro (N.S.)

2 Academy of Agricultural and Forestry Sciences, 61 Marasti Blvd, 011464 Bucharest, Romania; c.croitoru@sodinal.com

* Correspondence: gabriela.rapeanu@ugal.ro; Tel.: +40-336-130-177

Abstract: In this study, eggplant peel extract was used to obtain hydrogels. Two experimental variants were realized by varying the wall materials. Whey proteins isolate (WPI), citrus pectin (P), and sodium carboxymethylcellulose (CMCNa) were used as wall materials. The microcapsules were obtained by the gelation technique, followed by freeze-drying in order to obtain powders. Both experimental variants were analyzed in terms of phytochemical content, antioxidant activity, storage stability, and in vitro digestibility. Additionally, confocal microscopy was used to observe the encapsulation of the bioactive compounds from the eggplant peel extract into the selected matrices. The encapsulation efficiency of the powders varied from 64.67 ± 0.68% for variant 1 (V1) to 96.44 ± 3.43% for variant 2 (V2). Both powders presented high bioactive compound content with high antioxidant activity. V2 showed the highest stability within 28 days of storage, but also in the simulated digestive system.

Keywords: eggplant peels; bioactive compounds; hydrogels; anthocyanins

Citation: Condurache, N.-N.; Turturică, M.; Enachi, E.; Barbu, V.; Bahrim, G.-E.; Stănciuc, N.; Croitoru, C.; Râpeanu, G. Impact of Wall Materials on Physico-Chemical Properties and Stability of Eggplant Peels Anthocyanin Hydrogels. *Inventions* **2021**, *6*, 47. https://doi.org/10.3390/inventions6030047

Academic Editor: Farid Chemat

Received: 10 June 2021
Accepted: 28 June 2021
Published: 29 June 2021

1. Introduction

The agro-food industries generate annually overwhelming amounts of wastes by the processing of various vegetables and fruits [1]. Until a few years ago, food waste disposal was not a matter of concern. However, the increasing amounts of waste generated and the environmental issues caused by them have drawn attention. Currently, various organizations around the world are trying to find solutions to minimize food wastes with lower economic costs [2]. Due to the high content of proteins, lipids, sugars, fibers, pectin, or phytochemical compounds, the food wastes can be the ideal source from which value-added foods can be obtained for feeding the growing population. Their valorization in value-added food products represents an advantageous way of managing the wastes problems, turning them from wastes into by-products [1,2].

The fruit and vegetable processing industry generates by-products especially rich in natural antioxidants, such as polyphenols. These compounds are appreciated for their nutritional and functional properties. A regular intake of phenolic compounds appears to reduce the risk for developing coronary heart disease, hypertension, diabetes, obesity, gastrointestinal diseases, etc. [3]. Their antioxidant activity delays the formation of off-flavors and rancidity in food products, making them the perfect natural preservatives. They can also be used as natural pigments for food and beverage coloring or in cosmetics and nutraceuticals [4].

Aubergine, or eggplant, the name under which it is known worldwide, is a non-tuberous *Solanaceae* family vegetable. *Solanum melongena* is a tropical fruit with multiple shapes, sizes, and colors, and is very perishable yet very tasty. The most commercially important one is the purple variety (*Solanum melongena* L.) due to the anthocyanins located in the peels [5].

The anthocyanins are flavonoids with red, blue, and purple colors in fruits, vegetables, and flowers. Their basic structure is composed of a flavylium cation to which different sugars, hydroxyl or methoxyl groups are attached [6]. Of all the phenolic compounds, anthocyanins are the most sensitive. They easily degrade at higher pH, temperatures, or at prolonged exposure to light, oxygen, enzymes, etc. [7]. The anthocyanins are also sensitive to the digestive system conditions, especially when consumed separately from the original food matrix, with only 1–2% of the ingested amount being absorbed [8]. The latest researches propose encapsulation as a method of increasing the bioavailability and bioaccessibility of anthocyanins. When choosing the suitable encapsulation method, the properties of the bioactive compound, encapsulating material, and desired final product must be taken into account [9].

The present study aimed to obtain functional ingredients from eggplant peels bioactives for food or nutraceuticals applications. To extract the phenolic compounds from the eggplant peels, the ultrasound-assisted extraction (UAE) method was applied. The extract was characterized in terms of phytochemical content and antioxidant activity. Further, the extract was used for encapsulation by the gelation technique, followed by freeze-drying in order to obtain powders. Sodium carboxymethylcellulose (CMCNa), pectin (P), and whey protein isolate (WPI) were used as wall materials in different concentrations due to their ability to form hydrogels with three-dimensional networks. The two experimental variants resulting were analyzed in terms of encapsulation efficiency (EE), total anthocyanin content (TAC), total flavonoid content (TFC), total polyphenol content (TPC), and antioxidant activity. The microstructure of the powders was analyzed by confocal scanning laser microscopy (CLSM). The evolution of phytochemicals during the storage stability test was also monitored for 28 days. The in vitro release profile of the TAC and antioxidant activity under simulated gastric and intestinal juices was also performed.

2. Materials and Methods

2.1. Materials

Whey protein isolate (protein content of 95%) from Fonterra (New Zeeland). Ethanol 96% (EtOH) from Titolchimica (Italy). HPLC purity methanol (MeOH), 2,2-diphenyl1-picrylhydrazyl (DPPH), glacial acetic acid (CH_3COOH), sodium nitrite solution ($NaNO_2$), potassium chloride solution (KCl), sodium acetate solution (CH_3COONa), aluminum chloride ($AlCl_3$), sodium hydroxide (NaOH), sodium carbonate (Na_2CO_3), apple pectin, sodium carboxymethyl cellulose (CMCNa), 6-hydroxy-2,5,7,8-tetramethylchromane-2-carboxylic acid (Trolox), Folin–Ciocâlteu reagent, gallic acid, hydrochloric acid (HCl), sodium bicarbonate ($NaHCO_3$), Trizma buffer, the standards used for the HPLC analysis, delphinidin 3-*O*-glucoside, delphinidin 3-*O*-rutinoside, and cyanidin 3-*O*-rutinoside were obtained from Sigma-Aldrich Steinheim, Germany.

2.2. Methods

2.2.1. Biologically Active Compounds Extraction

The *Solanum melongena* L. autochthonous variety fruits were purchased from a local market in Galați, Romania. The purple outer layer of the fruit were peeled in uniform strips, washed with ultrapure water, dried with paper towels, and frozen. Subsequently, the peels were freeze-dried until 90% dry weight (dw), at −42 °C, under a pressure of 0.10 mBar, with a CHRIST Alpha 1-4 LD plus equipment (Germany) for 48 h.

The biologically active compounds were extracted using the UAE method described by Condurache et al. [10]. Briefly, 1 g of freeze-dried eggplant peels was mixed with 15 mL of a solvent mixture composed of EtOH 96% and glacial acetic acid, in a 4:1 ratio. The mixture was exposed for 15 min to ultrasounds at a 40 kHz frequency and 25 °C on a Smart sonic cleaner ultrasonic bath (MRC. LTD, Holon, Israel). Further, the extracts were centrifuged for 10 min at 14,000 rpm and 4 °C. The resulting supernatant was concentrated to dryness under reduced pressure at 40 °C (AVC 2-18, Christ, Shropshire, UK), and finally phytochemically analyzed.

2.2.2. Extract Characterization

The extract was characterized in terms of yield of extraction, TAC, TFC, TPC, and antioxidant activity, as described by Turturică et al. [11]. In brief, the TACs of 10 mg/mL eggplant peels' extract in ultrapure water were quantified using the pH differential method, and the results were expressed as mg delphinidin 3-*O*-glucoside (D3G)/g dw [11]. The TFC was analyzed using the colorimetric method based on the capacity of $AlCl_3$ to form stable complexes with the flavones or flavonols. The results were expressed as mg catechin equivalent (CE)/g dw [11]. The TPCs were quantified using the colorimetric method with the Folin–Ciocâlteu reagent, and the results were expressed as mg gallic acid equivalents (GAE)/g dw [11]. The free radical scavenging activity of the extract was measured using the DPPH, and the results were expressed as mM trolox equivalents (TE)/g dw [11]. The yield of extraction was calculated using formula (1) as described by Swer et al. [12]:

$$\text{EY} = \frac{extract\ weight}{eggplant\ peels\ weight} \times 100 \tag{1}$$

2.2.3. HPLC Analysis of the Anthocyanins

The chromatographic analysis of the anthocyanins found in eggplants, including separation, identification, and quantification, was made by the slightly modified method described by Turturica et al. [11]. The used HPLC system was a Thermo Finnigan Surveyor containing a diode array detector and Xcalibur software (Finnigan Surveyor LC, Thermo Scientific, Waltham, MA, USA). The volume used for the injection of the samples was 20 μL with a flow rate of 1.0 mL/min. Before the injection, the samples were filtered using 0.22 μm syringe filters. All standard compounds used in this study were acquired from Sigma-Aldrich (Darmstadt, Germany) and were of high purity (>95.0%).

2.2.4. Encapsulation of the Biologically Active Compounds from the Eggplant Peel Extract

The biologically active compounds from the eggplant peels' extract were encapsulated in hydrogels using the method described by Serrano-Cruz et al. [13] with slight modifications. The hydrogels are 3D networks made from polymeric chains cross-linked by physical or chemical bonds, having a high affinity for water [14]. In our study, two experimental variants were obtained using CMCNa, P, and WPI as wall material. For the first variant (V1), 1.5% P, 1.5% CMCNa, and 3% WPI were dispersed in ultrapure water during the agitation on a magnetic stirrer (IKA RCT Basic, Staufen Germany) for 6 h at 45 °C and 450 rpm. For the second variant (V2), 2.25% P, 2.25% CMCNa, and 1.5% WPI in ultrapure water were also hydrated for 6 h at 45 °C and 450 rpm on the heating magnetic stirrer (Figure 1). Both variants were allowed to stand at 4 °C overnight to ensure the full hydration of the wall materials, after which 25 mg/mL of extract was added in each mixture and homogenized for 2 h at 25 °C and 450 rpm. The measured pH of the mixtures was 2.5 due to the extract. The samples were frozen at −80 °C. Further, the mixtures were freeze-dried to obtain stable powders with a Alpha 1-4 LD plus equipment (CHRIST, Osterode am Harz, Germany) for 48 h, at −42 °C under a pressure of 0.10 mBar until a 94.37% dw for V1 and 96.31% dw for V2. Finally, the powders were collected and packed in plastic tubes with light protection and stored at 4 °C for later analyses.

V1	**V2**
100 mL ultrapure water +1.5% P +1.5% CMCNa +3% WPI	100 mL ultrapure water +2.25% P +2.25% CMCNa +1.5% WPI
↓	↓
Solubilization 6 h, 45 °C, 450 rpm	Solubilization 6 h, 45 °C, 450 rpm
↓	↓
Overnight rest, 4 °C, dark	Overnight rest, 4 °C, dark
↓	↓
+ 25 mg/mL eggplant peel extract	+25 mg/mL eggplant peel extract
↓	↓
Solubilization 2 h, 25 °C, 450 rpm	Solubilization 2 h, 25 °C, 450 rpm
↓	↓
Freezing −80 °C	Freezing −80 °C
↓	↓
Freeze-drying 48 h, −42 °C, 0.10 mBar pressure, 94.37% dw.	Freeze-drying 48 h, −42 °C, 0.10 mBar pressure, 96.31% dw.

Figure 1. The eggplant peel extract encapsulation scheme using the gelation technique.

2.2.5. Powder Characterization

The obtained powders were phytochemically analyzed in terms of initial TAC, TFC, TPC, and antioxidant activity [11]. The encapsulation efficiency (EE) of the anthocyanins was also calculated by measuring the TAC and the surface anthocyanins content (SAC) as described by Condurache et al. [15].

To measure the TAC, TFC, TPC, and antioxidant activity, 200 mg of each powder was mixed with 7 mL of methanol/glacial acetic acid/distilled water (25:4:21 *v*/*v*/*v*). The mixtures were vortexed for 1 min and sonicated for 30 min at 25 °C and 40 kHz using an ultrasonic bath (MRC. LTD, Holon, Israel). Further, the samples were centrifuged at 14,000 rpm for 10 min at 4 °C, and the supernatant was analyzed.

The EE represents the content of anthocyanins encapsulated in the matrix, and it was calculated as a percentage ratio between TAC and surface anthocyanin content (SAC). The SAC was measured by mixing 200 mg of each powder with 7 mL of ethanol/methanol (1:1 *v*/*v*), vortexing for 1 min, and centrifuging at 14,000 rpm for 10 min at 4 °C. The resulting supernatants were analyzed using the pH differential method [11]. The EE was calculated using Equation (2).

$$\%\ \mathrm{EE} = \frac{(\mathrm{TAC} - \mathrm{SAC})}{\mathrm{TAC}} \times 100 \quad (2)$$

TAC—Total Anthocyanin Content; SAC—Surface Anthocyanin Content.

2.2.6. Storage Stability

The encapsulated powders were evaluated regarding the TAC, TFC, TPC, and antioxidant activity during storage at room temperature in plastic tubes with light protection. The biologically active compounds were extracted as described in Section 2.2.5. Powder characterization and the phytochemicals were analyzed after 28 days of storage.

2.2.7. Confocal Laser Scanning Microscopy (CLSM)

The Confocal Laser Scanning Microscopy technique was used to observe the encapsulation process of the eggplant peel extracts' bioactives within the selected matrices. As an analysis method, confocal microscopy allows the morphology and structure observation without the fragmentation of the targeted microparticles. CLSM was performed using a Zeiss confocal laser system (LSM 710) equipped with a diode laser (405 nm), Ar-laser (458, 488, 514 nm), DPSS laser (diode pumped solid state e 561 nm) and HeNe-laser (633 nm). The obtained powders were fluorescently stained with Red Congo (40 μM), and the distribution of the bioactives into the powder matrix was observed using a Zeiss AxioObserver Z1 inverted microscope equipped with a 40x apochromatic objective (numerical aperture 1.4). Furthermore, the FS49, FS38, and FS15 filters were also used for the analysis. The 3D images were rendered, classified, and analyzed with the ZEN 2012 SP1 software (black edition).

2.2.8. Powders Behavior in Simulated Digestion

The simulated gastrointestinal digestion of the powders was performed according to Oancea et al. [16], at 37 °C and 150 rpm on an SI-300R orbital shaking incubator (Medline Scientific, Chalgrove, UK). The static model that simulates the digestion in the stomach was performed to measure the bioavailability of the anthocyanins and the antioxidant activity from the two variants of powders. The gastric digestion was performed for 2 h using simulated gastric fluid (SGF) that contained porcine pepsin (40 mg/mL in 0.1 M HCl) at a pH = 2.00. The intestinal digestion was performed using intestinal fluid (SIF) with 2 mg/mL pancreatin at Ph = 5.3. The release was calculated using Equation (3):

$$\% \text{ Release} = \frac{\text{initial concentration}}{\text{digested concentration}} \times 100 \quad (3)$$

2.2.9. Statistical Analysis of Data

The statistical analysis of data was performed using the Minitab 17 Statistics Software. The differences between the samples were assessed using the Tukey test with the One-way ANOVA method. All the experiments were completed in triplicates, and the results were expressed as average values with a standard deviation.

3. Results

3.1. Eggplant Peel Extract Characterization

Eggplant peels are rich in biologically active compounds, mainly in anthocyanins. In the present work, the bioactives from the eggplant peels were extracted using the UAE method, with ethanol as a solvent. This combination between the method, solvent, and acid had a 74.79% extraction yield. The obtained extract was analyzed in terms of phytochemical content and antioxidant activity, and the results are presented in Table 1. The extract highlighted a TAC of 0.35 ± 0.07 mg C3G/g dw, a TFC of 2.99 ± 0.12 mg CE/g dw, and a TPC of 12.79 ± 0.66 mg GAE/g dw. The DPPH scavenging capacity of the extract presented 193.14 ± 1.25 mM TE/g dw.

Table 1. The phytochemical content of the eggplant peel extract.

Phytochemical Content	TAC mg D3G/g dw	TFC mg CE/g dw	TPC mg GAE/g dw	Antioxidant Activity mM TE/g dw
Eggplant peel extract	0.35 ± 0.07	2.99 ± 0.12	12.79 ± 0.66	193.14 ± 1.25

TAC = total anthocyanin content; TFC = total flavonoid content; TPC = total polyphenol content; D3G = delphinidin 3-*O*-glucoside; CE = catehin equivalent; GAE = gallic acid equivalent; TE = trolox equivalent; dw = dry weight.

Hosseini et al. [17] presented a TAC of 0.43 ± 1.24 mg D3G/g fw and a TPC of 2.07 ± 4.20 mg GAE/g fw after the conventional extraction of bioactives from eggplant peels using water/ethanol/acetic acid in a 50:48:2 ratio. On the contrary, Horincar et al. [18] reported higher TAC, TFC, and TPC values than ours after using acidified ethanol for the UAE extraction of the eggplant peels bioactives. However, the antioxidant activity on DPPH free radical reported by them presented lower values than our extract. Jung et al. [19] extracted the phenolics from different parts of eggplant and reported a TPC content of

55.19 ± 1.30 mg GAE/g extract and TFC of 6.19 ± 0.28 mg CE/g extract. The differences between our results and the other studies' results are due to the phytochemical variability from the raw material and extraction conditions. Though, the obtained data from Table 1 confirm that the eggplant peels are a rich source of biologically active compounds, especially anthocyanins.

3.2. HPLC Analysis of the Anthocyanin

In order to achieve the characterization of the eggplant anthocyanin profile a chromatographic analysis was performed by using the HPLC technique (Figure 2). The identification and the quantification of anthocyanin were made depending on the retention time and by comparison with the available standards and the data existing already in the literature.

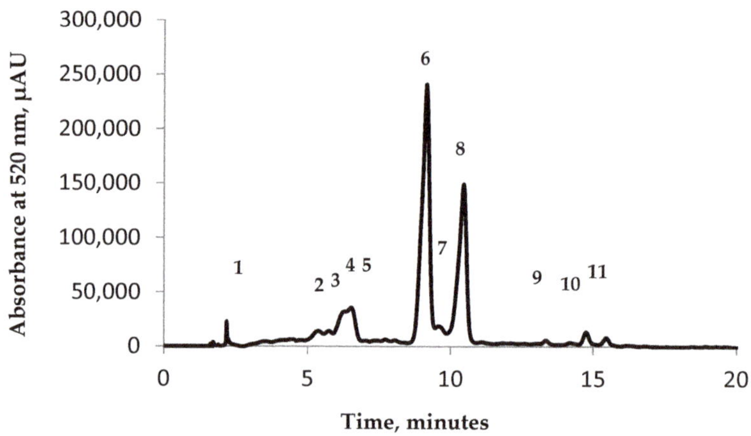

Figure 2. Chromatographic profile of eggplant peel extracts: Peak 1—delphinidin 3-*O*-rutinoside-5-glucoside; Peak 2—5: unidentified; Peak 6—delphinidin 3-*O*-glucoside; Peak 7—unidentified; Peak 8—delphinidin 3-*O*-rutinoside; Peak 9—unidentified; Peak 10—cyanidin 3-*O*-rutinoside; Peak 11–petunidin 3-*O*-rutinoside.

The anthocyanin identification was made at 520 nm, and the chromatographic analysis revealed the presence of 11 compounds: delphinidin 3-*O*-rutinoside-5-glucoside (Peak 1), delphinidin 3-*O*-glucoside (Peak 6), delphinidin 3-*O*-rutinoside (Peak 8), cyanidin 3-*O*-rutinoside (Peak 10), and petunidin 3-*O*-rutinoside (Peak 11), whereas peaks 2—5, 7, and 9 were unidentified.

The eggplant peel extract (Figure 2) revealed that delphinidin 3-*O*-glucoside is the major anthocyanin identified, having a concentration of 82.51%. These results are in agreement with those obtained by Azuma et al. [20]. They managed to identify delphinidin 3-*O*-rutinoside and nasunin as being the major compounds found in the eggplant extract. Delphinidin 3-*O*-rutinoside followed delphinidin 3-*O*-glucoside closely, with a concentration of 66.94%. The other identified anthocyanin compounds varied in terms of concentration, such as: delphinidin 3-*O*-rutinoside-5-glucoside—11.57%, cyanidin 3-*O*-rutinoside—5.08%, and petunidin 3-*O*-rutinoside—2.14%. In their studies, Ferarsa et al. [21], Dranca et al. [22], and Mauro et al. [23] managed to separate and identify only five anthocyanins from eggplant peel extracts. In our research conducted so far, we highlighted that the major anthocyanin found in eggplant is delphinidin 3-*O*-rutinoside. This study revealed a high concentration of delphinidin 3-*O*-glucoside. The difference between the two is explained by Mauro et al. [23], thereby passing to different stages of ripening, the delphinidin 3-*O*-rutinoside concentration in different eggplant cultivars show a significant decrease.

3.3. Encapsulation Efficiency and Powders Characterization

The encapsulation efficiency refers to the potential of wall material to entrap and hold the core material inside the capsule [24]. In our study, different wall material concentrations caused significant differences in the anthocyanin encapsulation efficiencies, as presented in Table 2 ($p < 0.05$). Thus, the encapsulation efficiency significantly increased with the polysaccharides concentration, ranging from 64.67 ± 0.67% for V1 to 96.44 ± 3.43% for V2 ($p < 0.05$). Our results are in agreement with other studies. Condurache et al. [15] reported encapsulation efficiencies of the eggplant peel anthocyanins in CMC, P, and WPI ranging from 69% to 77%. In this study, an increase in the encapsulation efficiency was also observed with the increase of the CMC concentration.

Table 2. Characterization of the encapsulated eggplant peels extract.

Phytochemical Content	TAC µg D3G/g dw	TFC mg CE/g dw	TPC mg GAE/g dw	Antioxidant Activity mM TE/g dw	Encapsulation Efficiency %
V1	50.41 ± 2.13 ^{a}	1.53 ± 0.06 ^{a}	8.03 ± 0.18 ^{a}	41.96 ± 0.28 ^{a}	64.67 ± 0.67 ^{a}
V2	94.94 ± 7.94 ^{b}	1.64 ± 0.14 ^{a}	7.22 ± 0.18 ^{b}	36.60 ± 0.83 ^{b}	96.44 ± 3.43 ^{b}

For each tested phytochemical and powder variant, values from the same column that do not share a letter are statistically different at $p < 0.05$ based on the Tukey method and 95% confidence. TAC—Total Anthocyanin Content; TFC—Total Flavonoid Content; TPC—Total Polyphenol Content.

Higher CMC and P concentrations also led to higher anthocyanin content entrapped in the powders. Thus, V2 presented a significantly higher TAC than V1 ($p < 0.05$). On the contrary, significantly higher TPC and antioxidant activity values were obtained for the variant with higher WPI concentration ($p < 0.05$). The TFC was not significantly different, regardless of the powder variant ($p > 0.05$). An opposite behavior was reported by Stanciuc et al. [25], who obtained higher polyphenol concentrations and antioxidant activities for higher pectin concentrations in grape skin bioactive-containing powders.

However, from Table 2, it can be observed that both powder variants showed high encapsulation efficiencies with high antioxidant activities. This leads us to conclude that the wall material combinations successfully encapsulated the phytochemicals from the eggplant peel extract.

3.4. Storage Stability of the Powders

The powders were stored at room temperature for 28 days and were characterized in terms of phytochemical content and antioxidant activity. The changes in TAC, TFC, TPC, and antioxidant activity of both powder variants during storage are shown in Figure 3.

The TAC and TFC of both powder variants did not significantly change during the 28 days of storage ($p > 0.05$). Instead, the TPC of V2 presented a significant increase ($p < 0.05$), while the TPC of V1 remained constant. Regarding the antioxidant activity, V1 showed significantly lower values after 28 days of storage, while V2 presented significantly higher values than the initial values ($p > 0.05$). Similar behavior was also reported by Moser et al. [26]. They suggested that the microencapsulation of grape juice combined with storage at low temperatures offered storage stability to the anthocyanins. On the contrary, Azarpazhooh et al. [27] reported a decrease in the TAC of the pomegranate peels bioactives microencapsulated powder as the period of storage increased.

However, from Figure 3, it can be observed that both variants showed low antioxidant activities and phytochemical content variation in time. This leads us to conclude that the used combination between CMC, P, and WPI successfully encapsulated the anthocyanins from eggplants peel extract, providing them good stability.

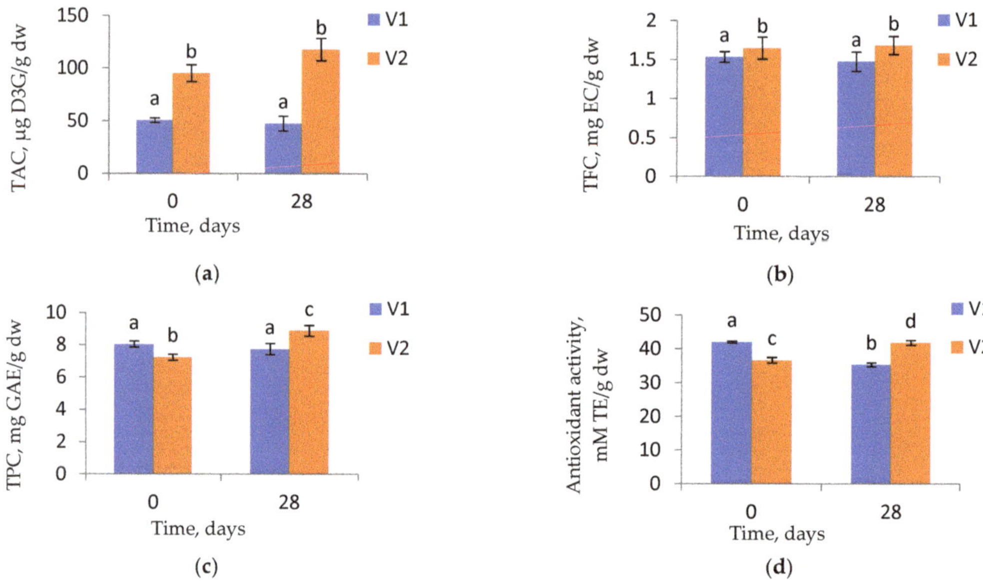

Figure 3. The TAC (**a**), TFC (**b**), TPC (**c**), and antioxidant activity (**d**) stability of the encapsulated powders after 28 days of storage at room temperature. For each tested phytochemical and powder variant, columns that do not share a letter are statistically different $p < 0.05$. TAC—Total Anthocyanin Content; TFC—Total Flavonoid Content; TPC—Total Polyphenol Content.

3.5. Morphological Structure of the Powders

Confocal microscopy is an excellent tool to investigate the morphostructural features of microencapsulated powders. Using the LSM 710 point-by-point scanning equipment, with the DPSS (561 nm) and HeNe (633 nm) lasers and the corresponding filters (FS38 WF and FS15 WF, respectively), the samples were analyzed in both their native form (to observe the autofluorescence emission of the phytopigments from the eggplant extracts) as well as after staining of the samples with Congo Red, which has an affinity for the protein component of the microencapsulating matrix. Thus, the bioactives from the eggplant exocarp, by encapsulation, generated a digitiform, laced appearance in the V1 variant (Figure 4a) or compact irregular scales in the V2 variant (Figure 4b). Interestingly, the same plant extract displayed different autofluorescent properties depending on the proportion of biopolymers used for the microencapsulating matrix, probably due to the transient bonds created. The higher percentage of WPI in the V1 variant determined a frameshift of the emission to the range of 640–680 nm, while in the presence of a matrix richer in polysaccharides (V2), the emission spectrum of the phytopigments in the extract appeared in the green-yellow domain (520–540 nm). Similar results were obtained by Chanoca et al. [28] in 2016 who used fluorescence lifetime microscopy to study the subcellular localization of anthocyanins in plant cells [28]. By labeling with Congo red (Figure 4c,d), the powders formed more or less homogeneous biofilms. In the WPI network (in red) predominant in the V1 sample matrix, a large diversity of bioactives (in green) was observed (Figure 4c). As the WPI content decreased and the content of carbohydrate polymers in the matrix increased, the powder became more hydrophilic, finer, and more homogeneous, and the fluorescent labeling with Congo red was weaker.

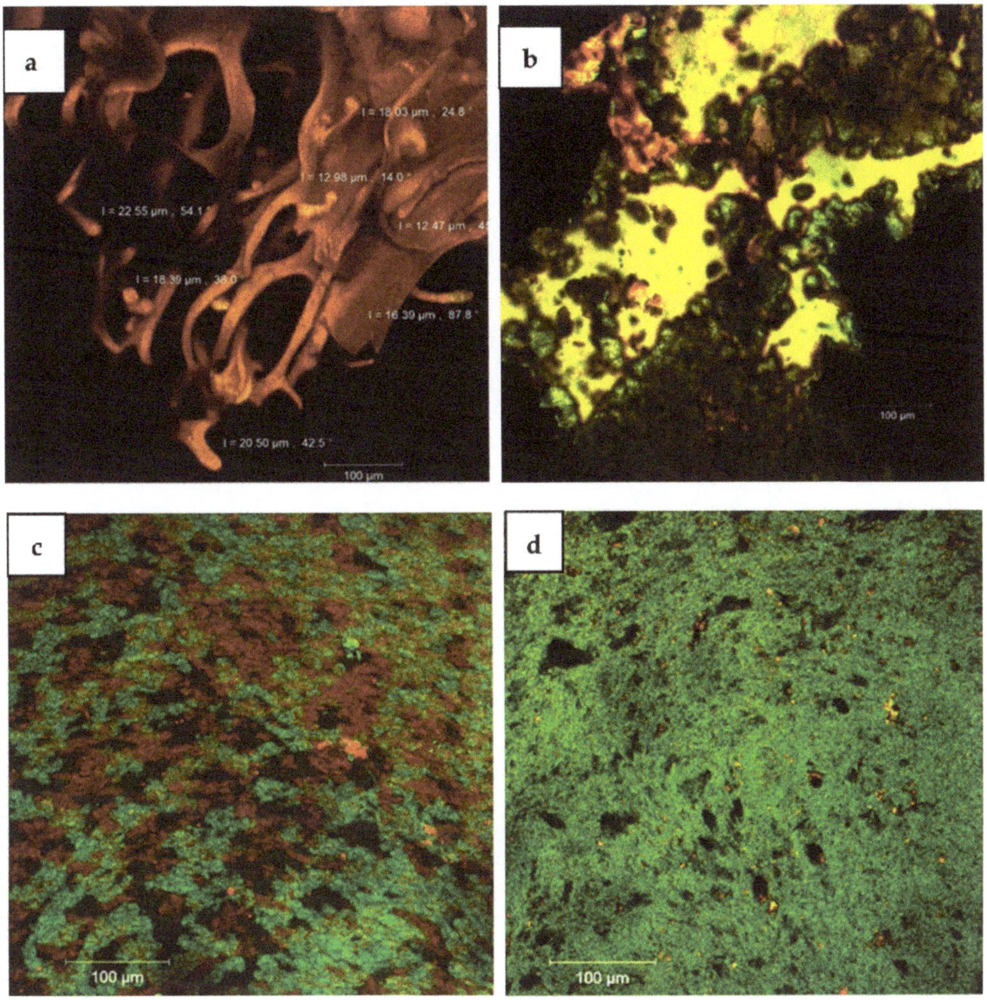

Figure 4. Confocal laser scanning images of the microencapsulated powders: V1 ((**a**)—native, (**c**)—stained) and V2 ((**b**)—native, (**d**)—stained).

3.6. Powders Behavior in Simulated Digestion

The anthocyanins have the lowest bioavailability among polyphenols, with only 1–2% of them being absorbed by the human body. Their absorption is affected by the physicochemical conditions from the gastrointestinal tract, such as pH, enzymes, and microbiota [29]. The combination between a top-down method such as gelation and a bottom-up method such as freeze-drying can represent a solution to assure an intelligent delivery system for the eggplant peel anthocyanins.

In our study, the encapsulated powders' behavior was studied under simulated gastric and intestinal conditions. The results obtained for the in vitro digestibility in simulated gastric fluid (SGF) showed that the encapsulation methods and the selected coating materials presented a protective effect on the anthocyanin release. In Figure 5a, we observed a remarkable anthocyanin content stability, with only 3% and 7% release for V1 and V2, respectively, after 120 min of gastric digestion. After the intestinal fluid was added, an accelerated anthocyanin release from the matrices was observed (Figure 5b). Thus, the anthocyanins from V1 completely released after only 60 min. For V2 instead, the anthocyanins started to gradually release in time. Thus, after 30 min of digestion, half of the anthocyanins were released from the matrix, reaching complete release after 120 min of digestion. It can be noticed a higher anthocyanin protective effect by the matrix containing higher CMCNa and P concentration.

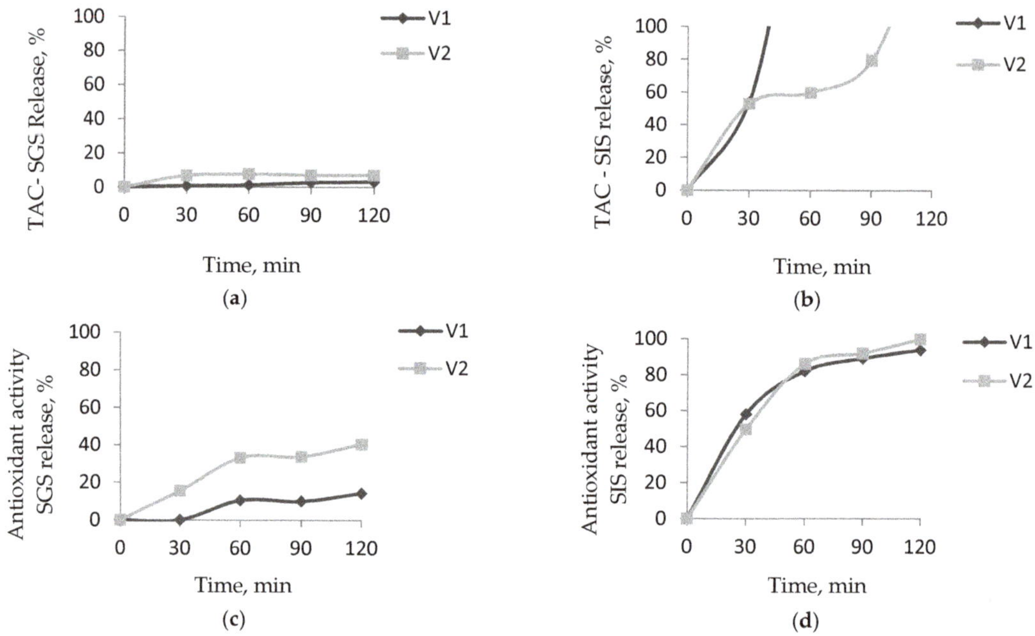

Figure 5. In vitro digestibility of encapsulated anthocyanin and antioxidant activity in simulated gastric fluid (**a**,**c**) and simulated intestinal fluid (**b**,**d**).

Condurache et al. [15] reported a high protective effect of the anthocyanins from eggplant peels encapsulated with proteins and polymers, with a maximum of 42% intestinal release. Instead, Huang and Zhou [30] reported that a complete release of the encapsulated anthocyanins from black rice occurred after 20 min of intestinal digestion.

In terms of antioxidant activity, there is a release of up to 14% for V1 and 40% for V2 after 120 min of simulated gastric digestion (Figure 5c). If in the anthocyanin's case, their complete release in simulated intestinal digestion took place after only 60 min for V1, in the case of antioxidant activity, a gradual release is observed for both variants (Figure 5d). Thereby, a 50% release is observed for both variants after the first 30 min of simulated intestinal digestion. Afterward, the release took place more slowly, reaching full release after 120 min of digestion. These results may be explained by the phenolic acids formed during the degradation of the anthocyanins in intestinal digestion [31].

4. Conclusions

This study focused on the obtaining of functional ingredients for food or nutraceuticals applications by the extraction and encapsulation of phenolic compounds from eggplant peels, mainly on the anthocyanins. Thereby, the anthocyanins were extracted using the ultrasound-assisted extraction method and phytochemically characterized. The major anthocyanin from the extract is delphinidin 3-*O*-glucoside. Further, the present study aimed to compare the impact of the two encapsulation matrices on the stability and controlled release of the anthocyanins. Thus, the gelation technique combined with the freeze-drying technique was used, and the wall materials tested in this study were the CMCNa, P, and WPI in different concentrations. It has been demonstrated that higher concentrations of CMCNa and P in the encapsulation matrix resulted in higher anthocyanin retention, encapsulation efficiency, and storage stability. The confocal microscopy revealed a digitiform, laced appearance in the V1 and compact irregular scales in the V2 generated by the eggplant peels' bioactives. As the content of polysaccharides increased in the matrix, the powder became more hydrophilic and homogeneous. The in vitro digestion study indicated that each type of matrix exhibited a different protection mechanism for the anthocyanin and the antioxidant activity of the encapsulated powders. A higher polysaccharide concentration in the matrix provides high anthocyanin stability in the gastric system and a gradual release in the intestinal system, which can lead to their better absorption.

Our results can be used in formulating new functional foods with high value added. In this regard, more research will be further addressed.

Author Contributions: Conceptualization, N.-N.C.; methodology, N.-N.C.; software, N.-N.C., E.E., V.B., and M.T.; validation, N.S. and G.R.; formal analysis, N.-N.C., E.E., V.B., and M.T.; investigation, N.-N.C.; resources, G.-E.B., G.R., and C.C.; data curation, N.-N.C.; writing—original draft preparation, N.-N.C.; writing—review and editing, G.R., C.C., and N.S.; visualization, G.-E.B.; supervision, N.S.; project administration, G.-E.B.; funding acquisition, G.-E.B. and G.R. All authors have read and agreed to the published version of the manuscript.

Funding: This work was supported by project number PN-III-P1-1.2-PCCDI-2017-0569-PRO-SPER (10PCCI) within the PNCDI program.

Institutional Review Board Statement: Not applicable.

Informed Consent Statement: Not applicable.

Data Availability Statement: The data that support the findings of this study are available from the corresponding author (G.R.) upon reasonable request.

Acknowledgments: This work was supported by the Integrated Center for Research, Expertise and Technological Transfer in Food Industry (Bioaliment TehnIA), which provided technical support.

Conflicts of Interest: The authors declare no conflict of interest.

References

1. Nayak, A.; Bhushan, B. An overview of the recent trends on the waste valorization techniques for food wastes. *J. Environ. Manag.* **2019**, *233*, 352–370. [CrossRef]
2. Galanakis, C.M. Food waste recovery: Prospects and opportunities. In *Sustainable Food Systems from Agriculture to Industry. Improving Production and Processing*; Galanakis, C.M., Ed.; Academic Press: Cambridge, MA, USA, 2018; pp. 401–419.
3. Shahidi, F.; Ambigaipalan, P. Phenolics and polyphenolics in foods, beverages and spices: Antioxidant activity and health effects—A review. *J. Funct. Foods* **2015**, *18*, 820–897. [CrossRef]
4. Baenas, N.; Abellán, A.; Rivera, S.; Moreno, D.A.; García-Viguera, C.; Domínguez-Perles, R. Foods and supplements. In *Polyphenols: Properties, Recovery, and Applications*; Galanakis, C.M., Ed.; Woodhead Publishing: Cambridge, MA, USA, 2018; pp. 327–362.
5. Concellón, A.; Añon, M.C.; Chaves, A.R. Effect of low temperature storage on physical and physiological characteristics of eggplant fruit (*Solanum melongena* L.). *LWT* **2007**, *40*, 389–396. [CrossRef]
6. Burton-Freeman, B.; Sandhu, A.; Edirisinghe, I. Anthocyanins. In *Nutraceuticals. Efficacy, Safety and Toxicity*; Gupta, R.C., Ed.; Academic Press: Cambridge, MA, USA, 2016; pp. 489–500.
7. Rodriguez-Amaya, D.B. Update on natural food pigments—A mini-review on carotenoids, anthocyanins, and betalains. *Food Res. Int.* **2019**, *124*, 200–205. [CrossRef]
8. Xiong, J.; Chan, Y.H.; Rathinasabapathy, T.; Grace, M.H.; Komarnytsky, S.; Lila, M.A. Enhanced stability of berry pomace polyphenols delivered in protein-polyphenol aggregate particles to an in vitro gastrointestinal digestion model. *Food Chem.* **2020**, *331*, 127279. [CrossRef]
9. Celli, G.B.; Ghanem, A.; Brooks, M.S.-L. Bioactive Encapsulated Powders for Functional Foods—A Review of Methods and Current Limitations. *Food Bioprocess Technol.* **2015**, *8*, 1825–1837. [CrossRef]
10. Condurache, N.-N.; Croitoru, C.; Enachi, E.; Bahrim, G.-E.; Stănciuc, N.; Râpeanu, G. Eggplant Peels as a Valuable Source of Anthocyanins: Extraction, Thermal Stability and Biological Activities. *Plants* **2021**, *10*, 577. [CrossRef] [PubMed]
11. Turturică, M.; Stănciuc, N.; Bahrim, G.; Râpeanu, G. Effect of thermal treatment on phenolic compounds from plum (prunus domestica) extracts—A kinetic study. *J. Food Eng.* **2016**, *171*, 200–207. [CrossRef]

12. Swer, T.L.; Mukhim, C.; Bashir, K.; Chauhan, K. Optimization of enzyme aided extraction of anthocyanins from *Prunus nepalensis* L. *LWT* **2018**, *91*, 382–390. [CrossRef]
13. Serrano-Cruz, M.R.; Villanueva-Carvajal, A.; Morales-Rosales, E.J.; Dávila, J.F.R.; Dominguez-Lopez, A. Controlled release and antioxidant activity of Roselle (*Hibiscus sabdariffa* L.) extract encapsulated in mixtures of carboxymethyl cellulose, whey protein, and pectin. *LWT* **2013**, *50*, 554–561. [CrossRef]
14. Batista, R.A.; Espitia, P.J.P.; Quintans, J.D.S.S.; Freitas, M.M.; Cerqueira, M.Â.; Teixeira, J.A.; Cardoso, J.C. Hydrogel as an alternative structure for food packaging systems. *Carbohydr. Polym.* **2019**, *205*, 106–116. [CrossRef]
15. Condurache, N.N.; Aprodu, I.; Crăciunescu, O.; Tatia, R.; Horincar, G.; Barbu, V.; Enachi, E.; Râpeanu, G.; Bahrim, G.E.; Oancea, A.; et al. Probing the Functionality of Bioactives from Eggplant Peel Extracts Through Extraction and Microencapsulation in Different Polymers and Whey Protein Hydrolysates. *Food Bioprocess Technol.* **2019**, *12*, 1316–1329. [CrossRef]
16. Oancea, A.-M.; Hasan, M.; Vasile, A.M.; Barbu, V.; Enachi, E.; Bahrim, G.; Râpeanu, G.; Silvi, S.; Stănciuc, N. Functional evaluation of microencapsulated anthocyanins from sour cherries skins extract in whey proteins isolate. *LWT* **2018**, *95*, 129–134. [CrossRef]
17. Hosseini, S.; Gharachorloo, M.; Ghiassi-Tarzi, B.; Ghavami, M. Evaluation the Organic Acids Ability for Extraction of Anthocyanins and Phenolic Compounds from different sources and Their Degradation Kinetics during Cold Storage. *Pol. J. Food Nutr. Sci.* **2016**, *66*, 261–269. [CrossRef]
18. Horincar, G.; Enachi, E.; Stănciuc, N.; Râpeanu, G.; de Jos, D.; University of Galati. Extraction and characterization of bioactive compounds from eggplant peel using ultrasound-assisted extraction. *Ann. Univ. Dunarea Jos Galati Fascicle VI Food Technol.* **2019**, *43*, 40–53. [CrossRef]
19. Jung, E.J.; Bae, M.S.; Eun-Kyung, J.; Jo, Y.J.; Lee, S.C. Antioxidant activity of different parts of eggplant. *Res. J. Med. Plant* **2011**, *5*, 4610–4615. [CrossRef]
20. Azuma, K.; Ohyama, A.; Ippoushi, K.; Ichiyanagi, T.; Takeuchi, A.; Saito, T.; Fukuoka, H. Structures and Antioxidant Activity of Anthocyanins in Many Accessions of Eggplant and Its Related Species. *J. Agric. Food Chem.* **2008**, *56*, 10154–10159. [CrossRef] [PubMed]
21. Ferarsa, S.; Zhang, W.; Moulai-Mostefa, N.; Ding, L.; Jaffrin, M.Y.; Grimi, N. Recovery of anthocyanins and other phenolic compounds from purple eggplant peels and pulps using ultrasonic-assisted extraction. *Food Bioprod. Process.* **2018**, *109*, 19–28. [CrossRef]
22. Dranca, F.; Oroian, M. Optimization of ultrasound-assisted extraction of total monomeric anthocyanin (TMA) and total phenolic content (TPC) from eggplant (*Solanum melongena* L.) peel. *Ultrason. Sonochem.* **2016**, *31*, 637–646. [CrossRef]
23. Mauro, R.P.; Agnello, M.; Rizzo, V.; Graziani, G.; Fogliano, V.; Leonardi, C.; Giuffrida, F. Recovery of eggplant field waste as a source of phytochemicals. *Sci. Hortic.* **2020**, *261*, 109023. [CrossRef]
24. Stoll, L.; Costa, T.M.H.; Jablonski, A.; Flores, S.; Rios, A.D.O. Microencapsulation of Anthocyanins with Different Wall Materials and Its Application in Active Biodegradable Films. *Food Bioprocess Technol.* **2016**, *9*, 172–181. [CrossRef]
25. Stănciuc, N.; Turturică, M.; Oancea, A.M.; Barbu, V.; Ioniţă, E.; Aprodu, I.; Râpeanu, G. Microencapsulation of Anthocyanins from Grape Skins by Whey Protein Isolates and Different Polymers. *Food Bioprocess Technol.* **2017**, *10*, 1715–1726. [CrossRef]
26. Moser, P.; Telis, V.R.N.; Neves, N.D.A.; García-Romero, E.; Gómez-Alonso, S.; Hermosín-Gutiérrez, I. Storage stability of phenolic compounds in powdered BRS Violeta grape juice microencapsulated with protein and maltodextrin blends. *Food Chem.* **2017**, *214*, 308–318. [CrossRef] [PubMed]
27. Azarpazhooh, E.; Sharayei, P.; Zomorodi, S.; Ramaswamy, H.S. Physicochemical and Phytochemical Characterization and Storage Stability of Freeze-dried Encapsulated Pomegranate Peel Anthocyanin and In Vitro Evaluation of Its Antioxidant Activity. *Food Bioprocess Technol.* **2018**, *12*, 199–210. [CrossRef]
28. Chanoca, A.; Burkel, B.; Kovinich, N.; Grotewold, E.; Eliceiri, K.W.; Otegui, M.S. Using fluorescence lifetime microscopy to study the subcellular localization of anthocyanins. *Plant J.* **2016**, *88*, 895–903. [CrossRef] [PubMed]
29. Tarone, A.G.; Cazarin, C.B.B.; Junior, M.R.M. Anthocyanins: New techniques and challenges in microencapsulation. *Food Res. Int.* **2020**, *133*, 109092. [CrossRef]
30. Huang, Y.; Zhou, W. Microencapsulation of anthocyanins through two-step emulsification and release characteristics during in vitro digestion. *Food Chem.* **2019**, *278*, 357–363. [CrossRef]
31. Kim, I.; Moon, J.K.; Hur, S.J.; Lee, J. Structural changes in mulberry (*Morus Microphylla.* Buckl) and chokeberry (*Aronia melanocarpa*) anthocyanins during simulated in vitro human digestion. *Food Chem.* **2020**, *318*, 126449. [CrossRef]

inventions

MDPI

Article

Multi-Criteria Analysis of the Mass Tourism Management Model Related to the Impact on the Local Community in Constanta City (Romania)

Catalin Anton *, Angela-Eliza Micu and Eugen Rusu

Doctoral School of Social Human Sciences, Dunarea de Jos University of Galati, 47 Domneasca Street, 800008 Galati, Romania; angelaelizamicu@yahoo.com (A.-E.M.); eugen.rusu@ugal.ro (E.R.)
* Correspondence: catanton@gmail.com

Abstract: Traditionally and socially, the tourism in Constanta is considered to be important to the local economy. Sun and beach locations are both a draw for locals and tourists to the city, on the Black Sea. However, vacation-oriented activities in the city only have a seasonal cycle. In this paper, we proposed to analyze the mass tourist activity in Constanta, taking into account economic, social, and environmental conditions. Additionally, we attempted to build a model based on the data available. The model was developed using a PESTEL analysis to determine the supportability factor of the indicators identified. We also set out to create a projection of the activities proposed for analysis by 2050. To create a model for coastal areas, the data used in this research must be accurate and consistent. Furthermore, correctly identifying indicators and their relationships is a critical step in conducting a thorough study. Last but not least, finding the calculation coefficient for the activity in question is critical, as collecting data from various activities might be challenging when trying to find a feasible model.

Keywords: mass tourism; management model; Constanta; PESTEL; Black Sea; sustainability

Citation: Anton, C.; Micu, A.-E.; Rusu, E. Multi-Criteria Analysis of the Mass Tourism Management Model Related to the Impact on the Local Community in Constanta City (Romania). *Inventions* **2021**, *6*, 46. https://doi.org/10.3390/inventions6030046

Academic Editor: Biswajit Sarkar

Received: 18 May 2021
Accepted: 22 June 2021
Published: 28 June 2021

1. Introduction

There is interdisciplinarity in regards to the science that has been carried out in coastal areas. Due to the interdependence of land and sea, it brings several threats to biodiversity and quality of life in this coastal area [1]. Understanding both the natural and the systems processes that keep them in harmony, is a means to understand the complementary nature of their relationship [2].

Integrative application of pressure appears to be the optimal solution for the coastal zone's longer-term growth.

It is particularly significant for the European Union, given that it includes areas of over 68,000 km of coastline with an integrated approach to management. Even with these significant disadvantages and weaknesses, the administration looks to be simpler with respect to land–sea governance because of a lack of medium-term approaches, and lacks medium data collected and analysis systems. While integrated coastal management (IZCM), climate planning (CPP), and urban/regional marine special planning (MSP) are tools that can be useful, they are discussed only at the strategic level and are not sufficiently implemented at the grassroots level. It is an indisputable fact that the increasing human impact has left a tangible and visible residual imprint on the Earth's climate. For scientific researchers, 'sustainable development', 'climate change', and 'environmental protection' have gained increasing acceptance in recent years, while 'environmental management' is more frequently seen today [3]

The city of Constanta, located in the southeastern part of Romania, on the Black Sea's coasts, has as its main goal the growth of tourism, with mass tourism as an option. This is evident in both the paper Integrated Strategy for Sustainable Tourism Development in

Constanta County, 2019–2028 [4], and the Master Plan for the Protection and Rehabilitation of the Coastal Area [5]. The basis for medium- and long-term city development planning are laid out in these strategic development publications. There are, however, studies that examine the impacts of mass tourism on a town which draw attention to the traffic and pollution issues that this activity causes [6]. This paper is structured as follows: Section 2 describes the Constanta area regarding the tourism activity, Section 3 presents the methods used in this research, followed by results, and the last section presents the conclusion.

2. Case Study

For this study, we focus on the Constanta area of Romania. Constanta is an open sea city, located in the east of Romania, as shown in Figure 1. Constanta is one of the most important cities due to national/international tourism to the Black Sea. Due to its physico-geographical characteristics, the Romanian coastline has benefited greatly from the largest sums of tourist attention in the first half of the 1960s and in terms of attendance amenities in this time period. Due to this arrangement, Constanta was found to be the county with the most structures and the most vacation options. Mamaia is one of the administrative districts of Constanta and one of the best beach resorts, with a 7 km-long beach which faces to the east [7].

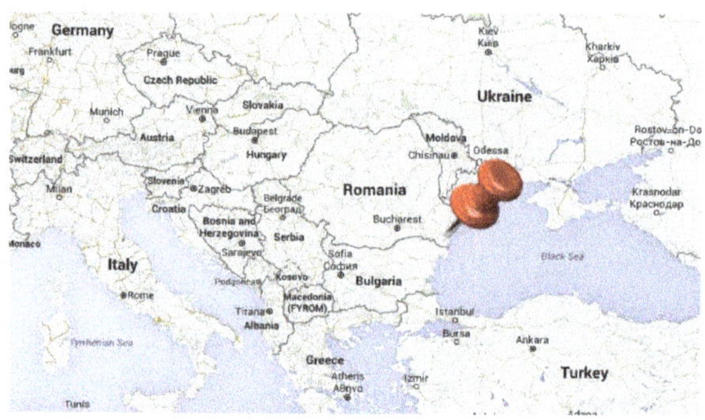

Figure 1. Location of Constanta City in Europe. Source: Google map processed image.

Even if tourism is a significant county-level activity, according to the town halls (Figure 2), the data indicate that the bulk of the operations are conducted on the shoreline of the Black Sea in Constanta County. Both the national tourism strategy [8] and the local strategic documents discussed in the preceding chapter specify a growth in tourism and accommodation capacity, as well as an improvement in the percentage of nights spent in accommodation units. This validates the concept of mass tourism as the only viable alternative for decision-makers. Simultaneously, one of the steps used by the authorities to enhance the number of tourists is to expand the surface of the beaches by constructing infrastructure. The formulation of infrastructure works in three phases is anticipated in the document Masterplan Protection and Rehabilitation of the Coastal Area, with the first phase already finished (in the period 2013–2015, with the surface of the beaches rising by over 120,000 square meters). Following that, we will look at the capacity of existing lodging in the area.

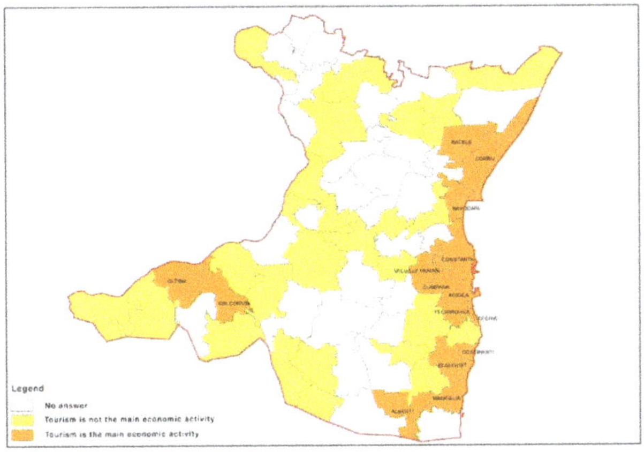

Figure 2. Localities in Constanta County where tourism is considered a main economic activity. Source: http://www.cjc.ro/dyn_doc/turism/Strategia_jud.Constanta_Faza2.pdf, accesed on 20 April 2021.

According to data from the National Statistical Institute, the total revenue generated by hotels, other lodging facilities, restaurants, and tourist agencies totaled RON 1,541,559,560 in 2017, accounting for 3.4 percent of the total revenue generated by all fields of activity in Constanta County. County-level tourism turnover increased by 56.7% between 2013 and 2017, but total business activity only slightly. If this type of business in 2013 held 2.2% of the market, it now has 3.4% of the market. (Table 1) [8–11].

Table 1. Turnover of active enterprises in Constanta County, from 2013 to 2017 (RON, 1 USD = 4.04 RON). Source: TEMPO-Online database (http://statistici.insse.ro:8077/tempo-online/#/pages/tables/insse-table) accessed on 10 June 2021.

	2013	2014	2015	2016	2017
Total domains of activity, of which:	44,640,521,503	45,607,349,710	44,166,254,796	42,177,088,057	44,672,535,846
HR domain	983,700,816	1,086,868,232	1,399,752,043	1,460,529,931	1,541,559,560

Tourist structure.

- Tourist reception structures with accommodation functions.
- Tourism benefits from tourist accommodation. This is why the resort's location mostly controls the capacity of tourist flows.

Due to its physico-geographical characteristics, the Romanian coastline has benefited greatly from the largest sums of tourist attention in the first half of the 1960s and in terms of attendance amenities in this time period. Due to this arrangement, Constanta was found to be the county with the most structures and the most vacation options [12].

According to the latest official data for 2018, 834 units (9.9% of the national total) were open in Constanta County with a total of 84,891 places (24.0% of the national total) (Table 2).

Table 2. Evolution of accommodation capacity at the level of Constanta County and for the total country. Source: TEMPO-Online database (http://statistici.insse.ro:8077/tempo-online/#/pages/tables/insse-table) accessed on 10 June 2021.

Years	Constanta County		Romania		% of Total Romania	
	Units	Places	Units	Places	Units	Places
2012	738	84,690	5821	301,109	12.7%	28.1%
2013	745	85,756	6009	305,707	12.4%	28.1%
2014	746	87,496	6130	311,288	12.2%	28.1%
2015	755	87,848	6821	328,313	11.1%	26.8%
2016	761	85,285	6946	328,888	11.0%	25.9%
2017	838	84,157	7905	343,720	10.6%	24.5%
2018	834	84,891	8449	353,308	9.9%	24.0%

A report by the National Statistical Institute shows the following reality in Constanta County for the past seven years:

- The number of lodging structures increased significantly in 2018, with 13.0 percent more units than in 2012.
- The number of available accommodations has remained relatively constant.
- When compared to 2012, operational accommodation capacity (number of places–days) decreased by 2.0 percent in 2018.

The situation is explained by the fact that the new structures, while more numerous, are still much smaller in size (pensions, villas) in comparison to the large units that have emerged from the tourist circuit for various reasons (camping, camps, hotels) [13–15]. Additionally, the reduced capacity in all vacation properties also means reduced availability of accommodation, and in particular, the large hotels in tourist destinations are made available for less time in operation year to year. As was researched in Constanta County, the average duration of accommodation structures was approximately 125.8 days in 2012, and decreased to 124 days in 2017. By comparison, the national average is about 255 days [16].

Data retrieved from the government tourism agencies are considerably different than that found in the National Statistical Institute itself. According to the December 2018 data, there are 1807 tourist reception structures with accommodation functions in Constanta County, totaling 122,262 places. In comparison to the rest of the country, Constanta County has 13.4 percent of the total number of accommodation structures and 27.2 percent of the total number of places to stay, making the county of Constanta, and particularly the seaside area of the county, the most important tourist destination in the country (Figures 3 and 4).

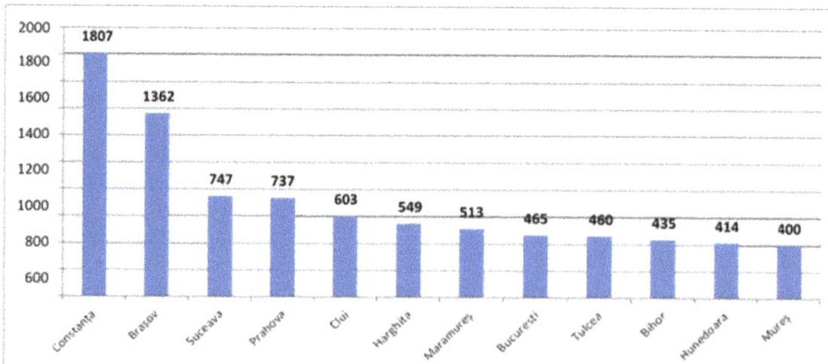

Figure 3. Number of tourist reception structures with accommodation functions, by county (December 2018). Source: http://turism.gov.ro/web/autorizare-turism/, accessed on 20 April 2021.

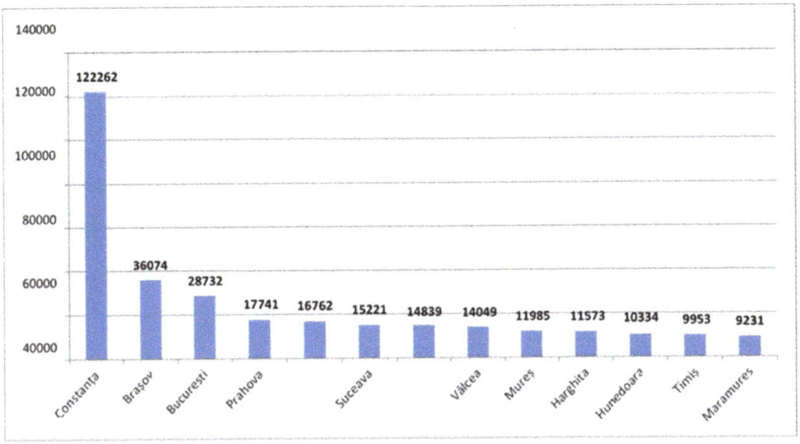

Figure 4. Number of accommodations, by county (December 2018). Source http://turism.gov.ro/web/autorizare-turism/, accessed on 20 April 2021.

In Table 3 is presented the accommodation capacity of Constanta County. The first main resorts on the Romanian Black Sea coast by number of places are: Mamaia (20.3%), Eforie Nord (14.6%), and Costineşti (13.0%).

Table 3. Accommodation capacity in Constanta County, by locality (December 2018). Source: http://turism.gov.ro/web/autorizare-turism/, accessed on 20 April 2021—processed data.

Localities	Accommodation Structures		Accommodation Places	
	Number	Percent	Number	Percent
Constanța	**309**	**17.1%**	**28.396**	**23.2%**
Constanța	87	4.8%	3.519	2.9%
Mamaia	222	12.3%	24.877	20.3%
Mangalia	**258**	**14.3%**	**32.904**	**26.9%**
Mangalia	29	1.6%	1.683	1.4%
Cap Aurora	7	0.4%	2.184	1.8%
Jupiter	36	2.0%	5.848	4.8%
Neptun-Olimp	88	4.9%	9441	7.7%
Saturn	40	2.2%	6.438	5.3%
Venus	58	3.2%	7.310	6.0%
Eforie	**431**	**23.9%**	**23.857**	**19.5%**
Eforie Nord	318	17.6%	17.795	14.6%
Eforie Sud	113	6.3%	6.062	5.0%
Costinești	**355**	**19.6%**	**15.850**	**13.0%**
Limanu	**127**	**7.0%**	**4.177**	**3.4%**
2 Mai	38	2.1%	1.047	0.9%
Vama Veche	88	4.9%	3.084	2.5%
Limanu	1	0.1%	46	0.0%
Năvodari	**226**	**12.5%**	**13.114**	**10.7%**
Techirghiol	**28**	**1.5%**	**1.390**	**1.1%**

Table 3. *Cont.*

Localities	Accommodation Structures		Accommodation Places	
	Number	Percent	Number	Percent
Other localities	**73**	**4.0%**	**2.574**	**2.1%**
Adamclisi	1	0.1%	30	0.0%
Agigea	10	0.6%	204	0.2%
Aliman	2	0.1%	28	0.0%
Băneasa	1	0.1%	20	0.0%
Cernavodă	4	0.2%	142	0.1%
Cobadin	1	0.1%	10	0.0%
Crucea	1	0.1%	54	0.0%
Cumpăna	1	0.1%	18	0.0%
Lipniţa	2	0.1%	42	0.0%
Lumina	1	0.1%	59	0.0%
Medgidia	6	0.3%	188	0.2%
Mihail Kogălniceanu	1	0.1%	36	0.0%
Oltina	1	0.1%	16	0.0%
Ostrov	3	0.2%	74	0.1%
Ovidiu	3	0.2%	63	0.1%
Saligny	1	0.1%	12	0.0%
Seimeni	1	0.1%	32	0.0%
23 August	20	1.1%	1.314	1.1%
Corbu	13	0.7%	232	0.2%
TOTAL	**1.807**	**100%**	**122.262**	**100%**

3. Materials and Methods

The multi-criteria model was applied to mass tourism in Constanta using the PESTEL technique. This model accounted for the impact of relevant indices on all three major factors: economic, social, and environmental. These processes can be applied to any residential or business destination, especially in a touristic or coastal area, identifying the activities and considering each location's unique characteristics [17–20].

It is important to run multicriteria analyses over time to identify possible changes in the factors. The more in-detailed the analysis you give, the more time you have to identify and reduce security risks that overlap with the various variables.

Indicators of activity in the area, as well as their impact on other economic, social, and environmental activities, were identified. The different perspectives were evaluated on the basis of the PESTEL model [21–25]; for Constanta, used three parameters of analysis:

Economic factor: In terms of agriculture and aquaculture, the analysis was carried out while taking into account local products and their market access. Despite being a predominantly agrarian country, Romania's trade balance still favors food imports, according to existing data at the Institute of Statistics. The occupancy of hotels and restaurants, as well as the economic use of the coastal beach, were the two indicators considered in this study. We also considered the activity of retailers in the studied area. In this chapter, we used data from the retail industry's turnover and profit from 2011 to 2019. Car traffic, use of local transportation, and electricity consumption are also important indicators of the stress that the community faces during the summer season. For real-time traffic monitoring, we used Google Maps' traffic function, and for relevant traffic data on the number of vehicles in a given area, we used TOMTOM's free data.

Social factor: This took into account the social environment and identified the trends that existed in the analyzed area. To examine this factor, two indicators were developed: employment and crime.

Environmental factor: This factor referred to the impact of other factors on the environment as well as the impact of ecological aspects. The environmental factor includes indicators for air, water (sea water and fresh water), waste (food waste and general waste), and protected areas.

Consideration of only the proposed indicators cannot lead to sustainable tourism activity. It is important to identify which of the variables add more or less pressure on some of the other variables. In fact, the supportability factor was applied, taking into account how the supportability of these activities would compare to each other. The tipping point occurs where one activity must stop before the other disappears. For activities that harmed one another, their credibility rating was reduced to zero if the percentage of total activity was small. Similarly, supportability has an effect on every other activity, and is inversely proportional to it. A rough estimate of the extent of the damage was based on existing data from government sources [26].

In the multi-criteria analysis, we used two scenarios: scenario number one, which refers to a moderate number of tourists, and scenario number two, which depicts the actual number of tourists. The relationship between the indicators and the number of tourists was established by using coefficients for the three months of the season under consideration, namely June, July, and August.

The coefficients added for the summer months were between 1.5 and 2.5, based on the number of tourists present and the population of the constant city.

In multi-criteria analysis, some of the indicators remained constant. This was due to the fact that mass tourism had little impact on them. In these cases, the indicators were labeled with the number 1, indicating that they were at the positive end of the analysis. For example, if we consider the consumption of local food products, because Romania is already an importer of food products, it follows that local producers with a limited stock of products exhaust their stocks regardless of tourist traffic. In general, retailers rely on suppliers who bring in imported goods to meet their food needs during the summer season. Furthermore, during the summer season, the workforce is typically relocated from one unit located in the city to another unit located on the coast, both units belonging to the same economic agent. The Statistical Institute's employment figures clearly show that the unemployment rate does not differ significantly between the seasonal and off-season months [27,28]. This demonstrates, on the one hand, the relocation of the workforce between units, and, on the other, the hiring of personnel from outside Constanta, with the newcomers having a negligible economic impact on the local community.

Returning to the method of calculating the indicators, this took the form of:

$$i_{mt} = i_{wt} \times c_{nt}/100 \quad (1)$$

where:

i_{mt} = *indicator that supports mass tourism.*

i_{wt} = *indicator that does not support mass tourism.*

c_{nt} = *coefficient for the number of tourists.*

Another equation was used to calculate the indicator for the retailers' indicator. This was:

$$i_{mt} = (Pn/Rn) \times c_{nt} \quad (2)$$

where:

Rn = *net result.*

Pn = *net profit.*

c_{nt} = *coefficient for the number of tourists.*

A linear regression was used to predict the values until 2050 for different factors of the analyzed model. The values were predicted based on the second variable in the simple linear regression. The predicted values are presented as a function of the other variable in the form of a straight line in the simple linear regression of this section. Linear regression is used to find the right line through the points that is best suited to the data. The best line is also known as a regression line.

4. Results

Based on those presented in the previous section, we analyzed the evolution of all three factors—economic, social, and environment—based on both scenarios—with mass tourism and without mass tourism—for summer time—months June, July, August—in the period 2011–2015. To view the future effects on these factors, we extended the analysis until the year 2050.

Figure 5 depicts the evolution of passenger transportation in the city of Constanta under scenarios 1 (no mass tourism) and 2 (with mass tourism). There is a noticeable difference in the efficiency of public passenger transportation during the months when tourists visit the coast. At this point, we can say that local policies should focus on encouraging residents to use public transportation more frequently, so that public transportation activity is more efficient during the off season. This indicator represents a situation in which mass tourism is a positive factor in the local community's development.

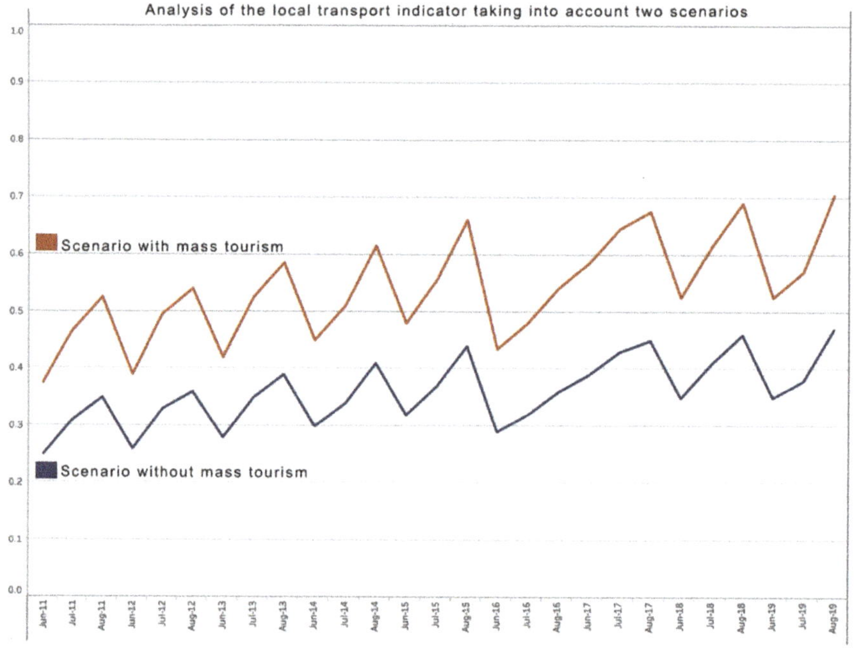

Figure 5. Local transport activity in the two scenarios.

By projecting local transportation to 2050 (Figure 6), it is possible to see that, based on the current data introduced into the system, the mass tourism hypothesis is more beneficial, reaching an optimal around 2039, whereas the assumption without mass tourism reaches an optimal of 0.84 out of 1 at the level of 2050. The trend is clear in both scenarios, and the optimization of this activity is at the disposal of the authorities, who can take measures to stimulate this indicator.

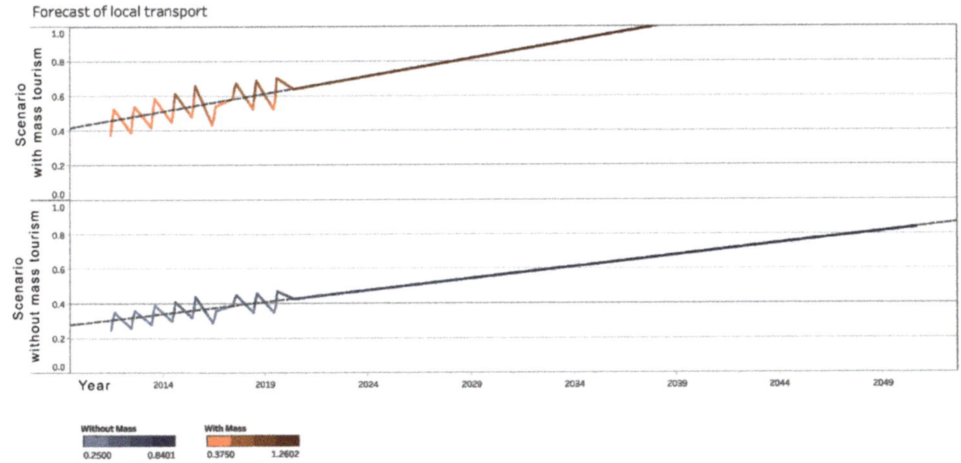

Figure 6. Forecast of local transport activity by 2050.

Similarly, we have expanded the analysis of economic factors (Figures 7 and 8), social factors (Figure 9), and environmental factors (Figure 10) to account for the two scenarios, one with and one without mass tourism.

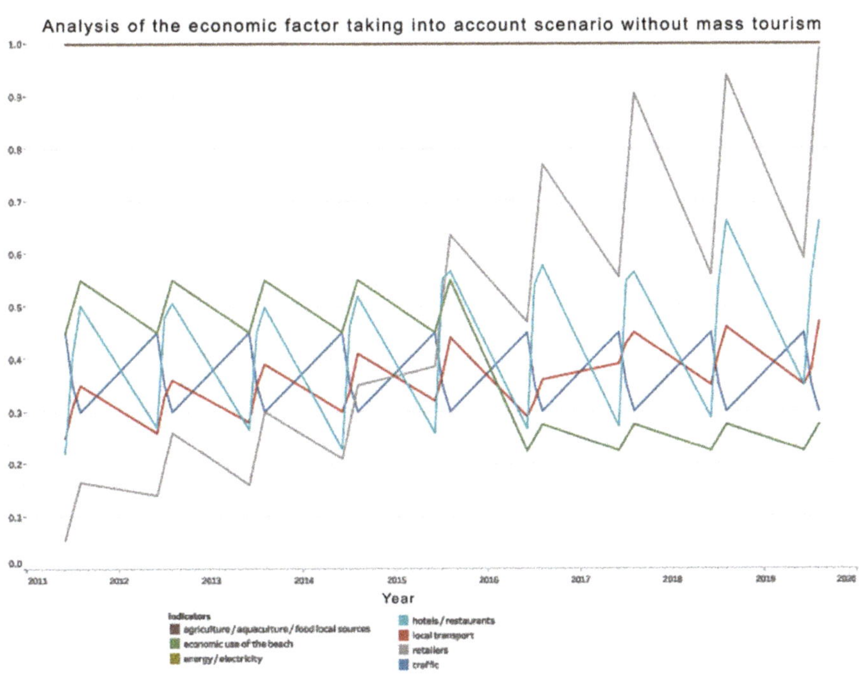

Figure 7. Analysis of economic factor indicators for the without mass tourism scenario.

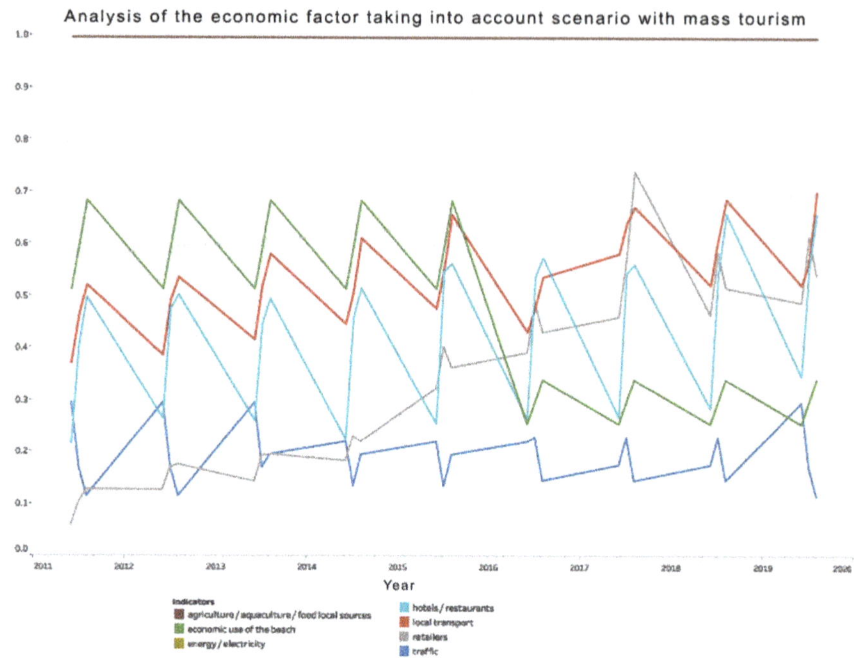

Figure 8. Analysis of economic factor indicators for the mass tourism scenario.

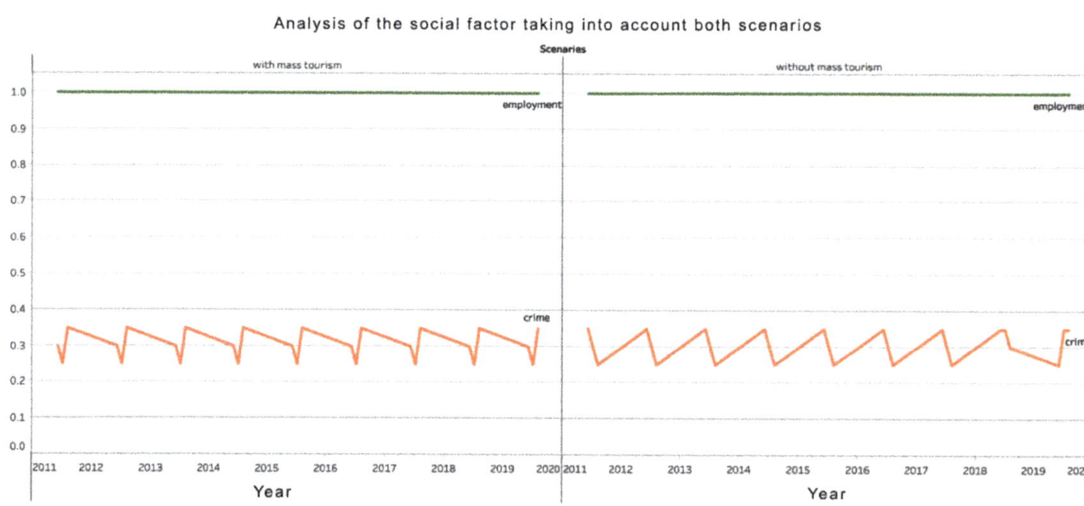

Figure 9. Analysis of social factor indicators for the two scenarios.

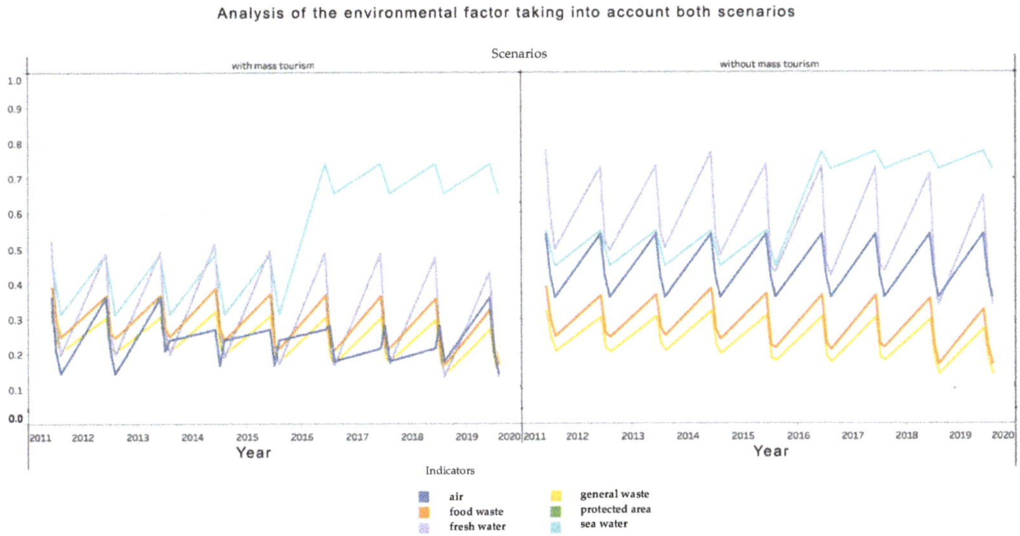

Figure 10. Analysis of environmental factor indicators for the two scenarios.

The evolution of indicators oscillates, with peaks occurring most frequently in August of each year, with this month being the most dynamic. In the case of the economic factor under consideration, several relevant indicators were considered, including the consumption of local products from agricultural or aquaculture sources, the activity of retailers in public food, the activity of the hotel and restaurant sector, local transportation and traffic, electricity consumption, and the economic use of the beach. This last indicator was added because, according to existing studies and strategies, the beach is the primary destination of seaside tourism, and the Romanian government has invested significantly in this goal. The best economic use of the beach was calculated based on the number of tourists and locals who use the beach's surface. The optimal area required for a beach tourist, which is 4 square meters, is established by both domestic legislation and international standards. In these circumstances, the optimal economic use of the beach for us was that the beach must be suitable in terms of area for the number of existing visitors, with a beach with an insufficient or excessively large area having a negative impact on tourist activity.

In the case of the social factor, we only looked at two indicators: labor force participation and criminality during the summer season. As previously stated, we discovered that the workforce used is especially the one relocated from the city between the units of the same operator, or the workforce that comes from other localities, by analyzing the data provided by the Statistical Institute, but also based on other studies and interviews with factors involved. In terms of criminality, it is high during the summer season but fairly consistent from year to year.

We considered the air quality, which was primarily related to the increased traffic on the coast during the summer season, the amount of waste produced (food and general), the quality of bathing water, the quantity and quality of drinking water, and the impact of tourists on protected areas. In terms of waste production, the data were incorporated into the county waste management plan, which shows a significant increase (on average 30 percent) in the amount of waste produced during the summer season, waste produced by tourists. The Statistics Institute also publishes an annual report on food waste, which shows how much food each novel wastes each year. In terms of bathing water quality, the Public Health Directorate's annual reports to the European Commission show that the bathing water is in the optimal parameters, with moderate, good, and very good status. The reduced use of the beach also helped to improve the quality of the bathing water.

However, it is clear that, while the statistics are positive, more detailed analyses on this parameter are required, given the numerous cases of pollution and discomfort reported in recent years. However, these are not the topic of this work. In terms of protected areas, the most significant protected area within a reasonable distance of Constanta is Lake Siutghiol, which is an avifaunistic protected area with significant ecological activity, particularly for migratory birds during the winter. Because the influence of tourist activity during the summer season is negligible, the indicator does not have extremely low values. We did not examine the degree of urbanization in the protected area in this paper because we believe that tourist activity during the summer does not have a significant impact on migratory bird wintering activity.

By achieving an average of the economic factor indicators (Figure 11) between 2011 and 2019, it can be seen that, while mass tourism was initially more beneficial, the situation changed in the middle of the period, and the activity without mass tourism became the better option for the community. A projection of this trend until 2050 (Figure 12) shows that both options are economically beneficial to the community, because the trend is growing; however, the growth in the option without mass tourism is more pronounced, reaching an optimal around 2048, whereas the optimal in the other option is expected after 2055.

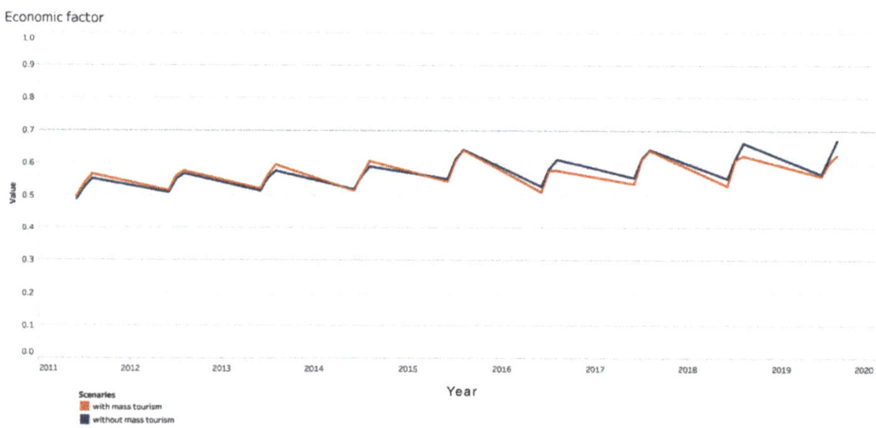

Figure 11. Analysis of economic factor average indicators for the two scenarios.

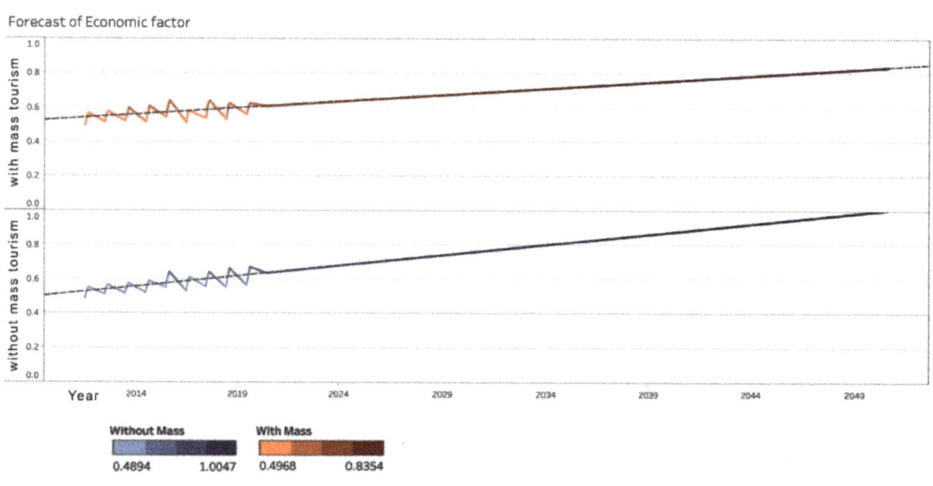

Figure 12. Economic factor forecast by 2050.

We conducted a similar analysis in terms of the social factor. If we focus on the intercalation of the two scenarios in the studied period, 2011–2019 (Figure 13), the situation becomes more interesting in terms of the forecasts of the two scenarios. The best option is also proven in the situation without mass tourism, with a higher value in 2050 than in the scenario with mass tourism. On the other hand, the trend is decreasing in terms of the option without mass tourism, while the trend is increasing in terms of the option with mass tourism (Figure 14). It is expected that in terms of community benefits, a choice will be made between the two options between 2050 and 2060. This social factor analysis provides some arguments for planners and public policymakers to choose one scenario over another based on the community's development directions. Simultaneously, the analysis can be improved by including additional social indicators such as education, access to culture, health, and so on.

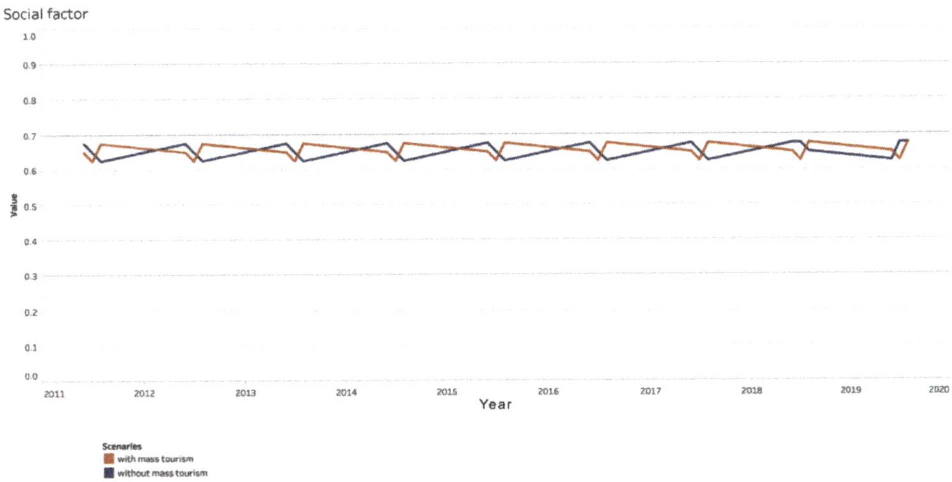

Figure 13. Analysis of social factor average indicators for the two scenarios.

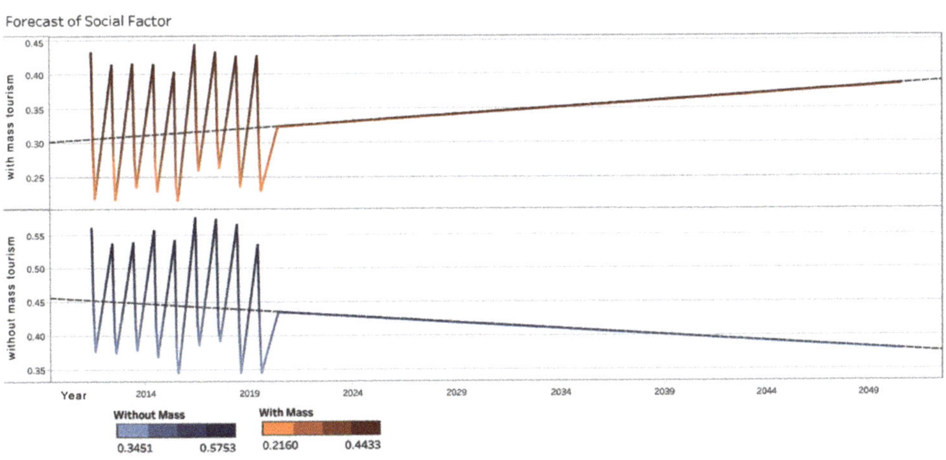

Figure 14. Forecast of the social factor until 2050.

In terms of the environment, an analysis of the averages of the proposed indicators for the period 2011–2019 (Figure 15) reveals that the scenario without mass tourism has values that are higher than the scenario with mass tourism. In terms of trends, the forecast until 2050 is particularly interesting, with the trend without mass tourism slightly decreasing and the trend with mass tourism slightly increasing (Figure 16). However, the main option remains one without mass tourism, and if current conditions continue, the option without mass tourism will be the best scenario for the community even after 2050.

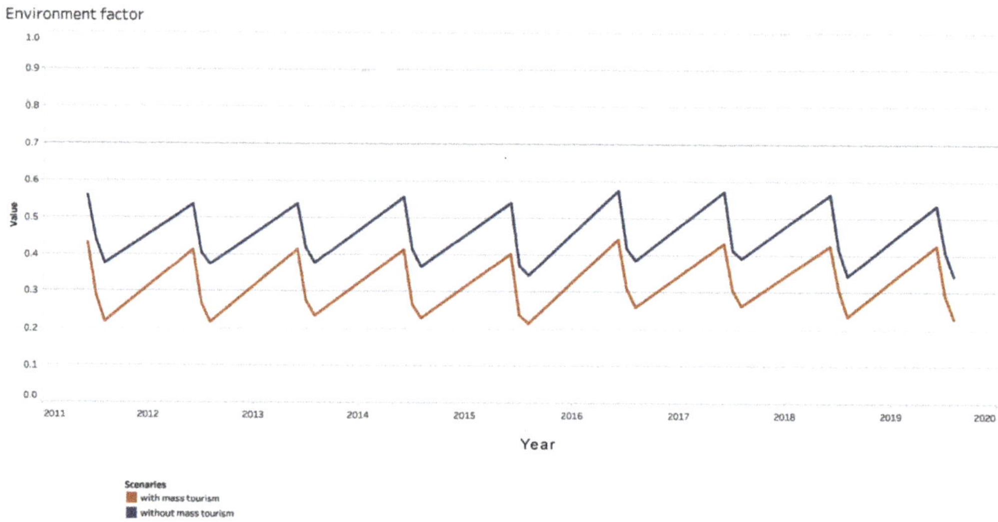

Figure 15. Analysis of environmental factor average indicators for the two scenarios.

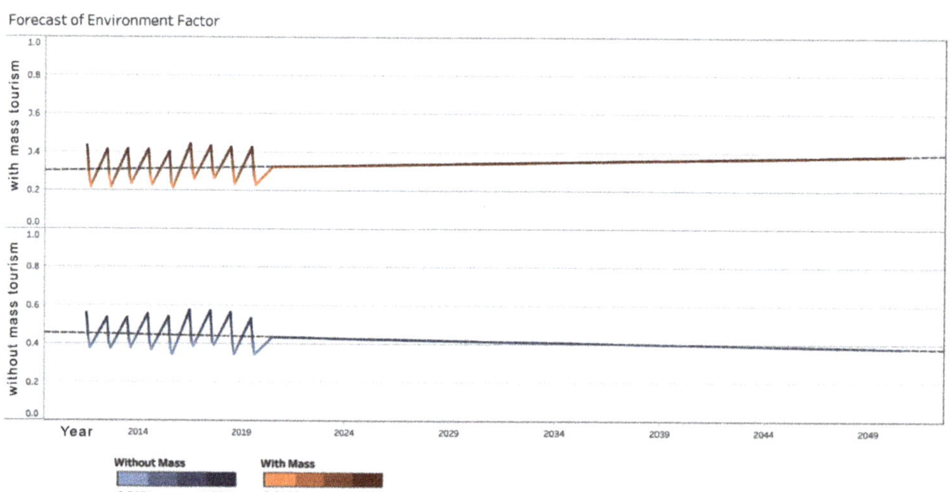

Figure 16. Average factor forecast by 2050.

The final part of the analysis was completed by calculating the average of the two scenarios for all of the proposed factors (economic, social, and environmental). As shown in Figure 17, the scenario without mass tourism outperforms the scenario with mass tourism over the period 2011–2019. This demonstrates that the previous option was not the best fit for the community. Regarding the projections of the two scenarios until 2050, the trends in both situations are increasing (Figure 18). The scenario without mass tourism has faster growth, but the difference between the two scenarios is not significant. This situation presents a challenge for policymakers seeking to improve local policies in accordance with the path they wish to take for community development.

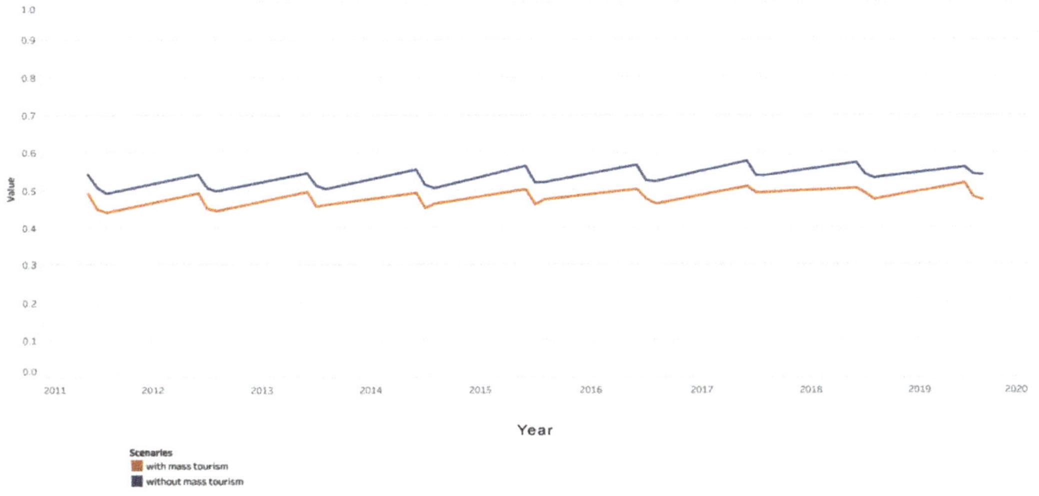

Figure 17. Analysis of indicator averages for the two scenarios.

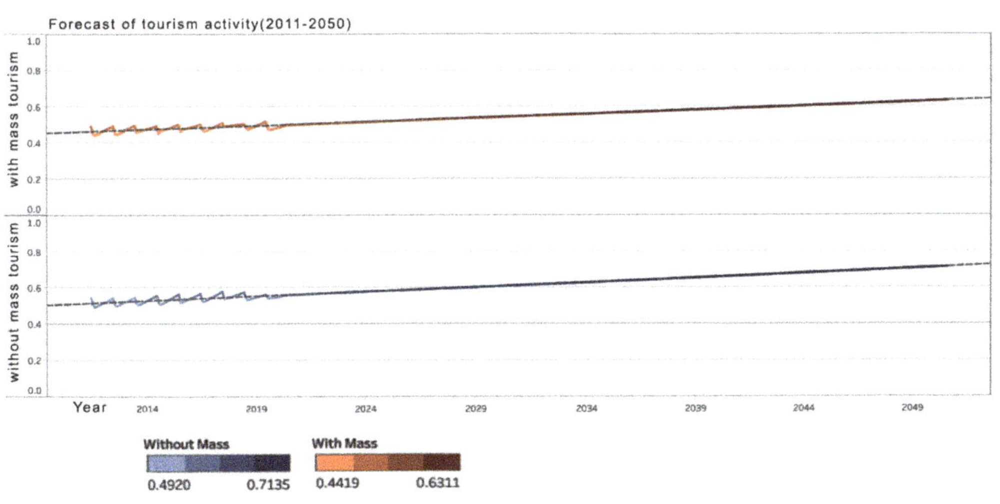

Figure 18. Forecast of indicator averages for the two scenarios for 2011–2050.

5. Conclusions

The multi-criteria analysis of the mass tourism management model in terms of its impact on the local community in Constanta is a complex tool that can aid in the development of local policies. Depending on the pressures exerted by various activities on other activities taking place in the community, the scenarios considered may or may not be viable. The accuracy and consistency of the data used in this analysis are critical. Furthermore, the correct identification of indicators and their interrelationships is an important process in carrying out a correct analysis. Last but not least, determining the calculation coefficient for the activity under consideration is critical, because bringing together data from various activities can be difficult when attempting to identify a viable model.

The PESTEL methodology was used to apply the multicriteria model to mass tourism in Constanța. The model's use demonstrated the impact of relevant actions on three primary factors: economic, social, and environmental. By recognizing existing activities and determining the relationships between these activities, such a model can be applied to any tourist or seaside location.

Multicriteria analysis must be performed throughout time in order to detect probable changes in components and provide more accurate model predictions.

Only considering the recommended indicators will not result in a long-term tourism activity. It is crucial to figure out which variables exert more or less pressure on the others. The affordability index must be determined for this. The equilibrium point must be determined, where one activity's pressure must cease before the other activity vanishes.

In the case studied in this paper, we assumed that the number of tourists visiting a location has a significant impact on the local community. Although it appears that mass tourism is beneficial from an economic standpoint, it can be observed that this type of tourism has a negative impact on some economic and environmental indicators, and the positive impact is often insignificant. This is not to say that a community cannot aim to use mass tourism as a form of tourism. It is just that in these cases, they must identify the high-pressure activities and take measures to relax the relationships between them.

One of the study's limitations was the data's availability and accuracy. Although most data are collected by public entities, they are not made available to the broader public. Furthermore, data gathering methods differ, and their accuracy in many circumstances is low. Another noteworthy limitation was the number of tourists visiting the area during the summer, even though the area is well known for the fact that many local people offer their own houses for accommodation. There are no relevant statistics on the number of tourists staying in these facilities. Another limitation of this study is that the activities in the coastal area are far more numerous and complex. As a result, in order to examine an effective situation, the interaction between these activities must be discovered, as well as the tolerance threshold between them. Last but not least, the study is time-limited in the sense that this analysis must be refreshed after a given period to determine whether the data entered are still relevant.

Author Contributions: Conceptualization, C.A. and A.-E.M.; methodology, A.-E.M.; validation, C.A., A.-E.M. and E.R.; formal analysis, C.A.; investigation, A.-E.M.; resources, A.-E.M. and E.R.; data curation, C.A.; writing—original draft preparation, C.A.; review and editing, E.R.; visualization, C.A.; supervision, A.-E.M. and E.R.; project administration, A.-E.M.; funding acquisition, E.R. All authors have read and agreed to the published version of the manuscript.

Funding: This work was carried out in the framework of the research project DREAM (Dynamics of the REsources and technological Advance in harvesting Marine renewable energy), supported by the Romanian Executive Agency for Higher Education, Research, Development, and Innovation Funding—UEFISCDI, grant number PN-III-P4-ID-PCE-2020-0008.

Institutional Review Board Statement: Not the case this study does not involve humans or animals.

Informed Consent Statement: Not the case this study does not involve humans or animals.

Data Availability Statement: The data that support the findings of this study are available in the public domain.

Acknowledgments: The results of this work will be also presented to the 9th edition of the Scientific Conference organized by the Doctoral Schools of "Dunărea de Jos" University of Galati (SCDS-UDJG) http://www.cssd-udjg.ugal.ro/ (accessed date 20 April 2021) that will be held on the 10 and 11 June 2021, in Galati, Romania.

Conflicts of Interest: The authors declare no conflict of interest.

References

1. Anton, C.; Gasparotti, C.; Anton, I.A.; Rusu, E. Implementation of a Coastal Management model in Kinvara Bay at the North Atlantic Ocean. *J. Mar. Sci. Eng.* **2019**, *1*, 71. [CrossRef]
2. Kindermann, G.; Anton, C.; Pender, K.; Cave, R. EU Life Feasability Study—Flag Funded Cuanbao Project. 2019; Unpublished Work.
3. Yüksel, I. Developing a Multi-Criteria Decision Making Model for PESTEL Analysis. *Int. J. Bus. Manag.* **2012**, *7*, 24. [CrossRef]
4. Institutul National de Cercetare-Dezvoltare in Turism. Strategia Integrată de Dezvoltare Durabilă a Turismului în Județul Constanța, 2019–2028. Bucuresti. 2019. Available online: http://www.cjc.ro/dyn_doc/turism/Strategia_jud.Constanta_Faza2.pdf (accessed on 20 April 2021).
5. Apele Romane ADMINISTRATIA MATIONALA. Masterplanul Protectia si Reabilitarea Zonei Costiere. *Constanta* **2012**, *1*. Available online: http://arhiva.rowater.ro/dadobrogea/Master%20Plan%20privind%20Protecia%20i%20Reabilitarea%20Zonei/Forms/viz%20MP1.aspx (accessed on 20 April 2021).
6. Akis, A. The effects of mass tourism: A case study from Manavgat (Antalya—Turkey). *Procedia Soc. Behav. Sci.* **2011**, *19*, 289–296. [CrossRef]
7. Rusu, E.; Soares, C.G. Wave energy assessments in the coastal environment of Portugal continental. In Proceedings of the International Conference on Offshore Mechanics and Arctic Engineering, Berlin, Germany, 9–13 June 2008; Volume 6.
8. Strategia Naţională a României Pentru Dezvoltarea Turismului 2019–2030. *Bucuresti* **2018**. Available online: http://b2b-strategy.ro/b2b/wp-content/uploads/Strategia-na%C8%9Bional%C4%83-a-Rom%C3%A2niei-pentru-dezvoltarea-turismului-Volumul-1.pdf (accessed on 10 June 2018).
9. Anton, I.A.; Rusu, L.; Anton, C. Nearshore wave dynamics at Mangalia beach simulated by spectral models. *J. Mar. Sci. Eng.* **2019**, *7*, 206. [CrossRef]
10. Rusu, E.; Măcuţă, S. Numerical modelling of longshore currents in marine environment. *Environ. Eng. Manag. J.* **2009**, *8*, 147–151. [CrossRef]
11. Rusu, E.; Conley, D.; Ferreira-Coelho, E. A hybrid framework for predicting waves and longshore currents. *J. Mar. Syst.* **2008**, *69*, 59–73. [CrossRef]
12. Ivan, A.; Gasparotti, C.; Rusu, E. Influence of the interactions between waves and currents on the navigation at the entrance of the Danube Delta. *J. Environ. Prot. Ecol.* **2012**, *13*, 1673–1682.
13. Anton, C.; Gasparotti, C.; Rusu, E. Identification of the economic pressure on environmental factors in the Romanian coastal zone-case study Eforie. In Proceedings of the 18th International Multidisciplinary Scientific GeoConference Surveying Geology and Mining Ecology Management (SGEM), Albena, Bulgaria, 2–8 July 2018; p. 461.
14. Anton, C.; Gasparotti, C.; Rusu, E.; Anton, I.A. Approach to the analysis and evaluation of the strategic intervention options in the Romanian coastal zone taking into account economic, social and environmental factors. In Proceedings of the International Multidisciplinary Scientific GeoConference Surveying Geology and Mining Ecology Management, SGEM 18 (5.3), Albena, Bulgaria, 2–8 July 2018; p. 67.
15. Anton, C.; Gasparotti, C.; Rusu, E. A challenge for the inland navigation—A connection between the Baltic and the Black seas. In Proceedings of the ICTTE Conference, Belgrad, Serbia, 2–8 July 2018.
16. Cheng, E.W.L.; Li, H. Application of ANP in process models: An example of strategic partnering. *Build. Environ.* **2007**, *42*, 278–287. [CrossRef]
17. Dare, C. The UK tour-operating industry: A competitive analysis. *J. Vacat. Mark.* **2006**, *6*, 357–367.
18. Saaty, T.L.; Takizawa, M. Dependence and independence: From linear hierarchies to nonlinear networks. *Eur. J. Oper. Res.* **1986**, *26*, 229–237. [CrossRef]
19. Shilei, L.; Yong, W. Target-oriented obstacle analysis by PESTEL modeling of energy efficiency retrofit for existing residential buildings in China's northern heating region. *Energy Policy* **2009**, *37*, 2098–2101. [CrossRef]
20. Shyur, H.J. COTS evaluation using modified TOPSIS and ANP. *Appl. Math. Comput.* **2006**, *177*, 251–259. [CrossRef]
21. Tzeng, G.H.; Chiang, C.H.; Li, C.W. Evaluating intertwined effects in e-learning programs: A novel hybrid MCDM model based on factor analysis and DEMATEL. *Expert Syst. Appl.* **2007**, *32*, 1028–1044. [CrossRef]
22. Vitkiene, E. Questions that have to be answered by a well-managed strategy of coastal recreation and tourism enterprises development. *TILTAI* **2009**, *4*, 117–124.
23. Wu, W.W. Choosing knowledge management strategies by using a combined ANP and DEMATEL approach. *Expert Syst. Appl.* **2008**, *35*, 828–835. [CrossRef]
24. Elomda, B.M.; Hefny, H.A.; Hassan, H.A. An extension of fuzzy decision maps for multi-criteria decision-making. *Egypt Inform. J.* **2013**, *14*, 147–155. [CrossRef]
25. Frynas, J.G.; Mellahi, K. *Global Strategic Management*; Oxford University Press: Oxford, UK, 2015.
26. Gray, S.A.; Gray, S.; Cox, L.J.; Henly-Shepard, S. Mental modeler: A fuzzy-logic cognitive mapping modeling tool for adaptive environmental management. In Proceedings of the 2013 46th Hawaii International Conference on System Sciences, Wailea, HI, USA, 7–10 January 2013.
27. Healey, N.M. The transition economic of central and eastern Europe: A political, economic, social and technological analysis. *Columbia J. World Bus.* **1994**, *29*, 62–70. [CrossRef]
28. Kosko, B. Fuzzy cognitive maps. *Int. J. Man Mach. Stud.* **1986**, *24*, 65–75. [CrossRef]

inventions

MDPI

Article

Assessment of the Wind Energy Potential along the Romanian Coastal Zone

Alina Girleanu *, Florin Onea and Eugen Rusu

Department of Mechanical Engineering, Faculty of Engineering, "Dunarea de Jos" University of Galati, 47 Domneasca Street, 800008 Galati, Romania; florin.onea@ugal.ro (F.O.); eugen.rusu@ugal.ro (E.R.)
* Correspondence: alina.girleanu@ugal.ro

Abstract: The present work aims to provide a comprehensive picture of the wind energy potential that characterizes the Romanian coastal environment using in situ measurements and reanalysis of wind data (ERA5) that cover a 42-year time interval (1979–2020). A total of 16 reference points (both land and offshore) equally distributed along the Romanian sector are used to evaluate the local wind energy potential, targeting in this way several sites where a renewable wind project could be established. Compared to the in situ measurements (land points), the ERA5 dataset underestimates the wind speed by at least 11.57%, this value increasing as we approach the coastline. From the analysis of the spatial maps, it is likely that the wind speed steadily increases from onshore to offshore, with a sharp variation near the coastline being reported. Furthermore, the assessment of some state-of-the-art offshore wind turbines was conducted using 12 systems defined by rated capacity ranging from 2 to 10 MW. Some scenarios were proposed to identify sustainable offshore wind projects to be implemented in the Romanian coastal zone based on these results.

Keywords: Romania; coastal area; wind energy; measurements; IEC classes; wind turbines

Citation: Girleanu, A.; Onea, F.; Rusu, E. Assessment of the Wind Energy Potential along the Romanian Coastal Zone. *Inventions* **2021**, *6*, 41. https://doi.org/10.3390/inventions6020041

Academic Editor: Pierluigi Siano

Received: 21 May 2021
Accepted: 3 June 2021
Published: 4 June 2021

1. Introduction

At the core of the energy transition towards renewables and planning for a post–COVID–19 economic rehabilitation, offshore wind resource is considered a possible pillar in reaching net–zero emissions of greenhouse gases by the end of 2050, a policy required by the European Green Deal. At the European level, offshore wind projects are already considered a competitive sector. According to [1], the European Union (EU) complete decarbonisation will require 450 GW of offshore wind capacity by 2050, representing 18 times more than the currently installed offshore wind capacity—25 GW and 5402 grid-connected wind turbines across 12 countries. The reason why offshore wind is more desirable than solar and even onshore wind power is its higher full-load and operating hours, its stability and its greater predictability. Moreover, offshore wind represents nowadays a very cost-effective alternative for power generation, taking into consideration the fact that the Levelized Cost of Electricity (LCOE) decreased from an average of 0.13 USD/kWh in 2018 to an average between 0.05–0.09 USD/kWh by 2030 and 0.03–0.07 USD/kWh by 2050 [2], falling in the low range of costs for fossil fuels and becoming more affordable than nuclear power. The increased economies of scale, together with the competitiveness of supply chains and further technological advances, will reduce the cost of harvesting wind energy. Nevertheless, for the time being, investing in onshore wind projects seems to be more economically viable, even though offshore designs generate more electricity [3].

There is a growing interest in developing wind projects in different coastal areas defined by resourceful wind energy [4–6]. Given the recent enhancements intended to reach the decarbonisation process, Romania must tap into its existing offshore wind power potential in the Black Sea. Currently, even though Romania is one of the leading countries in the south-eastern part of Europe in terms of onshore wind development, having an installed capacity of 3 GW, according to the National Energy and Climate Plan (NECP) 2021–2030 [7], there is inadequate support for the advancement of offshore capacities in the Black Sea. For example, almost 80% of the existing wind turbines in Romania are concentrated in the Dobrogea region, located near the sea. The most significant operational onshore wind farm in Europe is also established in Dobrogea, namely the Fantanele–Cogealac wind project, having an installed capacity of 600 MW [8].

Most of the latest published research regarding the Black Sea describes the wind conditions only from a meteorological point of view [9–11], probably due to the fact that the Black Sea coastal zone is considered less attractive than other enclosed or semi-enclosed seas, such as the Mediterranean or Baltic Sea [4–6]. Moreover, the performances of some commercial, state-of-the-art wind turbines were assessed by Onea and Rusu [12–14], considering several reference sites distributed along the Black Sea shoreline, at a water depth that, however, does not exceed 50 m. Additionally, several offshore sites that appear to be suitable for implementing a wind project in the Turkish coastal region were proposed in [15], while Raileanu et al. [16] assessed the performances of two wind turbines that operate offshore—Siemens SWT-3.6-120 and Senvion-6.2M-126.

Even though the topic of research appears similar to the one discussed in Onea and Rusu [3], the present study is characterized by significant differences. Firstly, the wind dataset and the number of values per day differ since ERA5 replaced the ERA–Interim database used in [3] with 24 values per day. Moreover, in [3], the assessment of the turbines' performances was carried out by adjusting the wind data values from 10 m to an operational hub height of 80 m, while in this case 12 turbines were evaluated by considering the ERA5 values directly reported at the height of 100 m above sea level. The selected wind turbines' rated capacity varies from 2 to 10 MW, including the existing operational systems and the brand-new generation of turbines expected to be implemented shortly. At the same time, it is essential to mention that in [17] the maximum rated power was 5 MW. Since the literature review concentrating on the Black Sea coastal area is considerably restrained, the purpose of this work is to identify the best performances of various wind turbines that might become operational in the Romanian coastal environment. From this perspective, this study will focus on the differences between the onshore and offshore wind power potential, which site is the most suitable for developing a renewable wind project and what type of wind turbine will be most appropriate to operate within the target area. This study can be considered an opportunity since there is a significantly growing interest both from investors and the Romanian authorities for developing offshore wind projects in this specific environment. In this context, the present work is defined by some elements of novelty, as follows: (a) The differences between the Romanian onshore and offshore wind resources are assessed by also including in-situ measurements for two sites located near the shoreline; (b) a long-term overview of the wind resources is provided (42 years), compared to only 20 years that were previously considered for this target area; (c) an extensive portfolio of turbines (12 different systems) is considered for evaluation to identify their expected performances.

2. Materials and Methods

2.1. Target Area

The Black Sea wind climate is mainly influenced by the action of polar, continental and marine-tropical air masses, from which the most noticeable impact is that of the polar air masses, especially during the wintertime, when shallow temperatures and extreme storms might occur. Notably, it is considered that the western side of the Black Sea is defined by more abundant wind resources, with 8 m/s representing the mean wind speed value recorded during the winter season [13]. Figure 1 reveals the differences in water depth across Romania's Exclusive Economic Zone (EEZ). The EEZ is characterized by an area of more than 22 thousand km^2 and the territorial surface of the water of 4084 km^2, associated with 5.30% of the entire Black Sea shoreline [3]. The thick black line indicates the boundaries of the Romanian EEZ. As shown in Figure 1, a meaningful part of Romania's EEZ is a deep-water area >50 m, which is more suitable for floating platforms than fixed systems. A closer look at the European offshore market [18] points out that, in 2019, the average distance to shore for developing projects was close to 60 km, which is just within the Romanian transition area from shallow to deep water [19]. Nevertheless, wind farms located more than 50 km from shore ought to be connected to the grid through high voltage direct current connections to reduce electrical losses significantly; such connections, though, demand more significant upfront investment costs [20]. Further details regarding the target area are presented in Figure 1, where 16 reference sites were defined for assessment. The sites considered for the study are equally distributed along the Romanian coastline and are located both onshore and offshore. Furthermore, the points located offshore were defined in water depths of 50–89 m, and the maximum distance from the shore of 118 km corresponds to Site B4. A more comprehensive picture of the selected points is outlined in Table 1. Considering these points, we are likely to achieve a general view of the wind power potential and the expected performances of a particular wind turbine that can be further extended to a more thorough investigation.

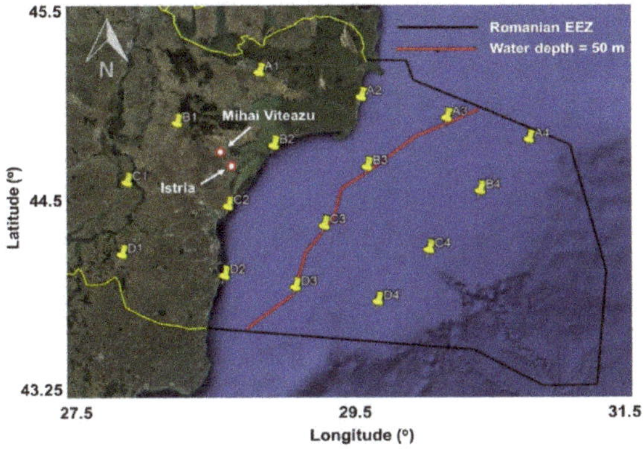

Figure 1. The Romanian coastal area, including the Exclusive Economic Zone (EEZ) boundaries and the reference sites considered. Map processed from Google Earth, 2021.

Table 1. The main characteristics of the sites considered onshore and offshore of the Romanian coastal environment. Data obtained from Google Earth 2021.

ID	Location	Long (°)	Lat (°)	Distance to Shore (km)	Height/Depth (m)
A1	Land	28.899	45.084	60	68
A2	Shoreline	29.635	44.952	0	0
A3	Offshore	30.247	44.837	49	−50
A4	Offshore	30.839	44.716	100	−81
B1	Land	28.309	44.800	57	185
B2	Shoreline	29.015	44.683	0	0
B3	Offshore	29.680	44.569	55	−61
B4	Offshore	30.478	44.438	118	−84
C1	Land	27.960	44.471	60	11
C2	Shoreline	28.693	44.351	0	0
C3	Offshore	29.377	44.250	56	−51
C4	Offshore	30.114	44.122	115	−89
D1	Land	27.942	44.081	60	145
D2	Shoreline	28.675	43.976	0	0
D3	Offshore	29.179	43.916	43	−51
D4	Offshore	29.755	43.841	90	−83

2.2. Wind Dataset

Several previous publications [21,22] have focused on assessing Romania's wind offshore resources. Against this background, the current study highlights several elements of novelty, including an analysis of a new wind database—ERA5—which is the latest product of the European Centre for Medium-Range Weather Forecasts (ECMWF) [23]. ERA5 is a replacement for the ERA–Interim database reanalysis that stopped being displayed in 2019 [24] and comprises vast amounts of historical observations into global estimates using state-of-the-art modelling and data assimilation systems. Moreover, wind speed values are reported at a hub height of 100 m ($U100$), obtained directly from the ERA5 data system. It is essential to underline that since the wind speed values at 100 m were directly provided, no further data processing was needed, such as the use of a logarithmic law to adjust the wind speed from a 10 m height above the sea level, where the velocity is much lower due to boundary layer mean velocity distribution. Additionally, in situ measurements were considered for two existing wind projects located onshore near the Dobrogea region (Mihai–Viteazu and Istria sites), considering the time interval from October 2006 and April 2009, which was used for comparison against ERA5 data.

2.3. Wind Turbines

Since there is yet no ongoing wind energy project in the Black Sea basin to emphasize the performances of a wind energy conversion system in the target area, some already operational wind turbines were considered for analysis. The main characteristics of these systems are presented in Table 2, systems that are currently considered for implementation in some European offshore wind projects. The average rated power is close to 8.20 MW, and more than 30% of the projects developed last year included larger systems. However, turbine orders placed online already reveal a trend towards the next generation in terms of size, with turbines ranging from 10 to 13 MW for upcoming projects starting with 2022 [25]. The size of new wind farms is estimated at 788 MW, while the distance to the shore and the water depth has increased to 52 km and 44 m, respectively [25]. The 12 selected turbines embrace a full spectrum of rated capacity, starting from 2 MW and ending with the AMSC wt10000dd SeaTitan—a production of American Superconductor, a manufacturer from the US—having the highest-rated power—10 MW. It is essential to mention that, although these systems can operate at different heights above sea level, their performances were evaluated at 100 m since, at this level, there is a higher wind potential. Figure 2 reveals the power curve of each system, identifiable throughout the cut-in, rated speed and cut-out thresholds, all operating at a hub height of 100 m.

Table 2. Technical details of the wind turbines considered in this work. Data processed from [26,27].

ID	Turbine	Power (MW)	Cut–in Speed (m/s)	Rated Speed (m/s)	Cut–out Speed (m/s)	Hub Height (m)
T1	V100–2.0	2.00	3.00	12	22	100
T2	V100–3.0	3.00	4.00	15	25	100
T3	SWT–2.3–93	2.30	4.00	13	25	100
T4	SWT–3.6–120	3.60	3.50	12	25	100
T5	SWT–4.0–130	4.00	5.00	12	25	100
T6	SWT–7.0–154	7.00	3.00	13	25	100
T7	REpower5M	5.075	3.50	14	25	100
T8	REpower6M	6.15	3.50	14	25	100
T9	Nordex N90/2500	2.50	3.00	13.50	25	100
T10	AREVA M5000–116	5.00	4.00	12.50	25	100
T11	Samsung S7.0–171	7.00	3.00	11.50	25	100
T12	AMSC wt10000dd SeaTitan	10.00	4.00	11.50	25	100

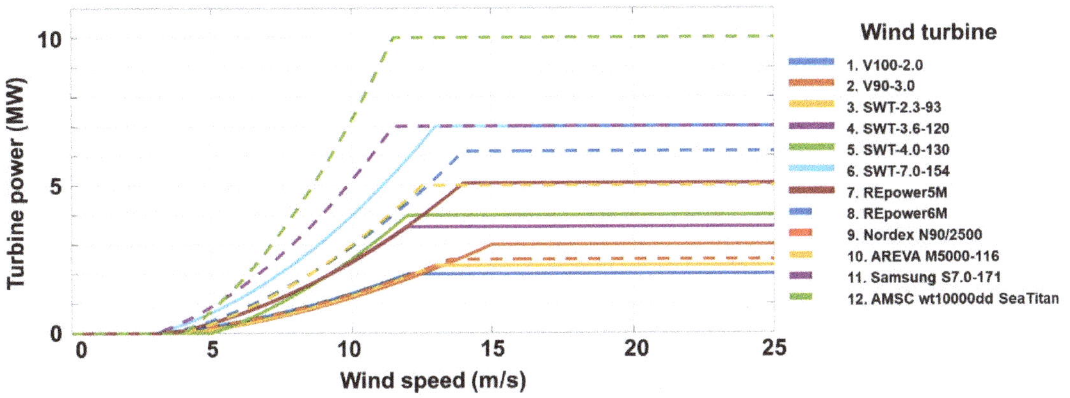

Figure 2. Representation of the wind turbine power curves considered for this study.

The Annual Electricity Production (*AEP*) of any wind turbine can be expressed as follows [28]:

$$AEP = T \times \int_{cut-in}^{cut-out} f(u)P(u)du \quad (1)$$

where *AEP* is displayed in MWh, *T* represents the average number of operational hours per year, $f(u)$ the Weibull probability density function, $P(u)$ the power curve of a wind turbine (also represented in Figure 2 for each of the 12 systems), and the cut-out and cut-in speeds are wind turbine technical terms.

The Weibull probability density function can be denoted as follows [29]:

$$f(u) = \left(\frac{k}{c}\right)\left(\frac{u}{c}\right)^{k-1} exp\left[-\frac{u}{c}^{k}\right] \quad (2)$$

where u represents the wind speed (expressed in m/s), k the shape parameter, and c the scale parameter (expressed in m/s).

One method, which can be used to estimate the reliability of a particular energy system converter, is to assess its capacity factor (C_f), which can be defined as follows [30]:

$$C_f = \frac{P_{turbine}}{P_{rated}} \times 100 \quad (3)$$

where $P_{turbine}$ represents the electric power expected to be generated by each turbine and P_{rated} represents the rated power of each generator.

3. Results

3.1. Analysis of the Wind Resources

ERA5 wind data is the principal data source used in this work; therefore, it is essential to understand and take into consideration the differences between alternative data sources, in this case, the in situ measurements. Based on the comparisons against some in situ measurements for two existing wind projects located onshore near the Dobrogea region, considering the time interval from October 2006 to April 2009, it appears that the ECMWF wind data underestimates the wind conditions, as illustrated in Figure 3. For the Mihai–Viteazu wind project, located at 39.70 m height, in the vicinity of reference sites B1 and B2 (see Figure 1), the ERA5 data undervalues the wind speed by at least 11.57%. Regarding the Istria site, located at a height of 42.60 m and closer to the coastline, the percentage is even higher, namely 13.52%.

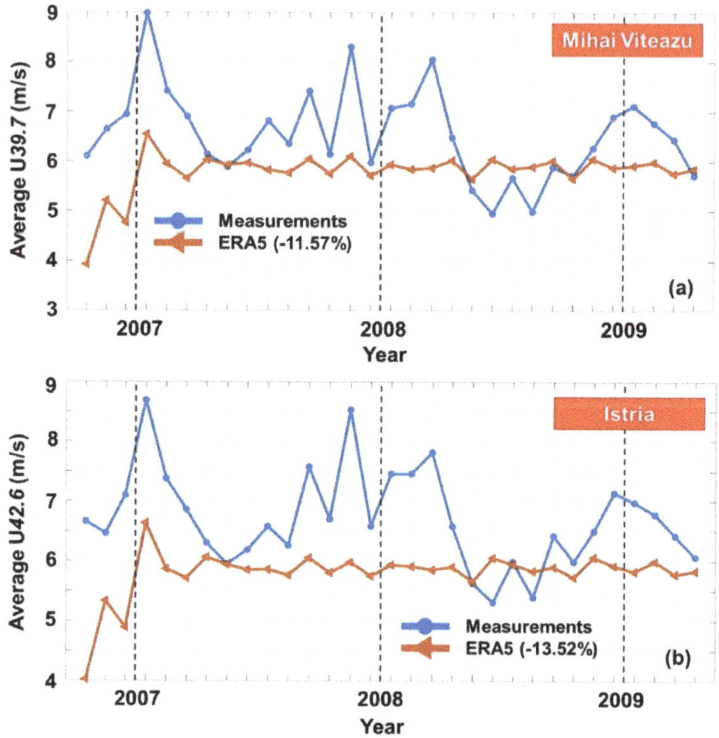

Figure 3. Wind speed comparison between ERA5 and in situ measurements, considering the time interval October 2006 to April 2009. Results for: (**a**) Mihai-Viteazu site, at 39.7 m height; (**b**) Istria site, at 42.6 m height. The in situ values were adapted from [30]. The percentage values indicate the average differences (in %) between ERA5 and in situ data, in this case ERA5 underestimating the wind speed.

Figure 4 and Table 3 show the Weibull probability distribution and characteristics, including the specific parameters (k and c), considering all 16 reference sites. It is important to mention that there is no significant variation between the sites, although the points determine a relatively extended area. The results are built upon the ERA5 values (January 1979 to December 2020) and provide insight into the distribution of the wind resources by intervals and each site's energy potential. All sites have wind conditions in the range of 6–20 m/s that will significantly increase the power production of an offshore wind turbine. The most moderate performances are expected for B2 and D2 sites, with wind speed peaks of approximately 4–5 m/s, while sites A3, A4, B4, and C3 exhibit the highest performances. Another noteworthy aspect is that none of the sites indicates wind conditions higher than 25 m/s, representing the cut-out threshold for most operational wind generators.

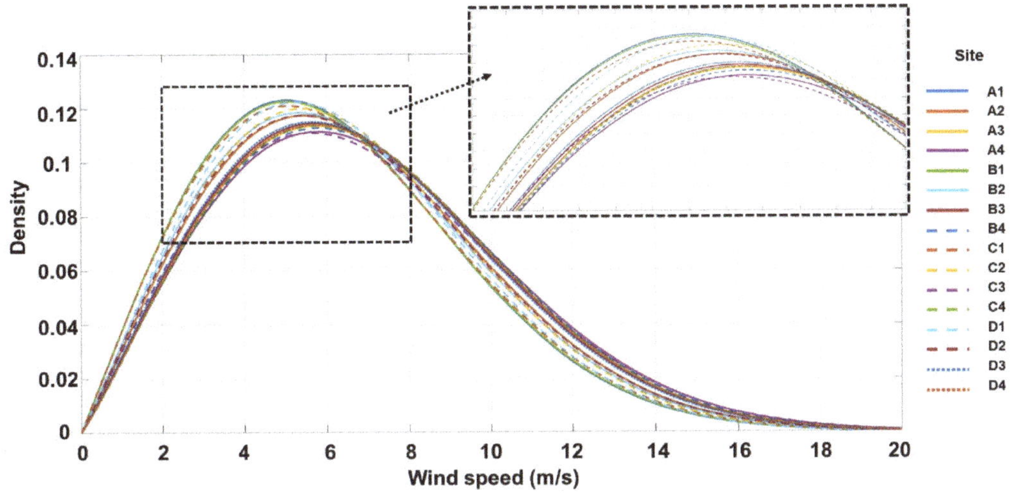

Figure 4. The Weibull distributions corresponding to the ERA5 values (January 1979 to December 2020).

Table 3. Weibull characteristics corresponding to locations considered. The results are reported for the *U*100 parameter for the time interval from January 1979 until December 2020.

Site	A1	A2	A3	A4	B1	B2	B3	B4
k (m/s)	7.04	7.49	7.81	7.92	7.06	7.38	7.75	7.87
c	2.03	2.08	2.10	2.09	2.02	2.06	2.10	2.10
Site	**C1**	**C2**	**C3**	**C4**	**D1**	**D2**	**D3**	**D4**
k (m/s)	7.12	7.38	7.82	7.82	7.20	7.46	7.70	7.80
c	2.01	2.10	2.04	2.11	2.08	2.06	2.10	2.11

Figure 5 provides the first representation of the wind energy distribution over the Romanian coastal zone, including the land. The results show that the wind speed increases from land to the shoreline and further to the offshore region. Moreover, Figure 5 illustrates the full-time and wintertime (December–January–February) distribution of the *U*100 parameter by wind classes (C1–C7), the higher classes (C4–C7) considered more attractive for developing future renewable projects [27]. The organization of wind capacities by classes is highlighted in Table 4, providing a clearer understanding of the classification. The results provided in Figure 5 are based upon the ERA5 values from January 1979 until December 2020. Regardless of the time interval analyzed, it can be noticed that C1 is the prevailing wind class present, with values of wind speed smaller than 6.10 m/s [31] and distribution of 50–60%, followed closely by the C7 wind class (characteristic of wind speeds higher than 9.70 m/s [31]) with a distribution of 12–20% (total time) and 14–24% (wintertime), respectively. As expected, the winter season is highly prone to more wind speed values close to C6 and C7 wind classes and less of C1 compared to the total time distribution.

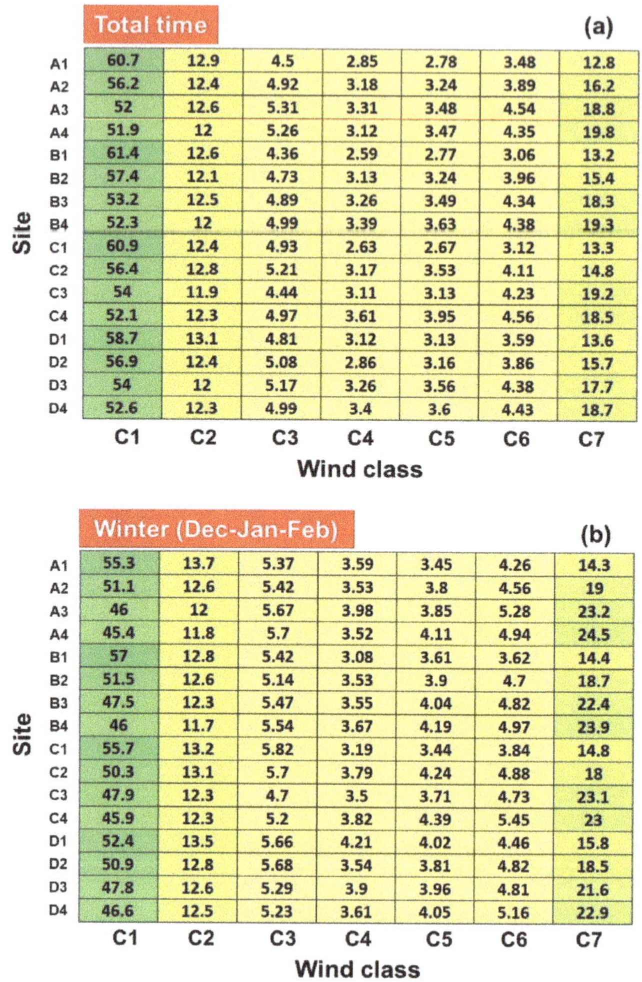

(a) Total time

Site	C1	C2	C3	C4	C5	C6	C7
A1	60.7	12.9	4.5	2.85	2.78	3.48	12.8
A2	56.2	12.4	4.92	3.18	3.24	3.89	16.2
A3	52	12.6	5.31	3.31	3.48	4.54	18.8
A4	51.9	12	5.26	3.12	3.47	4.35	19.8
B1	61.4	12.6	4.36	2.59	2.77	3.06	13.2
B2	57.4	12.1	4.73	3.13	3.24	3.96	15.4
B3	53.2	12.5	4.89	3.26	3.49	4.34	18.3
B4	52.3	12	4.99	3.39	3.63	4.38	19.3
C1	60.9	12.4	4.93	2.63	2.67	3.12	13.3
C2	56.4	12.8	5.21	3.17	3.53	4.11	14.8
C3	54	11.9	4.44	3.11	3.13	4.23	19.2
C4	52.1	12.3	4.97	3.61	3.95	4.56	18.5
D1	58.7	13.1	4.81	3.12	3.13	3.59	13.6
D2	56.9	12.4	5.08	2.86	3.16	3.86	15.7
D3	54	12	5.17	3.26	3.56	4.38	17.7
D4	52.6	12.3	4.99	3.4	3.6	4.43	18.7

(b) Winter (Dec-Jan-Feb)

Site	C1	C2	C3	C4	C5	C6	C7
A1	55.3	13.7	5.37	3.59	3.45	4.26	14.3
A2	51.1	12.6	5.42	3.53	3.8	4.56	19
A3	46	12	5.67	3.98	3.85	5.28	23.2
A4	45.4	11.8	5.7	3.52	4.11	4.94	24.5
B1	57	12.8	5.42	3.08	3.61	3.62	14.4
B2	51.5	12.6	5.14	3.53	3.9	4.7	18.7
B3	47.5	12.3	5.47	3.55	4.04	4.82	22.4
B4	46	11.7	5.54	3.67	4.19	4.97	23.9
C1	55.7	13.2	5.82	3.19	3.44	3.84	14.8
C2	50.3	13.1	5.7	3.79	4.24	4.88	18
C3	47.9	12.3	4.7	3.5	3.71	4.73	23.1
C4	45.9	12.3	5.2	3.82	4.39	5.45	23
D1	52.4	13.5	5.66	4.21	4.02	4.46	15.8
D2	50.9	12.8	5.68	3.54	3.81	4.82	18.5
D3	47.8	12.6	5.29	3.9	3.96	4.81	21.6
D4	46.6	12.5	5.23	3.61	4.05	5.16	22.9

Figure 5. Distribution of the *U*100 parameter by wind classes (January 1979 to December 2020), where: (**a**) Total time values; (**b**) winter values.

Table 4. Wind classification at 100 m, processed from Oh et al. [31].

Wind Class	100 m Reference Height	
	Power Density (W/m^2)	Wind Speed (m/s)
C1 (poor)	<260	<6.1
C2 (marginal)	260–420	6.1–7.1
C3 (fair)	420–560	7.1–7.8
C4 (good)	560–670	7.8–8.3
C5 (excellent)	670–820	8.3–8.9
C6 (outstanding)	820–1060	8.9–9.7
C7 (superb)	>1060	>9.7

Figure 6 presents the spatial distribution of the monthly and annual mean wind speed values (*U*100 parameter) based on the ERA5 data reported for 1979–2020, the reference sites A1, A4, D1, and D4, respectively. These four maps highlight that the wind speeds slightly increase as we approach the offshore area, although there are no significant differences. From the analysis of the parameter *U*100, it can be noticed that the wind speed values corresponding to the offshore sites are steadier and more resourceful, especially during the winter season (December–January–February), where there are constant values of 9–10 m/s, suggesting that during the most energetic season (winter), a wind project located offshore will obtain the best performances. The average wind speed value is a relevant parameter in this context since it indicates what type of wind turbine is most suitable for the analyzed coastal climate.

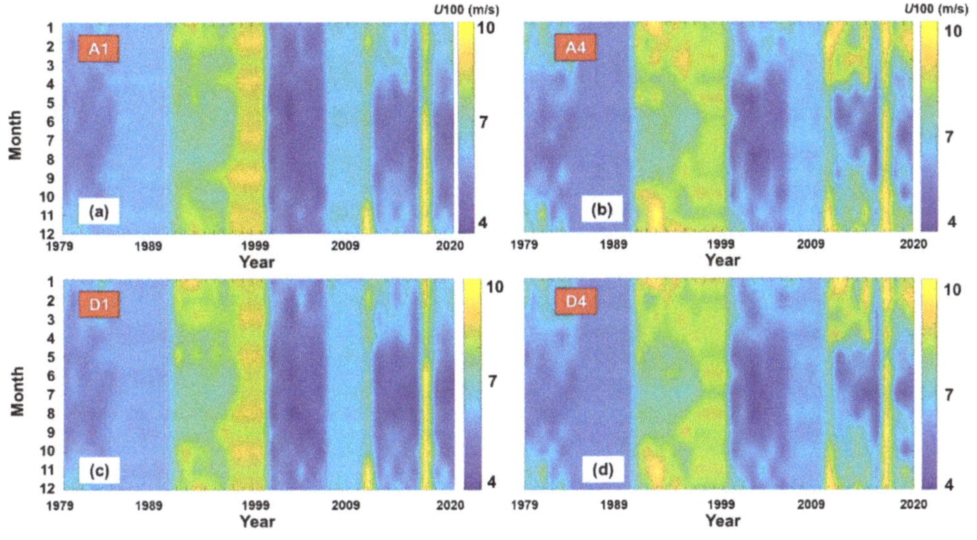

Figure 6. Monthly and annual distribution of the *U*100 (average values) reported between January 1979 and December 2020; (**a**) site A1; (**b**) site A4; (**c**) site D1; (**d**) site D4.

Another critical parameter is the wind direction, represented in Figure 7 for sites A3, B4, C3, D4, considering the 42-year time interval from January 1979 to December 2020, distributed by wind speeds. Each site has a different energy pattern, indicating for A3 a significant contribution coming from the offshore area from the north-eastern sector (7.34%) with wind speeds in the range 6–8 m/s. Compared to A3, the B4 site appears to have a higher share of wind resource from the southern sector, and D4 from the western one, especially wind speeds between 10–12 m/s. It seems that, stepping further into the offshore zone, the wind fields are oriented to the southwest direction. Moreover, the wind direction represents an essential viewpoint of the design of a wind farm, considering that the energy produced is significantly influenced by the 'wake effect'. According to [32], it is estimated that the total losses could reach 10–25% for offshore projects.

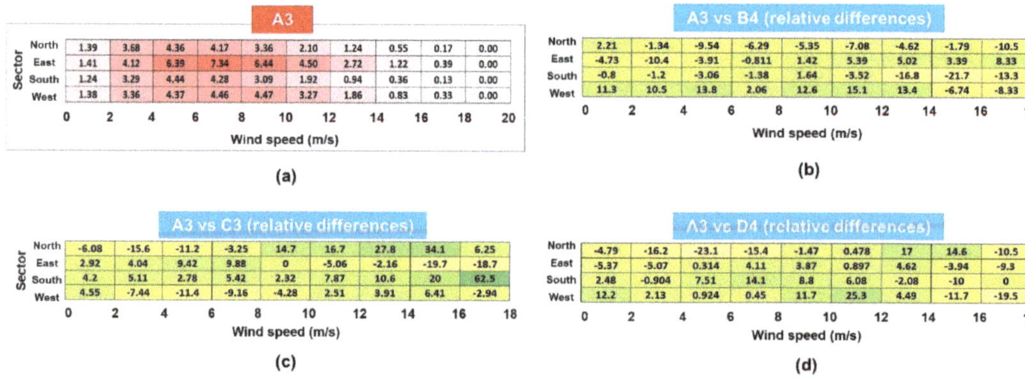

Figure 7. Distribution of the wind conditions (*U*100) by wind speed and directions, considering the 42-year time interval from January 1979 to December 2020. Results for: (**a**) A3-joint distribution; (**b–d**) differences (in %) reported between A3 and the sites B4, C3, and D4, respectively. A positive value indicates that, for A3, compared to the other sites, a higher share corresponds.

3.2. Evaluation of the Wind Turbine Performances

The first parameter analyzed is the operating capacity (in %), which indicates the percentage of the total time during which the specific system will perform at its total capacity. This parameter is computed based upon the cut-in speed—the initial point at which the wind turbine generates electricity; and the cut-out speed, which, for most significant offshore wind turbines, is set to 25 m/s, except for the V100–2.0 (T1), which has a cut-out speed of 22 m/s (see Table 2). When a system reaches the cut-out speed, it is set to shut down to prevent irreversible damage. Figure 8a illustrates that the most satisfying performances were obtained in the north-eastern part of the target area, at reference points A3 and A4, with a maximum value of 88.30% for the system Samsung S7.0-171, which has a cut-in speed of 3 m. As the cut-in speed increases from 3 to 4 m/s, a slight decrease in operating capacity is visible (80.10% for A3 and 80.30% for A4). The least attractive site, in this case, is represented by C2, regardless of the cut-in speed (77.10% when the cut-in speed is 4 m/s).

Secondly, the rated capacity (in %) highlights the maximum performance of each system, for which only the wind conditions located between the rated speed and cut-out values were assessed. In Figure 8b, the evolution of this parameter, which does not exceed 13.70%, is noticeable. The systems Samsung S7.0-171 (T11) and AMSC wt1000dd SeaTitan present the best performances (their rated capacity is >11.50 m/s) for the reference points A4 and C4 (13.70%), and A3 and B4 (13.60%). The values gradually decrease for B2 and D2, the rated capacity being in the range 5.60–6.30%, the lowest value being reported for C2 (5%).

The capacity factor (C_f), which is used to determine the efficiency of a particular generator, is presented in Figure 9a, based on the values corresponding to the 20-year time interval 2000–2019. Closer to shore, near the reference sites A2, B2, C2, and D2, a single turbine's capacity factor is in the extent of 16.30% (T2) and 34.50% (T11) and can reach up to 41.20% (T11) close to the 50 m contour line, near the A3 site. However, there is a slight improvement regarding the percentage (not even 1%) as the distance from the shore increases. For example, the C_f increases from 40.10% near reference site B3 to 40.80% near B4 for the system T11 and from 37.50% to 38.10% for T12, respectively. Nevertheless, given the fact that it has been considered that the wind measurements used for these calculations underestimate the wind power potential by 13.52%, Figure 9b illustrates the results by considering an adjustment of the *U*100 time series (each value) by 13.52%. Thus, a maximum value of 48.40% is registered for turbine T11 operating in B3 and 46.10% for T12. Regarding A3, where the highest values of C_f are located, the adjusted results report a

49.50% for T11 and 47.20% for T12. On the other hand, C2 and D2 sites represent the less attractive solutions for a wind project, regardless of the type of generator chosen.

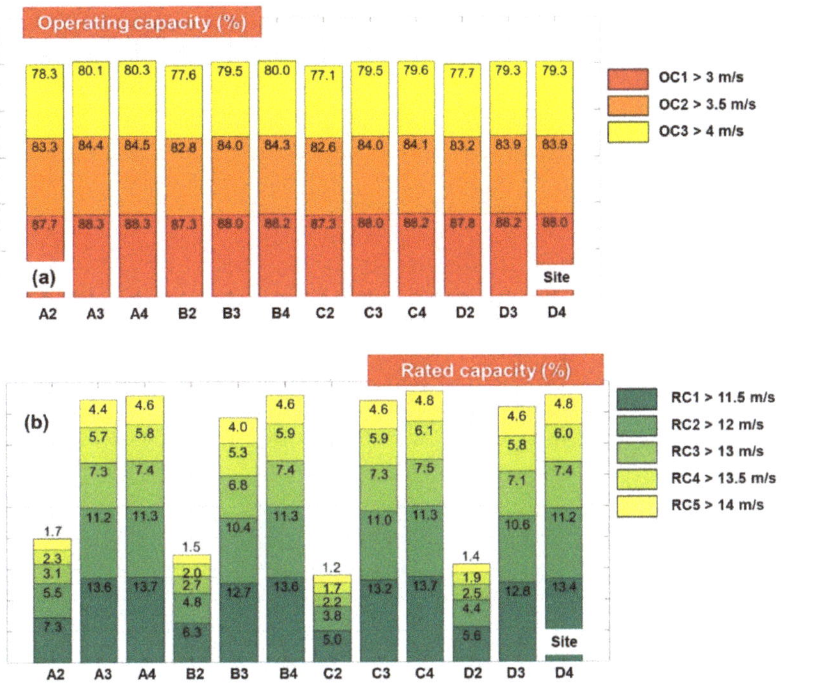

Figure 8. Performances of the selected wind turbines, for the time interval from January 2000 to December 2019. Results indicated for: (**a**) Operating capacity; (**b**) rated capacity. The OC and RC selection are based on the most frequent cut-in and rated speed values indicated by the current offshore wind turbines (see Table 2).

The estimation of the AEP is shown in Figure 10, the values corresponding to a single wind turbine. The analysis of the total time data reveals that, generally, high rated-power turbines, such as T12-AMSC wt10000dd SeaTitan, will not exceed an output of 33.80 GWh/year for the two sites that stand out regarding the wind energy potential—A3 and A4—based on the ECMWF data considered for the interval 2000–2019. However, when considering an adjustment of 13.52% of the *U*100 time series, the AEP index for T12 could reach up to 41.30 GWh/year for both sites as highlighted in Figure 10b (A3 is situated at a distance of 49 km from shore and A4 at 100 km). The second best performance was attributed to T11, indicating sites A3, A4, B4, and C4 with the following results: 25.20, 25.30, 25, and 24.80 GWh/year, respectively. Nevertheless, upon an adjustment of 13.52%, T11 could reach a substantial 30.30 GWh/year value.

Capacity factor (%) (a)

Turbine	A2	A3	A4	B2	B3	B4	C2	C3	C4	D2	D3	D4
T1	32.1	38.6	38.7	30.7	37.6	38.3	28.5	38	38	29.3	37.2	37.6
T2	18.9	24.4	24.5	18	23.6	24.3	16.3	24.1	24.2	16.9	23.6	23.9
T3	25.2	31.7	31.8	24	30.7	31.4	21.8	31.2	31.2	22.6	30.5	30.8
T4	30.7	37.5	37.5	29.4	36.4	37.1	27.1	36.8	36.8	27.9	36.1	36.4
T5	26	33.2	33.2	24.6	32.1	32.8	22.1	32.6	32.5	23	31.7	32.1
T6	27.7	34	34.1	26.5	33	33.7	24.5	33.5	33.5	25.2	32.8	33.2
T7	23	28.9	29	21.9	28	28.7	20.1	28.5	28.6	20.7	27.9	28.2
T8	23	28.9	29	21.9	28	28.7	20.1	28.5	28.6	20.7	27.9	28.2
T9	25.8	31.9	32	24.7	31	31.7	22.8	31.4	31.5	23.5	30.8	31.1
T10	27.1	33.9	33.9	25.9	32.8	33.5	23.6	33.3	33.3	24.4	32.5	32.9
T11	34.5	41.2	41.2	33.2	40.1	40.8	30.8	40.5	40.4	31.6	39.7	40
T12	31.6	38.6	38.6	30.2	37.5	38.1	27.7	37.9	37.8	28.5	37	37.4

Site

Capacity factor adjusted (%) $U100 + 13.52\%$ (b)

Turbine	A2	A3	A4	B2	B3	B4	C2	C3	C4	D2	D3	D4
T1	40.5	46.8	46.8	39.1	45.7	46.3	36.6	46.1	46	37.5	45.3	45.5
T2	25.6	32	32	24.5	31	31.7	22.4	31.4	31.5	23.2	30.8	31.1
T3	33.3	40.1	40.1	31.9	39	39.7	29.4	39.4	39.3	30.3	38.6	38.9
T4	39.3	45.9	45.9	37.9	44.8	45.4	35.4	45.2	45.1	36.2	44.3	44.6
T5	34.9	42.1	42	33.4	40.9	41.5	30.6	41.3	41.2	31.5	40.4	40.7
T6	35.6	42.2	42.2	34.3	41.1	41.8	31.9	41.5	41.4	32.8	40.7	41
T7	30.3	36.9	36.9	29.1	35.8	36.5	26.8	36.2	36.2	27.6	35.5	35.9
T8	30.3	36.9	36.9	29.1	35.8	36.5	26.8	36.2	36.2	27.6	35.5	35.9
T9	33.4	39.9	39.9	32.1	38.9	39.5	29.8	39.3	39.3	30.7	38.5	38.9
T10	35.6	42.4	42.4	34.2	41.2	41.9	31.6	41.7	41.6	32.5	40.8	41.1
T11	43.1	49.5	49.5	41.8	48.4	49	39.2	48.8	48.6	40.1	47.9	48.2
T12	40.5	47.2	47.2	39.1	46.1	46.7	36.4	46.4	46.3	37.3	45.6	45.8

Site

Figure 9. Capacity factor computed for different wind turbines and sites corresponding to the 20-year time interval 2000–2019. The results are indicated in terms of: (**a**) C_f—expected values based on ERA5 data; (**b**) adjusted C_f by considering an adjustment of the $U100$ time series (each value) by 13.52%.

According to the previous results, two types of wind turbines stand out considering both capacity factor and the annual electricity production, corresponding to the shoreline and offshore sites that are most suitable for developing wind projects. Based on the results available for the time interval from January 2000 until December 2019, the top five sites and the two systems that are the most assuring regarding the wind power potential in the Romanian coastal zone are listed in Table 5. As expected, the Samsung S7.0-171 (T11) has the highest capacity factor for the sites located offshore, namely A3, A4, B4, C3, and C4, with values of the C_f in the range of 40.40–41.20%. Moreover, the AMSC wt10000dd SeaTitan (T12) reported the highest AEP, ranging from 33.10 to 33.80 GWh for sites C4 and A3, respectively. All in all, A3 seems to be the most attractive solution since the other four sites are defined by a higher water depth (>50 m). However, the installation of floating wind turbines might be a viable solution in areas with water depths that do not allow the implementation of fixed systems.

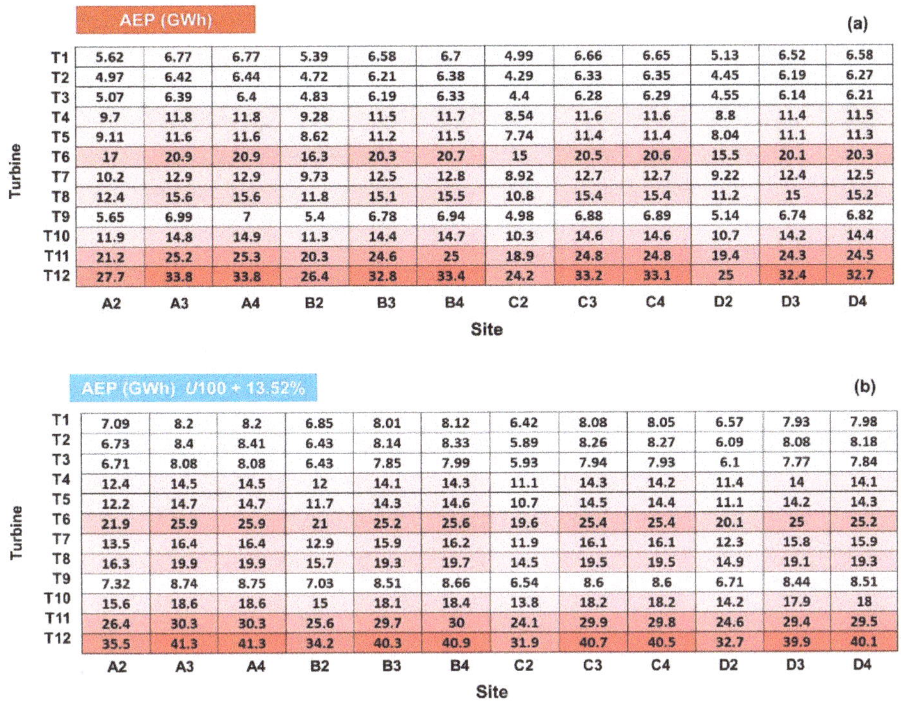

Figure 10. Annual Electricity Production (in GWh) expected to be provided by the wind turbines considered for the interval 2000–2019. Results indicated in terms of: (**a**) AEP—expected values based on ERA5 data; (**b**) adjusted AEP by considering an adjustment of the *U*100 time series (each value) by 13.52%.

Table 5. Top five capacity factor and AEP corresponding to the shoreline and offshore sites, excluding the points A1, B1, C1, and D1. Results available for the time interval from January 2000 until December 2019.

Site/Turbine	**A3—T11**	**A4—T11**	**B4—T11**	**C3—T11**	**C4—T11**
Capacity factor (%)	41.20	41.20	40.80	40.50	40.40
Site/Turbine	**A3—T12**	**A4—T12**	**B4—T12**	**C3—T12**	**C4—T12**
AEP (GWh)	33.80	33.80	33.40	33.20	33.10

Since there is currently no operational offshore wind project in the Black Sea zone, the purpose of the following section is to identify the expected AEP (in GWh) of a wind farm that replicates the Fantanele–Cogealac project (600 MW capacity) for the 12 reference sites considered. Thus, Figure 11 presents the equivalent number of necessary wind turbines to cover the Fantanele–Cogealac project and the estimated AEP generated. As expected, only 60 system AMSC wt10000dd SeaTitan will ensure A3 and A4 sites an AEP of 2028 GWh, while 86 turbine Samsung S7.0-171 will provide for the same sites 2167 GWh. As assumed, the number of turbines significantly increases as the rated power drops. Hence, for system T1, 300 turbines will be necessary to guarantee 2031 GWh for site A3. By studying the current European offshore sector, the tendency for the marine environment is to adopt large-scale wind turbines for cost-effectiveness purposes. Consequently, systems with low-rated capacities will no longer be considered an attractive solution.

Equivalent number of turbines required to cover the Fantanele-Cogealac project (600 MW)

Turbine \ Site	A2	A3	A4	B2	B3	B4	C2	C3	C4	D2	D3	D4
T1 (300)	1686	2031	2031	1617	1974	2010	1497	1998	1995	1539	1956	1974
T2 (200)	994	1284	1288	944	1242	1276	858	1266	1270	890	1238	1254
T3 (261)	1323	1668	1670	1261	1616	1652	1148	1639	1642	1188	1603	1621
T4 (167)	1620	1971	1971	1550	1921	1954	1426	1937	1937	1470	1904	1921
T5 (150)	1367	1740	1740	1293	1680	1725	1161	1710	1710	1206	1665	1695
T6 (86)	1462	1797	1797	1402	1746	1780	1290	1763	1772	1333	1729	1746
T7 (118)	1204	1522	1522	1148	1475	1510	1053	1499	1499	1088	1463	1475
T8 (98)	1215	1529	1529	1156	1480	1519	1058	1509	1509	1098	1470	1490
T9 (240)	1356	1678	1680	1296	1627	1666	1195	1651	1654	1234	1618	1637
T10 (120)	1428	1776	1788	1356	1728	1764	1236	1752	1752	1284	1704	1728
T11 (86)	1823	2167	2176	1746	2116	2150	1625	2133	2133	1668	2090	2107
T12 (60)	1662	2028	2028	1584	1968	2004	1452	1992	1986	1500	1944	1962

Figure 11. Expected AEP (in GWh) indicated by a wind farm that replicated the Cogealac–Fantanele project (600 MW capacity). Results corresponding to the interval 2000–2019, where, for example, T2 (200) means that from 200 wind turbines (V100–3.0) a total of 994 GWh will be generated for the site A2.

4. Discussion

According to the Statistical Review of World Energy [33], Romania's primary energy consumption and coal production had decreased from 2.80 EJ in 1989 to 0.20 EJ at the end of 2019 and from 2.60 EJ to 0.15 EJ, respectively. On the other hand, the electricity generation from renewable sources—hydroelectricity, solar, and wind—reveals an ascending trend, with the highest peak of 0.03 EJ recorded in 2017 [33]. Considering the time interval 2008–2018, however, the growth rate in wind power generation is more than 104% [33]. The foreseeable future of the energy market represents a constant topic of debate, especially in the EU countries, which constantly establish objectives and performance criteria that are currently stipulated in the European Green Deal, focusing on the near future, 2021–2050. Generally, the discussions concentrate on topics such as the effects of global warming, the strategies for ensuring energy security, and a sustainable future based upon net-zero greenhouse gas emissions. The most affordable way to support these policies is to identify renewable energy sources, such as those from the marine environment, where it is possible to develop large-scale projects. As discussed in this study, one of the most economical energy sources is offshore wind, where brand-new technical solutions have already been submitted, such as the emerging floating platforms that can operate in deep water.

The purpose of this work is to identify the most significant wind conditions in the Romanian coastal environment so that an offshore wind farm could become operational soon. The outcome of this research study is based on the ERA5 data, and from the assessment of the *U*100 index, it is noticeable that the wind assets are more attractive offshore than onshore, heedless of the period considered. Given that Fantanele–Cogealac, located in the Dobrogea region, is the largest onshore wind farm in Europe and represents approximately double the largest ever existing European wind farm (600 MW) is more than inspiring for the offshore wind market in Romania [34]. An important aspect that requires more detailed investigation is that offshore wind development across Europe, onshore grid access, and infrastructure pose serious challenges [20]. For example, any offshore wind farm developed in the Romanian waters will have to be connected to the grid in Dobrogea, where a significant part of the country's renewable power generation assets is already situated [20]. According to the NECP [7], some additional GW of renewable energy—primarily wind power—are expected to be produced in the already congested Dobrogea region.

In connection with some other related previous studies [3,21,22], this one embodies one of the first studies that evaluate the Romanian wind potential relying on the new ECMWF product—ERA 5. Even though it is used to estimate the resourcefulness of renewables, due to the relatively low resolution of the global model, it is reasonable to somehow underestimate the wind speed in some coastal zones (Mihai–Viteazu and Istria sites for the Black Sea—see Figure 3). Other constraints of the ERA5 datasets in the coastal areas of the North Sea are further highlighted in Kalverla et al. [35]. It is necessary to mention that most reanalysis datasets are provided at a standard height of 10 m [36], which is proper for evaluating the wind conditions from a meteorological point of view. A significant part of the previous studies evaluated the wind conditions in the Black Sea coastal zone at a height of 80 m ($U80$), where older-generation wind turbines used to operate [17]. The innovative feature of ERA5 is that it provides wind data directly at 100 m, admitting that this is a reference height for the offshore industry [7,37]; therefore, there is no need for further adjustments that could discredit the results.

On balance, considering the wind speed and water depth conditions, the study points out that the most suitable sites for the development of an offshore wind project appear to be located close to the 50 m contour line, thus bypassing any interference with shipping routes and port activities [38]. Furthermore, the figures indicated for the capacity factor of the most attractive offshore wind systems are in line with other values reported at the European level (around 40%). They could be even higher (almost 50%), acknowledging the underestimation of 13.52% of the wind speed data reported compared with the in situ measurements considered for this study (see Figure 3).

5. Conclusions

In the present work, the Romanian coastal zone's wind energy resources from the Black Sea region were evaluated by considering some environmental parameters (such as $U100$) and some relevant technical features of some state-of-the-art wind turbines that either operate or are currently implemented within several offshore wind projects. Based on the ERA5 wind datasets considering the 42-year time interval from January 1979 to December 2020, it was possible to outline the dominant energy patterns of the target area, thus revealing the most appropriate hot spots for developing renewable wind energy designs. The results reveal an upward trend concerning relevant wind conditions when approaching the offshore zone from the land, implying that an offshore project might become a certainty shortly.

During wintertime, which is the most energetic season, the average wind speed can frequently reach and overcome 9 m/s, values that are related to the International Electrotechnical Commission (IEC) classes II and III. The present work reveals that higher wind resources are characteristic in the northern part of the target area, where the capacity factor of the Samsung S7.0-171 system frequently registers a value of about 40% and can reach up to 41.20%, close to the 50 m depth contour line, near the A3 site, which is situated 49 km from the shore. Considering the proximity of some of the attractive sites for developing a wind project to the Port of Constanta and its facilities, these sites become compatible in terms of accessibility for both the development process and operation and maintenance activities. For example, opposed to fixed offshore turbines, floating platforms require constructing a different infrastructure for their installation, since elements are regularly assembled onshore and then towed to the offshore site. Moreover, floating turbines must be brought back to ports for some critical operations and maintenance activities. From this perspective, Romania's Port of Constanta can play a significant role in developing floating wind turbines in the Black Sea coastal zone. Furthermore, it could represent an enabling factor for tapping into uncharted territories represented by offshore fields, thus generating new business opportunities and jobs.

To obtain a comprehensive picture of the wind energy potential, an entire portfolio of offshore wind turbines had to be assessed to identify the most representative systems for this geographical region. Thus, another direction of this study was related to the evaluation of the performances provided by some wind turbines covering a full spectrum of rated capacities, in the range of 2–10 MW, being in line with other values reported from offshore wind farms at the European level.

To sum up, it has been proved that the Romanian coastal zone represents a viable solution for the development of an offshore wind project. The western part of the Black Sea basin has significant wind supplies that require precise quantification to authenticate the expected performances of the offshore wind turbines. As outlined in this study, the global reanalysis wind datasets underestimate the wind speed by at least 11.57%, and this value increases as we approach the coastline and go further in the offshore sector. Thus, the results achieved in terms of wind power potential and the evaluation of the wind turbine's parameters might not match the reality, the potential of the target area being much more substantial. Therefore, the present work should encourage researchers and investors to understand the limitations of the wind datasets by performing direct comparisons against in situ measurements.

Author Contributions: A.G. performed the literature review, wrote the final form of the manuscript and drew the conclusions. F.O. processed the wind data, carried out the statistical analysis and interpreted the results. E.R. guided the entire research and provided the needed resources. All authors have read and agreed to the published version of the manuscript.

Funding: This work was carried out in the framework of the research project DREAM (Dynamics of the REsources and technological Advance in harvesting Marine renewable energy), supported by the Romanian Executive Agency for Higher Education, Research, Development, and Innovation Funding—UEFISCDI, grant number PN–III–P4–ID–PCE–2020–0008.

Institutional Review Board Statement: Not applicable.

Data Availability Statement: The data used in this study are openly available. ERA5 data used in this study were obtained from the ECMWF data server.

Acknowledgments: The results of this work will also be presented to the 9th edition of the Scientific Conference organized by the Doctoral Schools of "Dunărea de Jos" University of Galati (SCDS-UDJG) http://www.cssd-udjg.ugal.ro/ (accessed on 3 June 2021) that will be held on 10th and 11th of June 2021, in Galati, Romania.

Conflicts of Interest: The authors declare no conflict of interest.

References

1. European Commission. A Clean Planet for All. A European Strategic Long–Term Vision for a Prosperous, Modern, Competitive and Climate Neutral Economy. In *Depth Analysis in Support of the Commission Communication*; European Commission: Brussels, Belgium, 2018.
2. Future of Wind. Deployment, Investment, Technology, Grid Integration and Socio–Economic Aspects. Available online: https://irena.org/-/media/Files/IRENA/Agency/Publication/2019/Oct/IRENA_Future_of_wind_2019 (accessed on 15 May 2021).
3. Onea, F.; Rusu, E. A Study on the Wind Energy Potential in the Romanian Coastal Environment. *J. Mar. Sci. Eng.* **2019**, *7*, 142. [CrossRef]
4. Raileanu, A.; Onea, F.; Rusu, E. An Overview of the Expected Shoreline Impact of the Marine Energy Farms Operating in Different Coastal Environments. *J. Mar. Sci. Eng.* **2020**, *8*, 228. [CrossRef]
5. Rusu, E. An Evaluation of the Wind Energy Dynamics in the Baltic Sea, Past and Future Projections. *Renew. Energy* **2020**, *160*, 350–362. [CrossRef]
6. Rusu, E.; Onea, F. An Assessment of the Wind and Wave Power Potential in the Island Environment. *Energy* **2019**, *175*, 830–846. [CrossRef]
7. NECP. *Planul National Integrat in Domeniul Energiei Si Schimbarilor Climatice 2021–2030*. COM, Romania. 2020. Available online: https://ec.europa.eu/energy/sites/ener/files/documents/ro_final_necp_main_ro.pdf (accessed on 25 May 2021).
8. Dragomir, G.; Serban, A.; Nastase, G.; Brezeanu, A.I. Wind Energy in Romania: A Review from 2009 to 2016. *Renew. Sustain. Energy Rev.* **2016**, *64*, 129–143. [CrossRef]
9. Rusu, L.; Raileanu, A.; Onea, F. A Comparative Analysis of the Wind and Wave Climate in the Black Sea along the Shipping Routes. *Water* **2018**, *10*, 924. [CrossRef]
10. Ganea, D.; Mereuta, E.; Rusu, L. Estimation of the Near Future Wind Power Potential in the Black Sea. *Energies* **2018**, *11*, 3198. [CrossRef]
11. Rusu, E. A 30–Year Projection of the Future Wind Energy Resources in the Coastal Environment of the Black Sea. *Renew. Energy* **2019**, *139*, 2283. [CrossRef]
12. Onea, F.; Rusu, E. Efficiency Assessments for Some State-of-the-Art Wind Turbines in the Coastal Environments of the Black and the Caspian Seas. *Energy Explor. Exploit.* **2016**, *34*, 217–234. [CrossRef]

13. Onea, F.; Rusu, E. Wind Energy Assessments along the Black Sea Basin. *Meteorol. Appl.* **2014**, *21*, 316–329. [CrossRef]
14. Onea, F.; Rusu, L. Evaluation of Some State-Of-The-Art Wind Technologies in the Nearshore of the Black Sea. *Energies* **2018**, *11*, 2452. [CrossRef]
15. Argin, M.; Yerci, V. The Assessment of Offshore Wind Power Potential of Turkey. In Proceedings of the 2015 9th International Conference on Electrical and Electronics Engineering (ELECO), Bursa, Turkey, 26–28 November 2015; pp. 966–970.
16. Raileanu, A.B.; Onea, F.; Rusu, E. Evaluation of the Offshore Wind Resources in the European Seas Based on Satellite Measurements. In Proceedings of the International Multidisciplinary Scientific Geo Conference SGEM, Albena, Bulgaria, 16–25 June 2015; pp. 227–234.
17. Onea, F.; Rusu, L. An Evaluation of the Wind Energy in the North-West of the Black Sea. *Int. J. Green Energy* **2014**, *11*, 465–487. [CrossRef]
18. IEA. Offshore Wind Outlook 2019, Paris. Available online: https://www.Iea.Org/Reports/Offshore-Wind-Outlook-2019 (accessed on 15 May 2021).
19. WindEurope. Offshore Wind in Europe—Key Trends and Statistics. 2019. Available online: https://www.iea.org/reports/offshore-wind-outlook-2019 (accessed on 15 May 2021).
20. *Romania's Offshore Wind Energy Resources: Natural Potential, Regulatory Framework, and Development Prospects*; EPG: Bucharest, Romania, 2020.
21. Onea, F.; Rusu, L. Offshore Wind Energy and the Romanian Energy Future. *E3S Web Conf. ICACER* **2019**, *103*, 001004. [CrossRef]
22. Onea, F.; Rusu, L. Assessment of the Romanian Onshore and Offshore Wind Energy Potential. *E3S Web Conf. REEE* **2019**, *122*, 01003. [CrossRef]
23. Girleanu, A.; Onea, F.; Rusu, E. The Efficiency and Coastal Protection Provided by a Floating Wind Farm Operating in the Romanian Nearshore. In Proceedings of the 6th International Conference on Clean Energy Research, ICACER 2021, Barcelona, Spain, 15–17 April 2021.
24. Hersbach, H.; Bell, B.; Berrisford, P.; Horanyi, A.; Sabater, J.M.; Nicolas, J.; Radu, R.; Schepers, D.; Simmons, A.; Soci, C.; et al. Global reanalysis: Goodbye ERA–Interim, hello ERA5. *ECMWF Newsl.* **2019**, *159*, 17–24.
25. WindEurope. Offshore Wind in Europe—Key Trends and Statistics. 2020. Available online: https://windeurope.org/intelligence-platform/product/offshore-wind-in-europe-key-trends-and-statistics-2020/ (accessed on 15 May 2021).
26. Welcome to Wind-Turbine-Models. Available online: https://en.wind-turbine-models.com/ (accessed on 1 June 2021).
27. Wen, Y.; Kamranzad, B.; Lin, P. Assessment of Long-Term Offshore Wind Energy Potential in the South and Southeast Coasts of China Based on a 55-Year Dataset. *Energy* **2021**, *224*, 120225. [CrossRef]
28. Salvacao, N.; Guesdes Soares, C. Wind Resource Assessment Offshore the Atlantic Iberian Coast with the WRF Model. *Energy* **2018**, *145*, 276–287. [CrossRef]
29. Al–Nasaar, W.K.; Neelamani, S.; Al-Salem, K.A.; Al–Dashti, H.A. Feasibility of Offshore Wind Energy as a Model. *Energy* **2019**, *169*, 783–796. [CrossRef]
30. *Project Description—Wind Parks Mihai Viteazu I–IV (80 MW)*; Nek Umwelttechnik AG: Constanta, Romania, 2010.
31. Oh, K.Y.; Kim, J.Y.; Lee, J.K.; Ryu, M.S.; Lee, J.S. An Assessment of Wind Energy Potential at the Demonstration Offshore Wind Farm in Korea. *Energy* **2020**, *46*, 555–563. [CrossRef]
32. Xie, K.; Yang, H.; Hu, B.; Li, C. Optimal Layout of a Wind Farm Considering Multiple Wind Directions. In Proceedings of the 2014 International Conference on Probabilistic Methods Applied to Power Systems (PMAPS), Durham, UK, 7–10 July 2014.
33. Statistical Review of World Energy. Energy Economics. Available online: https://www.bp.com/en/global/corporate/energy-economics/statistical-review-of-world-energy.html (accessed on 19 March 2021).
34. Girleanu, A.; Rusu, E.; Onea, F. An Insight into the Energy Market in the Context of Climate Change and the European Gre Green Deal. In Proceedings of the International Multidisciplinary Scientific GeoConference SGEM, Albena, Bulgaria, 2021; under review.
35. Kalverla, P.C.; Holtslag, A.A.M.; Ronda, R.J.; Steeneveld, G.J. Quality of Wind Characteristics in Recent Wind Atlases over the North Sea. *Meteorol. Soc.* **2020**, *146*, 1498–1515. [CrossRef]
36. Onea, F.; Rusu, L. Long–Term Analysis of the Black Sea Weather Windows. *J. Mar. Sci. Eng.* **2019**, *7*, 303. [CrossRef]
37. ERA5: How Are the 100m Winds Calculated, Copernicus User Support Forum—ECMWF Confluence Wiki. Available online: https://confluence.ecmwf.int/ (accessed on 29 April 2021).
38. Raileanu, A.; Onea, F.; Rusu, E. Implementation of Offshore Wind Turbines to Reduce Air Pollution in Coastal Areas—A Case Study of Constanta Harbour in the Black Sea. *J. Mar. Sci. Eng.* **2020**, *7*, 550. [CrossRef]

inventions

MDPI

Article

Roller Profiling for Generating the Screw of a Pump with Progressive Cavities

Nicuşor Baroiu, Georgiana-Alexandra Moroşanu *, Virgil-Gabriel Teodor and Nicolae Oancea

Faculty of Engineering, Dunarea de Jos University of Galati, 47 Domnească Street, 800008 Galati, Romania; nicusor.baroiu@ugal.ro (N.B.); virgil.teodor@ugal.ro (V.-G.T.); nicolae.oancea@ugal.ro (N.O.)
* Correspondence: alexandra.costin@ugal.ro; Tel.: +40-7-6598-5755

Abstract: Progressive cavity pumps are used in industry for the circulation of high viscosity fluids, such as crude oil and petroleum products, sewage sludge, oils, salt water, and wastewater. Also known as single screw pumps, these pumps are composed of a single rotor which has the shape of a rounded screw, which moves inside a rubber stator. The stator has an double helical internal surface which, together with the helical surface of the rotor, creates a cavity that moves along the rotor. The movement effect of the cavity inside the stator is the movement of the fluid with a constant flow and high pressure. In this paper, an algorithm for profiling the rollers for generating the helical surface of the pump rotor with progressive cavities is proposed. These rollers are constituted as tools for the plastic deformation of the blank (in case the pump rotor is obtained by volumetric deformation) or for its superficial hardening.

Keywords: incremental pumps; profiling of generating tools; helical surfaces

Citation: Baroiu, N.; Moroşanu, G.-A.; Teodor, V.-G.; Oancea, N. Roller Profiling for Generating the Screw of a Pump with Progressive Cavities. *Inventions* **2021**, *6*, 34. https://doi.org/10.3390/inventions6020034

Academic Editors: Eugen Rusu and Gabriela Rapeanu

Received: 20 April 2021
Accepted: 12 May 2021
Published: 14 May 2021

1. Introduction

Pumps with progressive cavities allow the circulation of fluids with high viscosity at a constant flow and high pressure [1–3]. Among the most used progressive cavity pumps are single screw pumps, consisting of an external helical rotor that rotates eccentrically inside a double internal helical stator [4,5]. The active surface of the rotor is a cylindrical helical surface of constant pitch, and the stator, whose active surface is a double helical surface, is also cylindrical and with constant pitch. The stator is usually made of rubber [6,7], and the rotor is made of steel, often superficially hardened. The exploitation activity of the pumps with progressive cavities has highlighted the fact that traditional pumps with a rubber stator, which surrounds the rotor, do not allow the existence of an interstice between them, which limits their activity until the stator wears out [8,9]. Alternatively, a new class of pumps that do not use elastomers has been developed, in which the rotor and the stator are made of metal, and are used in drilling activities in high temperature wells [10–13]. This eliminates the wear and greatly increases the lifetime of the pump for use with viscous, high density, and abrasive liquids, and with liquids containing material in suspension. To ensure proper operation, there must be a space between the rotor and the stator to ensure the leakage of the fluid. If this space is not well defined in the rotor and stator design stage, the efficiency of the pump will be seriously affected. Consequently, dimensional designs and optimizations of the rotor and stator, respectively, using analytical and dynamic simulations of the fluid flow through these pumps, have previously been explored [14–18].

The productive functioning of pumps with progressive cavities depends on proper design of the rotor profiles, respecting the technical conditions of their form. The cross-section form of helical pump rotors as ensembles of profiles associated with the rolling centrodes are determined based on the fundamental theorems of the enveloping surfaces (curves) and calculated based on Olivier's first theorem, the Gohman general theorem, or the Willis theorem (normals method) [19,20].

The geometric characteristics of the rotor, namely, the helical surface with long length and rounded profile, determine that it can be processed by volumetric deformation, using tools in the form of profiled rollers with axes parallel to the axis of the processing blank. The algorithm by which the tool profile can be determined contains the following steps: defining the analytical equations in the own reference system of the piece of the generated profile; defining the absolute movements of the piece and the blank during the generating process; defining the relative movements between the piece and the tool; calculation of the trajectories of the points on the piece profile that execute a relative movement towards the tool; defining the condition which allows the determination of these points, called the enveloping condition—these points belong to the trajectory's family and to the enveloping of that family; and association of the enveloping condition at the equations of the trajectory's family, a connection that allows the determination of the parametric equations of the tool profile [21,22]. The tool's profiling for generating surfaces of the rotors and the stators of the pump can start either from a physical model or from 3D models obtained by rapid prototyping. In practical activity, the determination of the rotor and the stator form can be undertaken with the help of measuring machines in high precision coordinates when the axial profiles of the generating tools in numerical form are known [23]. By comparison, these axial and graphical profiles can be determined using facilities of programming and graphical representation environments, such as the CATIA software application [24].

2. Materials and Methods

2.1. Generating Principle

The processing scheme involves the generating of a cylindrical helical surface of constant pitch, using helical deforming tools, with axes parallel to the axis of the deformed screw.

Figure 1 shows the two rolling centrodes: C_1—the centrode of the blank, having the radius R_{rp}, and C_2—the centrode of the tool of radius R_{rs}.

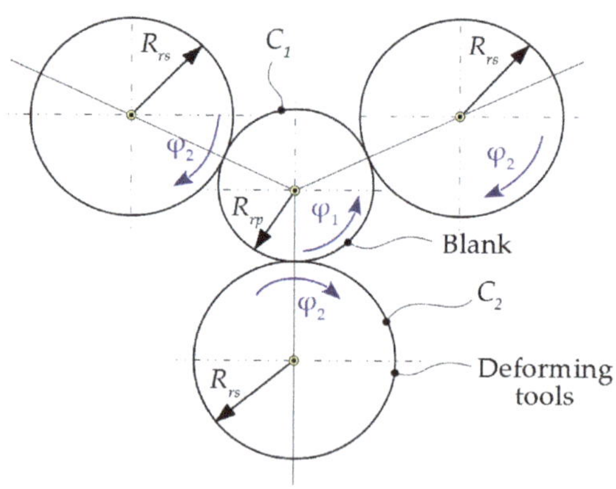

Figure 1. Generating principle. Rolling centrodes: C_1; C_2. C_1—the centrode of the blank, having the radius R_{rp}, and C_2—the centrode of the tool of radius R_{rs}.

Note that although three tools are used in the generating process to ensure the balance of large radial loads during processing, it is sufficient to consider a single tool for the tool profiling. The active profile is identical for all three tools being processed simultaneously. In the following, reference will be made to a single generating tool.

The rolling condition between the C_1 and C_2 centrodes has the form:

$$R_{rp} \cdot \varphi_1 = R_{rs} \cdot \varphi_2, \quad (1)$$

where, noting $i = \frac{R_{rp}}{R_{rs}}$, is obtained:

$$\varphi_2 = i \cdot \varphi_1. \quad (2)$$

Because the plastic deformation process occurs in the cross-section of the blank, the study of the enwrapping between the two helical surfaces can be performed as for a plane enveloping process in a common plane, perpendicular to the two axes of the tool, with respect to the blank.

To ensure the generation of the active surface of the pump screw, it is necessary that the two helical surfaces, the tool and the blank, have opposite directions and have the same inclination to their own axes on the rolling circles.

Noting the inclination angle of the helix to its own axis with ω, the above condition has the mathematical form given by Equation (3):

$$tg\omega = \frac{p_{et}}{R_{rt}} = \frac{p_{ep}}{R_{rp}}, \quad (3)$$

where p_{et} represents the helical parameter of the tool and p_{ep} is the helical parameter of the piece; also see Figure 2.

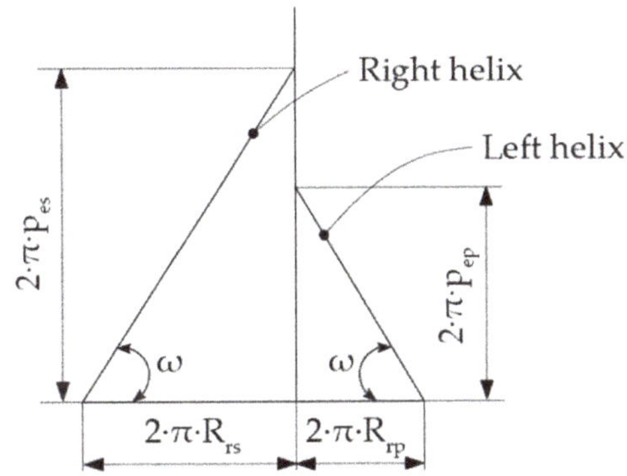

Figure 2. The slope angle of the two helical surfaces.

2.2. The Equations of the Active Surface of the Rotor

2.2.1. Cross-Section of the Screw

The cross-section of the active surface of the rotor is a circle of radius R, eccentric to the C_1 centrode, with e size.

The reference systems used for the active surface determination are presented in Figure 3.

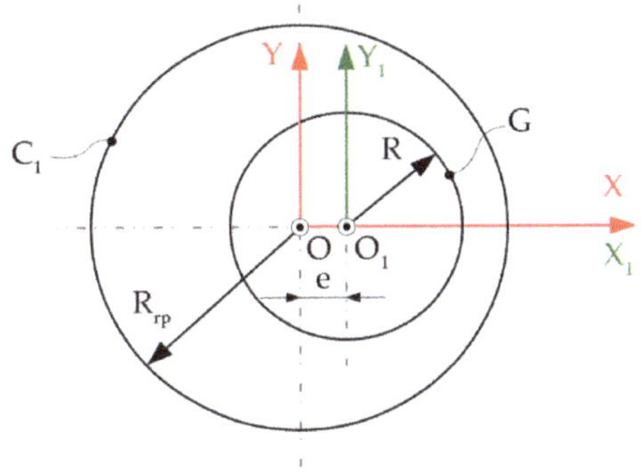

Figure 3. Reference systems used for active surface determination.

To solve the profiling problem of the generating tool, the following reference systems are used: xyz is a fixed reference system, with its origin in the center of the blank centrode and the z-axis superimposed on the rotational axis of the blank; $x_1y_1z_1$ is a fixed reference system, with its origin in the center of the tool centrode and the z_1 axis superimposed on the rotational axis of the tool; XYZ is a mobile reference system, joined with the blank—this system performs a rotational movement with the blank around the $Z \equiv z$ axis; $\xi\eta\zeta$ is a mobile reference system, joined with the tool—the axis $y \equiv z_1$ and the system has a rotational movement around it.

The generating profile of the active helical surface of the screw, the surface Σ, is given by the equations:

$$G : \left| \begin{array}{c} X(\beta) = R \cdot \cos\beta + e; \\ Y(\beta) = R \cdot \sin\beta; \\ Z(\beta) = 0. \end{array} \right. \tag{4}$$

2.2.2. Helical Surface

The helical movement of the G-profile, the Z-axis, and the p_{ep} helical parameter, generates the Σ surface:

$$\omega_3(\varphi_1) \cdot X + C, C = \begin{pmatrix} 0 \\ 0 \\ p_{ep} \cdot \varphi_1 \end{pmatrix}. \tag{5}$$

By development, the following is obtained:

$$\Sigma : \left| \begin{array}{c} x(\beta) = R \cdot \cos\beta \cdot \cos\varphi_1 + R \cdot \sin\beta \cdot \sin\varphi_1 + e \cdot \cos\varphi_1; \\ y(\beta) = -R \cdot \sin\beta \cdot \sin\varphi_1 + R \cdot \sin\beta \cdot \cos\varphi_1 - e \cdot \sin\varphi_1; \\ z(\beta) = p_{ep} \cdot \varphi_1. \end{array} \right. \tag{6}$$

2.3. *Determination of the Virtual Pole Position*

The determination of the virtual pole position is made from the condition that the normal to the profile of the piece intersects the centrode associated with it.

The profile of the piece has the equations given by (4), so its direction will be:

$$\vec{n} = \begin{vmatrix} \vec{i} & \vec{j} \\ \dot{X}_\beta & \dot{Y}_\beta \end{vmatrix} = \dot{Y}_\beta \cdot \vec{i} - \dot{X}_\beta \cdot \vec{j}, \tag{7}$$

with $\dot{X}_\beta = -R \cdot \sin\beta$ and $\dot{Y}_\beta = R \cdot \cos\beta$, resulting in:

$$\vec{n} = R \cdot \cos\beta \cdot \vec{i} + R \cdot \sin\beta \cdot \vec{j}. \tag{8}$$

In the current point, the normal to the profile can be written as:

$$\vec{N} = \left(X_i + \lambda \cdot \dot{Y}_\beta\right) \cdot \vec{i} + \left(Y_i + \lambda \cdot \dot{X}_\beta\right) \cdot \vec{j}, \tag{9}$$

with variable scalar λ.

By development, the following is obtained:

$$\vec{N} = (R \cdot \cos\beta + e + \lambda \cdot R \cdot \cos\beta) \cdot \vec{i} + (R \cdot \sin\beta + \lambda \cdot R \cdot \sin\beta) \cdot \vec{j}. \tag{10}$$

The C_1 centrode has the equations:

$$C_1 : \left| \begin{array}{l} X = R_{rp} \cdot \cos\varphi_1; \\ Y = R_{rp} \cdot \sin\varphi_1. \end{array} \right. \tag{11}$$

According to the virtual pole method [17,18], it is located at the intersection of the normal at the G profile at the current point with the centrode associated with the piece. The intersection condition becomes:

$$\begin{cases} R\cdot\cos\beta+e+\lambda\cdot R\cdot\cos\beta=R_{rp}\cdot\cos\varphi_1; \\ R\cdot\sin\beta+\lambda\cdot R\cdot\sin\beta=R_{rp}\cdot\sin\varphi_1. \end{cases} \tag{12}$$

By removing the parameter λ from Equation (12), the following is obtained:

$$\lambda=\frac{R_{rp}\cdot\cos\varphi_1-R\cdot\cos\beta-e}{R\cdot\cos\beta}=\frac{R_{rp}\cdot\sin\varphi_1-R\cdot\sin\beta}{R\cdot\sin\beta} \tag{13}$$

Or:

$$\begin{gathered} R_{rp}\cdot\cos\varphi_1\cdot\sin\beta-R\cdot\cos\beta\cdot\sin\beta-e\cdot\sin\beta=R_{rp}\cdot\sin\varphi_1\cdot\cos\beta-R\cdot\sin\beta\cdot\cos\beta\Rightarrow \\ \Rightarrow e\cdot\sin\beta=R_{rp}\cdot(\cos\varphi_1\cdot\sin\beta-\sin\varphi_1\cdot\cos\beta)\Rightarrow e\cdot\sin\beta=-R_{rp}\cdot\sin(\varphi_1-\beta)\Rightarrow \\ \Rightarrow \sin(\varphi_1-\beta)=\frac{-e\cdot\sin\beta}{R_{rp}}\Rightarrow\varphi_1=\beta+\arcsin\left(\frac{-e\cdot\sin\beta}{R_{rp}}\right). \end{gathered} \tag{14}$$

2.4. Contact Curve Determination

The determination of the contact curve equations is undertaken by applying the absolute movement of the piece (the angle rotation φ_1 around the origin O) to the current point.

When the current point passes through the entire G profile, the curve obtained represents the contact curve.

If the current point is denoted with T, its coordinates in the own system of the piece will be:

$$T:\left|\begin{array}{l} X_T=R\cdot\cos\beta+e; \\ Y_T=R\cdot\sin\beta; \\ Z_T=P_{ep}\cdot\beta. \end{array}\right. \tag{15}$$

When β goes through the entire domain of values, from 0 to $2\cdot\pi$, the current point goes through the entire contact curve.

At some point, the coordinates of the current point in the fixed system xyz will be given by:

$$x=\omega_3(\varphi_1)\cdot X_T \tag{16}$$

which can be developed as:

$$T\left|\begin{array}{l} x_T=X_T\cdot\cos\varphi_1+Y_T\cdot\sin\varphi_1; \\ y_T=-X_T\cdot\sin\varphi_1+Y_T\cdot\cos\varphi_1; \\ z_T=P_{ep}\cdot\varphi_1. \end{array}\right. \tag{17}$$

2.5. The Determination of the Cross-Section of the Active Surface

The determination of the cross-sectional profile of the active surface of the helical tool is undertaken by applying a movement corresponding to the absolute movement of the tool to each point determined on the contact curve.

The determination of the cross-section can be made if the problem is treated as a plane enwrapping problem, in a plane perpendicular to the rotational axes of the two elements, the X-axis of the blank, and the ζ axis of the tool.

The absolute movement of the tool is given by the transformation:

$$x_1=\omega_3^T(\varphi_2)\cdot\xi\Rightarrow\xi=\omega_3(\varphi_2)\cdot x_1. \tag{18}$$

Between the movement of the tool and that of the blank, Equation (2) is respected by the rolling condition.

Knowing the coordinates of the points on the contact curve in the *xyz* system, in the absolute movement of the $\xi\eta\zeta$ system, these are given by the following transformations:

$$x_1 = x + B, \tag{19}$$

where: $B = \begin{pmatrix} -A_{12} \\ 0 \\ 0 \end{pmatrix}$; $A_{12} = R_{rp} + R_{rt}$.

$$\xi = \omega_3(\varphi_2) \cdot x_1 = \omega_3(\varphi_2) \cdot [x + B] + C, \tag{20}$$

where: $C = \begin{pmatrix} 0 \\ 0 \\ p_{et} \cdot \varphi_2 \end{pmatrix}$.

By development, the following is obtained:

$$S: \left| \begin{array}{c} \xi_T = (x_T - A_{12}) \cdot \cos\varphi_2 + y_T \cdot \sin\varphi_2; \\ \eta_T = -(x_T - A_{12}) \cdot \sin\varphi_2 + y_T \cdot \cos\varphi_2; \\ \zeta_T = p_{et} \cdot \varphi_2. \end{array} \right. \tag{21}$$

The cross-section of the tool is obtained by intersecting the active surface with a plane perpendicular to the ζ axis, $\zeta = H$:

$$S.T. \left| \begin{array}{c} \xi_T = (x_T - A_{12}) \cdot \cos\varphi_2 + y_T \cdot \sin\varphi_2; \\ \eta_T = -(x_T - A_{12}) \cdot \sin\varphi_2 + y_T \cdot \cos\varphi_2. \end{array} \right. \tag{22}$$

3. Results

A numeric application was developed for the following values: R_{rp} = 30 mm; R_{rs} = 60 mm; R = 25 mm; $e = -5$ mm; $p_p = 2 \cdot \pi \cdot p_{ep} = 60$, the piece's pitch.

The numerical application was undertaken using the GNU Octave software program, a computer application for performing numerical calculations that is generally compatible with the MATLAB program.

The values table of the tool's profile and the contact curve are presented in Table 1 and graphically in Figure 4.

Table 1. The values table of the tool's profile and the contact curve.

No.	Tool's Profile			No.	Contact Curve		
	ξ [mm]	μ [mm]	ζ [mm]		x [mm]	y [mm]	z [mm]
1.	−70.000	0.000	0.000	1.	20.000	0.000	0.000
2.	−69.517	7.385	1.943	2.	20.093	0.287	1.943
3.	−68.093	14.581	3.878	3.	20.366	0.550	3.878
⋮	⋮	⋮	⋮	⋮	⋮	⋮	⋮
37.	70.000	0.000	60.000	37.	20.000	0.000	60.000
38.	69.517	−7.385	61.943	38.	20.093	0.287	61.943
39.	68.093	−14.581	63.878	39.	20.366	0.550	63.878
⋮	⋮	⋮	⋮	⋮	⋮	⋮	⋮
71.	−68.093	−14.581	116.120	71.	20.366	−0.550	116.120
72.	−69.517	−7.385	118.060	72.	20.093	−0.287	118.060
73.	−70.000	−0.000	120.000	73.	20.000	0.000	120.000

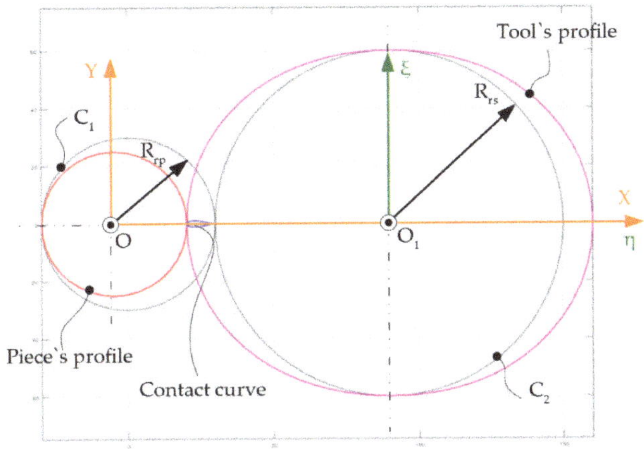

Figure 4. The tool's profile, the contact curve, and the profile of the piece.

Figure 4 shows an axial view of the piece profile, tool profile, and the contact curve (treated as a plane problem). In reality, the enwrapping surfaces are helical (three-dimensional) surfaces and Table 1 shows the 3D coordinates of the points belonging to the contact curve in the two reference systems: the tool system ($\xi\eta\zeta$) and the piece system (XYZ).

4. Conclusions

This paper presented an algorithm for profiling the rollers for processing the helical surfaces of a pump rotor with progressive cavities.

The algorithm is based on the virtual pole method, which is a variant of the normal theorem.

The new form of this theorem allows the determination of the enwrapping equations of a family of profiles generated in the rolling movement of two centrodes, eliminating the need to write the equations of relative movements between the generating profile and the profile to be generated. This eliminates an important source of possible errors and greatly simplifies the process of calculating the profile of the generating tool.

In future work, the aim is to apply the "virtual pole" method for the study of surfaces obtained by computer-aided inspection, using a three-dimensional measurement technique, namely, the reverse engineering concept. In other words, the surfaces will be scanned, using equipment based on a 3D optical scanner, without physical contact, existing within the Dunarea de Jos University of Galati, Faculty of Engineering. This device is an Atos Core, produced by GOM GmbH (Schmitzstraße 2, 38122, Braunschweig, Germany), which uses a modern and innovative technology for the fully automated inspection and measurement of the pieces. The scanning accuracy with the GOM 3D scanner is very high, almost entirely covering the surface of the pieces and providing all the details of the measured piece with great precision. The GOM 3D scanning system also ensures the identification of the manufacturing problems, based on the possibility of comparing the theoretical and real geometries of the obtained pieces. Therefore, using this scanning system, the geometric generation errors of surfaces can be estimated based on the "virtual pole" theorem.

Author Contributions: Conceptualization, N.B. and N.O.; methodology, V.-G.T.; software, V.-G.T.; validation, N.B., V.-G.T. and N.O.; formal analysis, N.O.; investigation, N.B.; resources, N.B.; data curation, V.-G.T.; writing—original draft preparation, G.-A.M.; writing—review and editing, G.-A.M.; visualization, N.B.; supervision, N.O.; project administration, N.B.; funding acquisition, G.-A.M. All authors have read and agreed to the published version of the manuscript.

Funding: Doctoral Schools of Dunarea de Jos University of Galati (SCDS-UDJG) http://www.cssd-udjg.ugal.ro/ (accessed on 13 May 2021).

Institutional Review Board Statement: Not applicable.

Informed Consent Statement: Not applicable.

Acknowledgments: The results of this work will be also presented to the 9th edition of the Scientific Conference organized by the Doctoral Schools of Dunarea de Jos University of Galati (SCDS-UDJG) http://www.cssd-udjg.ugal.ro/ (accessed on 13 May 2021) that will be held on 10 and 11 of June 2021, in Galati, Romania.

Conflicts of Interest: The authors declare no conflict of interest.

References

1. Lehman, M. Progressing cavity pumps in oil and gas production. *World Pumps* **2004**, *2004*, 20–22. [CrossRef]
2. Dunn, L. Progressing Cavity Pumping Systems Overview with a Focus on Coal Seam Gas Applications. Available online: https://fdocuments.in/document/progressing-cavity-pumping-systems-overview-with-a-focus-on-.html (accessed on 19 April 2021).
3. Chaudhary, P. Review on Optimization of PC-Pumps Installation in CBM Reservoirs and Troubleshooting Problems Faced By Them during Dewatering Imperial. *J. Interdiscip. Res.* **2016**, *2*, 1150–1155.
4. Progressive cavity pumps, S.C.; Confind, S.R.L. Câmpina, România. Available online: http://www.confind.ro/pompe_cavitati_progresive.html (accessed on 19 April 2021).
5. Wittrisch, C.; Cholet, H. *Progressing Cavity Pumps Oil Well Production Artificial Lift*; Editions Technip: Paris, France, 2013.
6. Liang, Y.; Cao, G.; Shi, G.; Wang, G.; Li, J.; Zhao, Y. Progressing cavity pump anti-scaling techniques in alkaline-surfactant-polymer flooding in the Daqing Oilfield. *Pet. Explor. Dev.* **2011**, *38*, 483–490. [CrossRef]
7. Al Naboulsi, M.I.; Antonescu, N.N.; Dinita, A.; Morosanu, M. Tribological Characterization of Some Elastomers Used at Progressive Cavity and Piston Pumps. In Proceedings of the MATEC Web of Conferences 2020, 318, 7th International Conference of Materials and Manufacturing Engineering (ICMMEN), Thessaloniki, Greece, 2–3 July 2020.
8. Wang, H.; Wang, S.; Lv, H. The effects of temperature on the mechanical and tribological properties of progressing cavity pump NBR stator rubber. *J. Mater. Sci.* **2016**, *22*, 308–312. [CrossRef]
9. Merey, S. Comparison of Sucker Rod Pump and Progressive Cavity Pump Performances in Batı Raman Heavy Oil Field of Turkey. *Celal Bayar Univ. J. Sci.* **2020**, *16*, 191–199.
10. Arredondo, M.; Caballero, D.; Morety, R.; Delgado, A.; Ortegano, B.J. *All Metal PCP Experiences in Orinoco Belt*; SPE Artificial Lift Conference & Exhibition-North America: Houston, TX, USA, 2014.
11. Bybee, K. First Metal-PCP SAGD Field Test Shows Promise for Heavy-Oil Hot Production. *J. Pet. Technol.* **2015**, *60*, 70–73. [CrossRef]
12. Li, M.; Jiang, D.; Guo, H.; Meng, F.; Yu, M.; Du, W.; Zhang, G. Study on clearance optimization of all-metal screw pumps: Experiment and simulation. *Mechanika* **2017**, *23*, 735–742. [CrossRef]
13. Zhang, H.; Wu, X.; An, Y. Study on Speed and Clearance Optimization of All-Metal Progressing Cavity Pumps: Experiment and Simulation. *ACS Omega* **2020**, *5*, 19533–19540. [CrossRef] [PubMed]
14. Zhou, D.; Yuan, H. Design of Progressive Cavity Pump Wells. In *SPE Progressing Cavity Pumps Conference*; Society of Petroleum Engineers: Houston, TX, USA, 2008. [CrossRef]
15. Ceballos, J.B.; Vivas, O.A. Mathematical model of controllers for progressive cavity pumps. *Rev. UIS Ing.* **2019**, *18*, 17–30. [CrossRef]
16. Nguyen, T.; Tu, H.; Al-Safran, E.; Saasen, A. Simulation of single-phase liquid flow in progressing cavity pump. *J. Pet. Sci. Eng.* **2016**, *147*, 617–623. [CrossRef]
17. SH Lee, S.H.; Kwak, H.S.; Han, G.B.; Kim, C. Design of Gerotor Oil Pump with 2-Expanded Cardioids Lobe Shape for Noise Reduction. *Energies* **2019**, *12*, 1126.
18. Om, I.L.; Ryo, S.I.; Kim, C.Y. Calculation for cross section and area efficiency of progressing cavity pump with hypotrochoidal multilobe using the differential geometric envelope approach. *J. Pet. Sci. Eng.* **2018**, *171*, 211–217. [CrossRef]
19. Litvin, F.L. *Theory of Gearing, Reference Publication 1212*; NASA, Scientific and Technical Information Division: Washington, DC, USA, 1984.
20. Oancea, N. *Surfaces Generation through Winding, Vol. I-III*; Galati University Press: Galati, Romania, 2004.
21. Costin, G.A.; Teodor, V.G.; Oancea, N. Virtual Pole Method-Alternative Method for Profiling Tools Which Generate by Enwrapping. *Ann. Dunarea Jos Univeristy* **2019**, *5*, 31–34.
22. Teodor, V.G.; Costin, G.A. Virtual Pole Method-Alternative Method for Profiling Rack Tools. *Ann. Dunarea Jos Univeristy* **2019**, *5*, 5–8.
23. Popa, C.L.; Teodor, V.G.; Baroiu, N.; Oancea, N. Side Mill Tool Profiling for Generation of Helical Surfaces Determined by Reverse Engineering. *Appl. Mech. Mater. Eng. Solut. Technol. Manuf.* **2014**, *657*, 28–32. [CrossRef]
24. Berbinschi, S.; Teodor, V.G.; Baroiu, N.; Oancea, N. Profiling Methodology for Side Mill Tools for Generation of Helical Compresor Rotor Using Reverse Engineering. *Ann. Dunarea Jos Univ. Galati Fascicle V Technol. Mach. Build.* **2011**, *2*, 111–116.

inventions

Review

Review on Study Methods for Reciprocally Enwrapping Surfaces

Georgiana Alexandra Moroşanu, Nicuşor Baroiu, Virgil Gabriel Teodor *, Viorel Păunoiu and Nicolae Oancea

Faculty of Engineering, Dunarea de Jos University of Galati, 47 Domnească Street, 800008 Galati, Romania; alexandra.costin@ugal.ro (G.A.M.); nicusor.baroiu@ugal.ro (N.B.); viorel.paunoiu@ugal.ro (V.P.); nicolae.oancea@ugal.ro (N.O.)

* Correspondence: virgil.teodor@ugal.ro; Tel.: +40-7-4525-3196

Abstract: The processing of the piece surfaces involves the transformation of the blank into a piece, the defining characteristics of the surfaces (the form, dimensions, deviations, etc.), depending to a large extent on the parameters of the generated processes of the surfaces. Generating surfaces by enveloping represents a complex problem, this mode of generating being a constant concern of researchers due to its advantages. A very important issue is the cutting edge profiling of the future tool, which will generate the wanted profile of the piece. Throughout time, researchers have studied the problems of enveloping surfaces and stated theorems related to them, based on which practical solutions have been conceive to determine the generating and generated surfaces form. The paper presents an overview on study methods for reciprocally enwrapping surfaces, as well as methods for profiling tools for generating helical surfaces by the kinematic method and by the method of decomposing the helical movement. This review can represent a basis for future research articles and projects.

Keywords: enwrapping surfaces; profiling; geometrical methods

Citation: Moroşanu, G.A.; Baroiu, N.; Teodor, V.G.; Păunoiu, V.; Oancea, N. Review on Study Methods for Reciprocally Enwrapping Surfaces. *Inventions* **2022**, *7*, 10. https://doi.org/10.3390/inventions7010010

Academic Editor: Eugen Rusu

Received: 9 December 2021
Accepted: 29 December 2021
Published: 7 January 2022

1. Introduction

The processing of the surfaces of the pieces assume transforming the blank into a piece, having the shape and dimensions prescribed in the design stage.

According to well-known authors such as Radzevich [1], the purpose of machining processes is to obtain the functional surfaces of the pieces. The defining characteristics of the surfaces of the processed pieces, such as their shape, dimensions, deviations, roughness, etc., largely depend on the parameters of the surface generating processes [1].

The generating of surfaces by enwrapping is a problem of utmost importance, both in the case of cutting processes and in the case of plastic deformation processes. This mode of generating has been a continuing concern of researchers and technologists due to the advantages of this mode of generating. Among these advantages are: high productivity and increased machining accuracy, due to the fact that the profile errors of the tool are greatly reduced on the profile of the generated piece [1,2].

According to Cioară [3], a piece can be obtained by various manufacturing processes, in accordance with different requirements such as: dimensional accuracy, surface quality or productivity. The shape of that part is characterized by the generating and directing curves, which can be obtained by copying, kinematically or can be materialized by the cutting edges of the tools.

In his work, Cioară [3] presents the cases of generating the surfaces characterized by the director obtained kinematically, in the form of the envelope of a family of curves.

Since the 19th century, there have been concerns about improving the profiles of the active surfaces of the gears. George Biddell Airy [4] performed research in this field, starting from previous studies conducted by Euler in 1760.

Concerns about changing the shape of teeth in order to improve their bearing capacity are still relevant, as evidenced by research [5–8].

Profiling the tools that generate by enveloping often requires an analytical representation of the surfaces of the pieces. In addition, the analytical representation of surfaces is used for processing on computer numerical control (CNC) machines and for computer-aided inspection processes (CAD, CAM) [9,10].

The generating of surfaces by enwrapping involves their processing by cutting, in which case, obviously, the quality of the surfaces obtained depends mainly on the geometry of the cutting tool.

A particularly important problem is the profiling of the edge of the future tool, which will generate the desired profile of the piece. This implies the existence of adequate methodologies, which allow us to solve the problem of profiling the tools that generate by enwrapping.

Over time, researchers have studied the aspects of surface enwrapping and stated theorems related to them, based on which practical solutions have been devised to determine the shape of generating and generated surfaces [1,2,11,12].

Initially, the issue of generating surfaces by enveloping had a graphical approach, due to the difficulties related to the analytical calculation, often very laborious, necessary to solve such problems.

The geometrical methods used in this regard are: the Camus theorem [13], a method based on the fundamental properties of enwrapping curves [5], the Poncelet method (1827) and the Rouleaux method (1842).

Subsequently, based on geometric methods, analytical methods for tool profiling were developed, which generate by enwrapping, in particular, by the rolling method [2,12].

According to the definition accepted in the theory of enwrapping surfaces, rolling is the relative movement of two curves that are constantly in contact.

It can be of two types: rolling without slipping and rolling with slipping. In the case of rolling without slipping, the relative speeds between the two curves have the direction of the common tangent. In the case of rolling with slipping, the curves have the same direction of the common tangent at the point of contact and the relative speeds are of opposite directions.

Theodore Olivier (1842) developed two fundamental theorems, which bear his name, for the case of generating enveloping surfaces with linear contact and, respectively, with point contact [2,14].

Subsequently, these theorems are reviewed by the mathematician Gohman (1896), who approaches a method that simplifies the way of analyzing the contact of enwrapping surfaces with linear contact and point contact [2,14].

In 1843, the normal theorem, also known as the Willis theorem, was developed for reciprocally enwrapping profiles associated with rolling centrodes [15].

The Nikolaev method (1950) is also another method specific to the profiling of surfaces, reciprocally enveloping, cylindrical or revolutionary [2].

Based on this fundamental method, Guo Q. et al. [16] developed an analytical method for calculating the envelope of the tool profile at processing on a numerically controlled five-axis machine.

In paper [17], based on the normal to the generating surface and the relative velocity vector between the tool and the blank, the analytical expression of the enwrapping of the cutting edge positions of the tool is obtained, during its movement according to the programmed trajectory. The method is validated by exemplifying it when processing on a machine tool with five axes numerically controlled and there is a wide discussion on the influence of geometric errors of the tool on the generated surface.

Complementary methods, such as the minimum distance method (1992), method of substituting circles (1998), the tangent method (2000) and the planar trajectory method (2002), are methods that use the direct way of expressing the enwrapping conditions and treat problems specific to the generating of reciprocally enveloping surfaces [18].

The development of industrial systems, as well as the increasing use of computer numerical controlled (CNC) machine tools, have allowed the generating of complex-shaped surfaces, which use circular or rectilinear generating tools. They describe complex trajectories, involving the development of applications for the study of enveloping surfaces made through graphic design programs, such as CATIA, AutoCAD, SolidEdge, etc. [16,18–36].

Albu [18] studied the possibility of simulating the milling processing with the front cylinder tool of the helical surfaces of a conveyor screw. In the respective paper, the possibility of obtaining a conical helical surface of variable pitch is analyzed.

Teodor V.G., Popa I. and Oancea N. (2010) created an algorithm to replace the plane generators of surfaces discreetly known by Bezier curves [37,38].

As a result of the development of computing systems and the evolution of graphic design media, graphic methods could be imagined and developed to solve the problems of tool profiling, which generate by enwrapping thus providing solutions to problems related to the study of reciprocally enveloping surfaces [21,37,39–44].

A number of authors have used the theory of enwrapping surfaces to study the effect of deviations of the tool from the programmed trajectory on the geometric shape of the generated surface [9,16,45–49].

2. Materials and Methods

2.1. Olivier Method

This method is based on Olivier's two fundamental theorems, namely:

- Theorem I—generating of surfaces with linear contact for conjugated surfaces that depend on a single parameter [11,14];
- Theorem II—generating of surfaces with point contact for conjugated surfaces that depend on two independent parameters [2,11].

In the case of Olivier's first theorem, in a straight orthogonal system, a family of surfaces that depends on a parameter is described by equations of form:

$$(\Sigma)_{\varphi}: \left| \begin{array}{l} X = X(u,v,\alpha); \\ Y = Y(u,v,\alpha); \\ Z = Z(u,v,\alpha), \end{array} \right. \tag{1}$$

or, in a vectorial expression:

$$\vec{R}(u,v,\alpha) = X(u,v,\alpha)\cdot\vec{i} + Y(u,v,\alpha)\cdot\vec{j} + Z(u,v,\alpha)\cdot\vec{k} \tag{2}$$

where, u and v are independent variable parameters of the current point belonging to the Σ surface and α is the parameter that determines the family of surfaces [11,14].

Observation: The characteristic curve on a surface Σ, Figure 1, which performs a movement in relation to a fixed part, depending on a single variable parameter, is the geometric place of the points on the surface, where the normal to it is perpendicular to the velocity vector, $\vec{v}$, in the absolute movement of the surface in relation to the fixed part (fixed reference system) [50].

Olivier's first theorem can be expressed, for a value of the variable parameter Σ, depending on the variable parameters u and v of the surface, by the analytical forms [50]:

$$\begin{array}{c} \varphi(u,v,\alpha) = 0; \\ \varphi'_{\alpha} = 0, \end{array} \tag{3}$$

or by a vectorial form equation:

$$C_{\alpha}: \left| \begin{array}{l} \vec{r} = \vec{R}(u,v,\alpha); \\ \vec{R}_u \cdot \vec{R}_v \cdot \vec{R}_{\alpha} = 0. \end{array} \right. \tag{4}$$

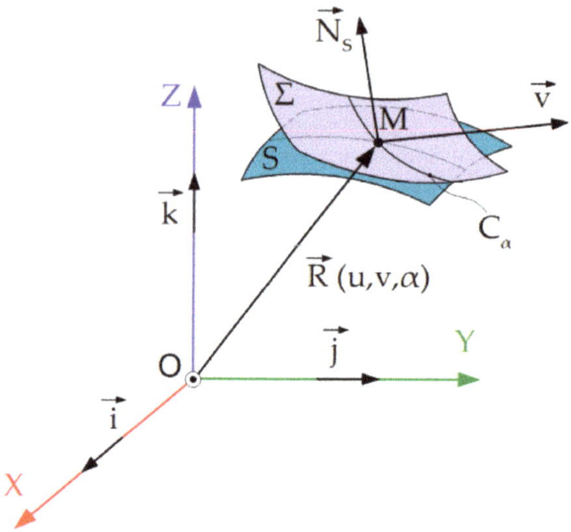

Figure 1. Characteristic curve.

In the case of a family of surfaces described by two independent parameters α and β, it will have the parametric equations of the form [2]:

$$(\Sigma)_{\alpha,\beta}: \left| \begin{array}{l} x = x(u,v,\alpha,\beta); \\ y = y(u,v,\alpha,\beta); \\ z = z(u,v,\alpha,\beta), \end{array} \right. \tag{5}$$

or:

$$\vec{R}(u,v,\alpha,\beta) = x(u,v,\alpha,\beta)\cdot\vec{i} + y(u,v,\alpha,\beta)\cdot\vec{j} + z(u,v,\alpha,\beta)\cdot\vec{k}. \tag{6}$$

According to Olivier's second theorem [2], if the function $\vec{R}(u,v,\alpha,\beta)$ admits partial derivatives after all arguments, in a domain D, of variation of the parameters u and v, for a defined domain of the independent parameters α and β, then the envelope of the family $(\Sigma)_{\alpha,\beta}$ can be determined from the system of equations:

$$S: \left| \begin{array}{l} \vec{r} = \vec{R}(u,v,\alpha,\beta); \\ \vec{R}_u \cdot \vec{R}_v \cdot \vec{R}_\alpha = 0; \\ \vec{R}_u \cdot \vec{R}_v \cdot \vec{R}_\beta = 0, \end{array} \right. \tag{7}$$

where $\vec{R}_u$, $\vec{R}_v$, $\vec{R}_\alpha$ and $\vec{R}_\beta$ are partial derivatives of the $\vec{R}$ vector.

2.2. Gohman Theorem

Olivier's theorems allowed the mathematician Gohman (1896) to develop a method for determining the characteristic curve on a moving surface.

The theorem states that: "a point on the S surface will belong to the characteristic curve only if, at that point, the normal to the S surface is perpendicular to the velocity vector $\vec{v}$, in the absolute movement executed by the surface" [2,51].

Figure 2 shows the S surface and the two reference systems.

The S surface, in the reference system XYZ, can be expressed as:

$$S: F(X,Y,Z) = 0. \tag{8}$$

The S surface executes a movement, depending on a single parameter τ. In this movement, the S surface describes the family of surfaces $(S)_\tau$ in the fixed system xyz in the form:

$$(S)_\tau : F(x, y, z, \tau) = 0. \quad (9)$$

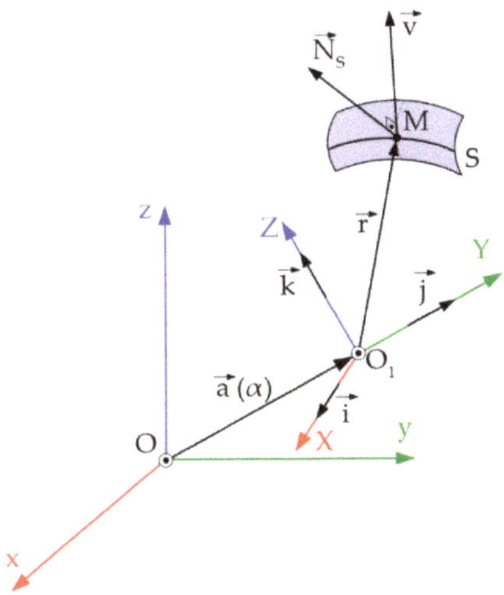

Figure 2. Reference systems: xyz—fixed reference system, XYZ—mobile reference system.

Gohman's theorem can define the characteristic curve C given by the system of equations:

$$C : \begin{cases} F(x, y, z, \tau) = 0; \\ F'_\tau = F_X \frac{dx}{d\tau} + F_Y \frac{dy}{d\tau} + F_Z \frac{dz}{d\tau} = 0; \\ \tau = \text{constant}. \end{cases} \quad (10)$$

The equation $F'_\tau = F_X \frac{dx}{d\tau} + F_Y \frac{dy}{d\tau} + F_Z \frac{dz}{d\tau}$ can be seen as the scalar product of two vectors, $F_X \cdot \vec{i} + F_Y \cdot \vec{j} + F_Z \cdot \vec{k}$, obviously representing the normal at the S surface, $\vec{N}_S$ and the vector $\vec{v} = \frac{dX}{d\tau} \cdot \vec{i} + \frac{dY}{d\tau} \cdot \vec{j} + \frac{dZ}{d\tau} \cdot \vec{k}$, representing the absolute velocity (relative to the fixed system) in the movement of the surface.

Therefore, the condition $F'_\tau = 0$ can be written as $\vec{N}_S \cdot \vec{v} = 0$, which allows the enunciation of Gohman's theorem.

The Gohman method can easily solve the problem of generating surfaces (surface vortexes) associated with a couple of rolling centrodes (the case of rack tool generating, gear shaped cutter tool generating or the generating of cylindrical helical surfaces with rotary cutter tools) [9].

Based on Gohman's theorem, Dooner [15] applies the theory of enwrapping surfaces to cylindrical, conical and hypoid gears.

Dooner and Griffis present the application of the first gear law to the study of worm and hypoid gears, in which, in general, normal at the flank of the tooth does not intersect the instantaneous rotation axis [52]. In the same paper, the second law of gearing is applied for the study of hyperboloid gears, noting that: "the spiral profile of the flank of the tooth, in these gears, is generated so as to allow any pressure angle with the conjugate profile" [52].

2.3. *Normals Method (Willis Theorem)*

The "gearing" theorem, also known as the "normals method" or "Willis theorem" (1843) [2,51] or sometimes as "Euler–Savary theorem" [52], can be stated as follows: "two profiles associated with rolling centrodes, which transmit rotational movement between two parallel axes, are reciprocally enveloping if, in the process of rolling the centrodes, the contact points of the profiles admit a common normal, which passes through the gearing pole".

Two enwrapping profiles are considered, Figure 3, having the two centrodes C_1 and C_2 tangent at the contact point, P. The centrodes rotate around the axes O_1 and O_2 having the velocities expressed by the parameters ω_1 and ω_2.

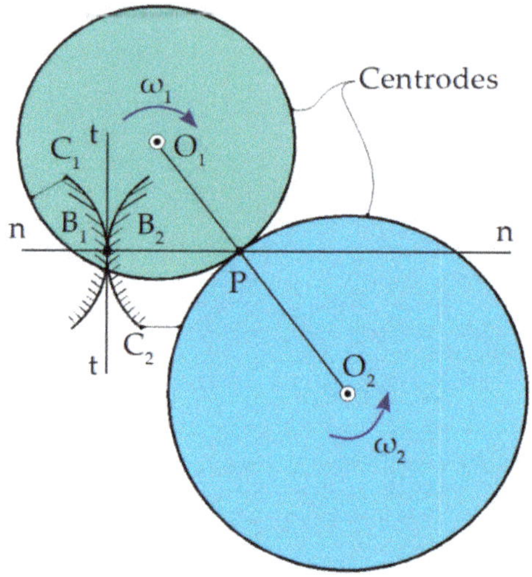

Figure 3. Conjugated profiles.

In these circumstances, the transmission ratio may be established:

$$i = \frac{\omega_2}{\omega_1}. \tag{11}$$

The relative velocity of points B_1 and B_2 will have the direction of the common tangent to the C_1 and C_2 curves at the contact point (line *t-t*). As a result, the instantaneous radius of rotation $PB_1 = PB_2$ has the direction of the common normal at the two conjugate profiles (line *n-n*).

2.4. Minimum Distance Method

Minimum distance method is an analytical method for the study of enwrapping surfaces based on a specific theorem [2,12]: "the envelope of a family of plane curves that performs a movement joined with a couple of rolling centrodes is the geometrical locus of the points belonging to the family for which, in the different rolling positions, the distance to the gearing pole is minimal", Figure 4.

The following are defined:

- Centrode associated with the vortex of the blank profiles—C_1 (radius circle R_{rp});
- Centrode associated with the flank of the generating rack—C_2 (line superimposed on the η axis);
- Reference systems: xy—fixed reference system; XY—mobile reference system, joined with the blank; and $\xi\eta$—mobile reference system, joined with the generating rack.

If the parametric equations of the C_Σ profile are defined:

$$C: \left| \begin{array}{l} X = X(u); \\ Y = Y(u) \end{array} \right. \tag{12}$$

and the relative movement in relation to the rack system, $\xi\eta$, in form:

$$\xi = \omega_3^T(\varphi) \cdot X - a; \tag{13}$$

$$a = \left\| \begin{array}{c} -R_{rp} \\ -R_{rp} \cdot \varphi \end{array} \right\|, \tag{14}$$

then, the family of profiles, C_Σ, in the system $\xi\eta$, is described by the transformation:

$$(C_\Sigma)_\varphi = \left| \begin{array}{l} \xi = X(u) \cdot \cos\varphi - Y(u) \cdot \sin\varphi + R_{rp}; \\ \eta = X(u) \cdot \sin\varphi + Y(u) \cdot \cos\varphi + R_{rp} \cdot \varphi. \end{array} \right. \tag{15}$$

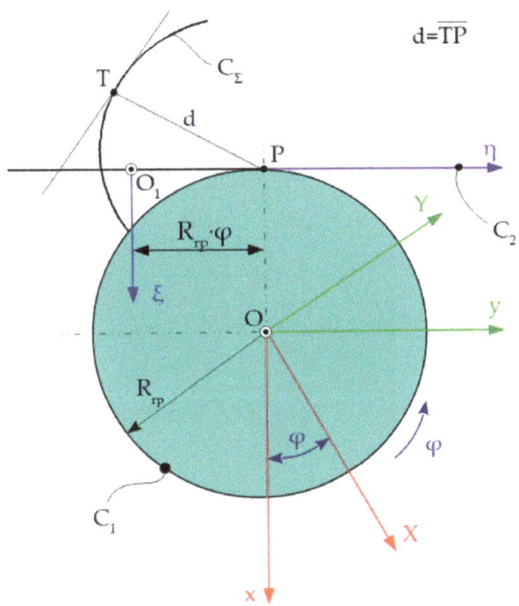

Figure 4. Minimum distance method applied to rack tool generating [12].

The distance from the current point of the family of curves $(C_\Sigma)_\varphi$ to the gearing pole P, in the system $\xi\eta$ of the rack is:

$$d = \sqrt{\xi^2 + (\eta - R_{rp})^2}. \tag{16}$$

The minimum condition of distance d becomes:

$$\xi \cdot \xi'_u + (\eta - R_{rp}) \cdot \eta'_u = 0, \tag{17}$$

representing the enwrapping condition according to this method.

The set of Equations (15) and (17) represents, in the $\xi\eta$ system, the enveloping surface of the family $(C_\Sigma)_{\varphi_u}$—the profile of the rack conjugated to the family $(C_\Sigma)_\varphi$.

2.5. Method of Substituting Circles

This method is based on a new principle, namely the family of substituting surfaces (curves) [2,12,51,53] and can be stated as follows: "the envelope of a curve associated with a couple of rolling centrode is the family of substituting circles transposed, in the rolling movement, on the conjugate centrode", Figure 5.

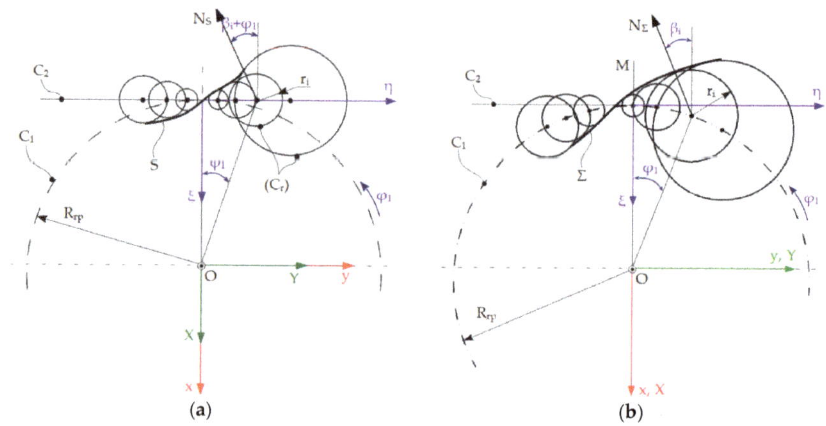

Figure 5. Family of substituting circles associated: (**a**) with a centrode; (**b**) Σ profile.

In the XY system, the family of substituting circles has the equations:

$$(C)_r : \left| \begin{array}{l} X = -R_{rp}\cos\varphi_1 - r_i\cos\beta_i; \\ Y = -R_{rp}\sin\varphi_1 - r_i\sin\beta_i. \end{array} \right. \quad (18)$$

The enwrapping condition is, in this case:

$$Y'(u) \cdot \sin\varphi_1 - X'(u) \cdot \cos\varphi_1 = \frac{X(u) \cdot X'(u) + Y(u) \cdot Y'(u)}{R_{rp}}, \quad (19)$$

with $tg\beta_i = -\frac{X'(u)}{Y'(u)}$ and $r_i = -\left[R_{rp}\cos(\varphi_1 + \beta_i) + X(u)\cos\beta_i + Y(u)\sin\beta_i\right]$.

The equations of the envelope S result:

$$\begin{array}{c} \xi = -r_i \cdot \cos(\varphi_1 + \beta_i); \\ \eta = -r_i \cdot \sin(\varphi_1 + \beta_i) + R_{rp} \cdot \varphi_1. \end{array} \quad (20)$$

The method of substituting circles has applicability in solving the problem of profiling the tools that generate by enwrapping and allows a suggestive graphic expression, thus preventing the avoidance of major errors. Applications of the method are presented in [54,55].

2.6. Trajectory Method

Trajectory method is a complementary method which is based on determining the enwrapping condition by identifying the trajectories of the points in the tool space, the trajectories tangent to the profile to be generated [12,56].

The trajectory method can be expressed as follows: "the enveloping of a profile associated with a centrode, belongs to a couple of rolling centrodes, is the enveloping of the trajectories described by the point of this profile in the space associated with the rolling centrode" [12].

Figure 6 shows a detail of the tool profile and the plane trajectories of the points belonging to the piece's profile in the tool's space.

The parametric equations of the characteristic curve, C_Σ, are given by Equation (12), u being a variable parameter.

Therefore, the rolling condition is:

$$\lambda = R_r \cdot \varphi. \tag{21}$$

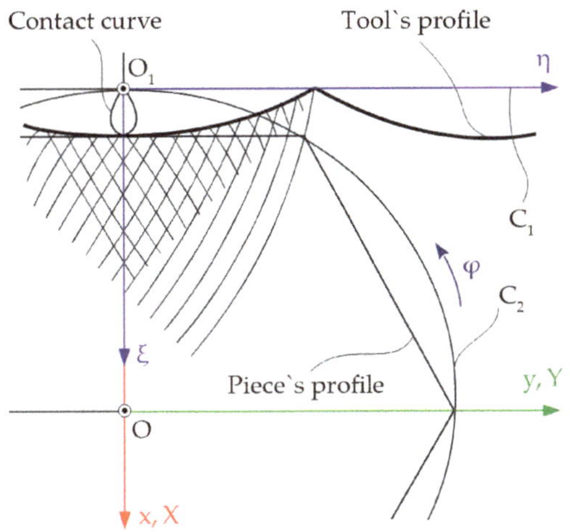

Figure 6. Tool's profile and the in-plane trajectories of piece [12].

The family of trajectories is expressed by the form:

$$(C_\Sigma)_\varphi : \left| \begin{array}{l} \xi = \xi(u, \varphi); \\ \eta = \eta(u, \varphi). \end{array} \right. \tag{22}$$

Relation (22), along with the condition:

$$\frac{\xi'_u}{\xi'_\varphi} = \frac{\eta'_u}{\eta'_\varphi}, \tag{23}$$

allows us to determine the profile of the envelope, C_S, which represents the enwrapping curve of the in-plane trajectories of the points belonging to the curve C_Σ, in the rolling movement of the centrode [2,12].

2.7. Comparative Study on Specific Forms of Enwrapping Condition

In the case of analytical methods, in order to determine the profiles of the tools that generate by enwrapping, by the rolling method, a typical algorithm is followed, consisting of the following steps:

(1) Establishing the parametric equations of the profile to be generated;
(2) Determining the form of the rolling condition, which determines that the lengths traversed by the instantaneous centers of rotation, on each of the two centrodes, are equal;
(3) Determining the absolute movements of the piece and the tool;
(4) Based on the absolute movements, the relative movements between the tool and the piece are determined;
(5) Considering the fixed tool, the family of curves generated by the profile of the piece is determined, during the relative movement that it has towards the tool;
(6) Determining the specific form of the enwrapping condition;
(7) Associating to the family of curves, determined in step 5, the enwrapping condition obtained in step 6, the profile of the generating tool is determined; practically, the en-

wrapping condition allows that, from the points belonging to the family of curves, only those belonging to the envelope, and, therefore, to the tool profile, to be selected.

(8) The parametric equations of the gearing curve can be obtained; this represents the geometric locus, in the fixed space, in which the tangency between the two reciprocally enveloping profiles takes place, that of the piece-known and that of the tool-determined.

2.7.1. Rack Tool Profiling—Gohman's Theorem

The rolling condition, according to the Gohman's Theorem [11,41,57,58], see Figure 7, of the centrode C_1 (circle of radius R_{rp}), associated with the profile Σ and of the centrode C_2, associated with the rack tool S is given by the Equation (21).

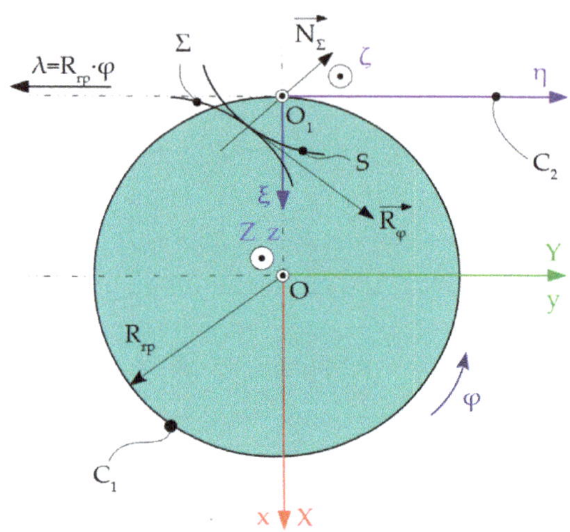

Figure 7. Reference systems for generating with rack tool, *xyz* is the fixed reference system, *XYZ*—mobile reference system, joined with the Σ profile, *ξηζ*—mobile reference system, joined with the rack tool.

The reference systems, in this case, are presented in Figure 7.

The relative movement of the existing point in the mobile reference system joined with the Σ profile towards to the mobile system joined with the rack tool is given by Equations (13) and (14).

The parametric equations of the Σ profile are defined, depending on the variable parameter *u*:

$$\Sigma : \begin{vmatrix} X = X(u); \\ Y = Y(u). \end{vmatrix} \tag{24}$$

The envelope of the surface family $(\Sigma)_\varphi$ in the reference system $\xi\eta\zeta$ is determined, which represents the flank of the generating rack, from (13) and (14):

$$(\Sigma)_\varphi : \begin{vmatrix} \xi = \xi(u,\varphi); \\ \eta = \eta(u,\varphi). \end{vmatrix} \tag{25}$$

According to Gohman's theorem, for profiling rack tools, the enwrapping condition will be [2]:

$$\vec{N}_\Sigma \cdot \vec{R}_\varphi = 0, \tag{26}$$

where $\vec{N}_\Sigma$ represents the normal at the Σ profile:

$$\vec{N}_\Sigma = N_x \cdot \vec{i} + N_y \cdot \vec{j} \tag{27}$$

and $\vec{R}_{\varphi}$:

$$\vec{R}_{\varphi} = [Y(u,v) - R_r \sin\varphi] \cdot N_X - [X(u,v) + R_r \cos\varphi] \cdot N_y = 0. \tag{28}$$

2.7.2. Gear Shaped Cutter Tools Profiling—Gohman's Theorem

In the case of gear shaped cutter tools [11,28,55,59,60], the rolling without sliding movement is at the contact between two rotating cylindrical surfaces. The radii of the two surfaces are denoted as follows: R_{rp}—for blank and R_{rs}—for tool. Figure 8 also shows the centrodes associated with the tool and the blank, C_2, respectively, C_1.

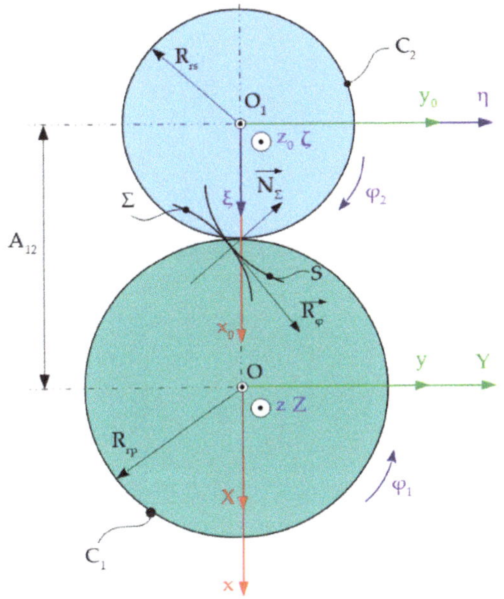

Figure 8. Generating with gear shaped cutter tool.

From the rolling condition, the transmission ratio expressed by the form is defined:

$$i = \frac{\varphi_2}{\varphi_1} = \frac{R_{rp}}{R_{rs}}, \tag{29}$$

with φ_1 and φ_2 angular parameters belonging to the two rotational movements.

The parametric equations in the mobile reference system XYZ that define the Σ profile are similar to Equation (24).

The relative movements of the mobile reference systems are:

$$\xi = \omega_3(-\varphi_2)\left[\omega_3^T(\varphi_1) \cdot X - a\right],\ a = \begin{pmatrix} -A_{12} \\ 0 \end{pmatrix}. \tag{30}$$

The family of surfaces, $(\Sigma)_{\varphi_1}$, in the mobile reference system $\xi\eta\zeta$ is given by the equation:

$$(\Sigma)_{\varphi_1} : \left| \begin{array}{l} \xi = X(u)\cos(\varphi_1+\varphi_2) - Y(u)\sin(\varphi_1+\varphi_2) + A_{12}\cos\varphi_2; \\ \eta = X(u)\sin(\varphi_1+\varphi_2) + Y(u)\cos(\varphi_1+\varphi_2) + A_{12}\sin\varphi_2, \end{array} \right. \tag{31}$$

A_{12} representing the distance measured between the z and z_0 axes.

The enwrapping condition, in the case of gear shaped cutter tools, has the form:

$$[(1+i)\cdot Y(u,v)-i\cdot A_{12}\cdot \sin\varphi_1]\cdot N_X-[(1+i)\cdot X(u,v)+i\cdot A_{12}\cdot \cos\varphi_1]\cdot N_Y=0. \quad (32)$$

2.7.3. Rotary Cutter Tools Profiling—Gohman's Theorem

In the case of generating with the rotary cutter tool [11,12,41,54], Figure 9, the plane of the blank will roll on the tool cylinder of radius R_{rs}. The centrodes are represented: C_1—the centrode associated with the blank, and C_2—the centrode of the rotary cutter.

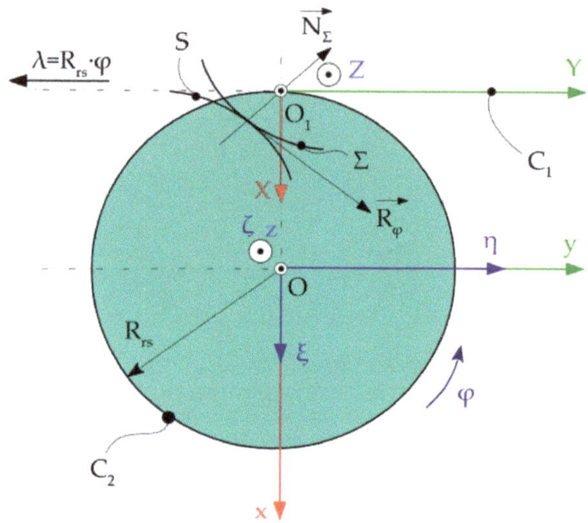

Figure 9. Generating with rotary cutter tool.

The rolling condition is:

$$\lambda = R_{rs}\cdot\varphi. \quad (33)$$

The relative movement of the blank relative to the $\xi\eta$ system is:

$$\xi=\omega_3(\varphi)\cdot[X+a],\ a=\left\|\begin{array}{c}-R_{rs}\\ -R_{rs}\cdot\varphi\end{array}\right\|. \quad (34)$$

The parametric equations in the mobile reference system are of the form (24). The enwrapping condition becomes, in this case:

$$[-Y(u)+R_{rs}\cdot\varphi]\cdot N_X+X(u)\cdot N_Y=0. \quad (35)$$

2.7.4. Rack Tool Profiling—Minimum Distance Method

Knowing the curve C_Σ of the profile to be generated by the equation of type (24), with u variable parameter, as well as the relative movement in the mobile reference system of the generating rack $\xi\eta$, (13), the family of profiles can be determined with the Equation (15) [12,61], with R_{rp}—rolling radius of the blank centrode.

Therefore, the enwrapping condition according to the minimum distance method is give by (17).

2.7.5. Gear Shaped Cutter Tool Profiling—Minimum Distance Method

Figure 10 shows the minimum distance method applied at generating with the gear shaped cutter tool [12,61].

The relative movements of the mobile reference systems are given by Equation (30). Equation (30) represents the relative movement of the blank in the reference system of the tool, the relative movement of the tool towards the piece being expressed in the form:

$$X = \omega_3(\varphi_1)\left[\omega_3^T(-\varphi_2)\cdot\xi + a\right]. \tag{36}$$

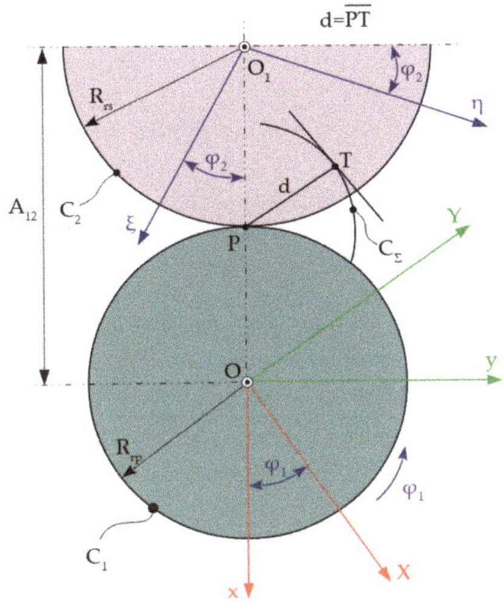

Figure 10. The minimum distance method applied at generating with the gear shaped cutter tool [2].

The enwrapping condition, according to the minimum distance method, is given by the relation:

$$(\xi - R_{rs}\cdot\cos\varphi_2)\cdot\xi'_u + (\eta - R_{rs}\cdot\sin\varphi_2)\cdot\eta'_u = 0. \tag{37}$$

2.7.6. Rotary Cutter Tool Profiling—Minimum Distance Method

Figure 11 shows the profiling of the rotary cutter tool [12,61], according to the minimum distance method.

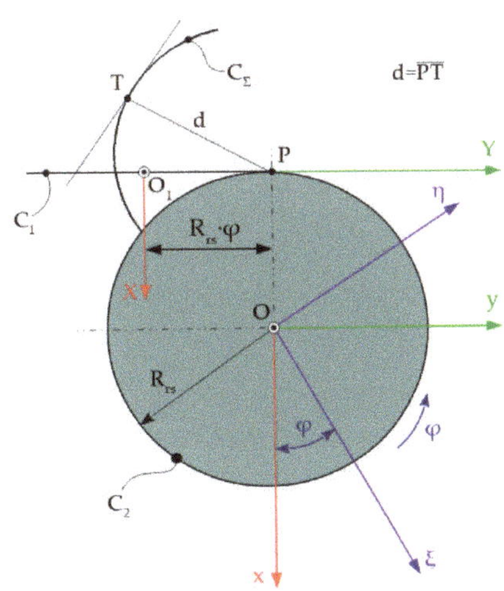

Figure 11. The minimum distance method applied at generating with rotary cutter tool [2].

The relative movement of the piece in the tool space is given by Equation (34).

The enwrapping condition is:

$$(\xi + R_{rs} \cdot \cos\varphi) \cdot \xi'_u + (\eta - R_{rs} \cdot \sin\varphi) \cdot \eta'_u = 0 \tag{38}$$

By combining Equations (34) and (38), the profile of the rotary cutter tool is obtained.

To ensure the generation of the active surface of the pump screw [62,63], it is necessary that the two helical surfaces, the tool and the blank, have opposite directions and have the same inclination to their own axes on the rolling circles.

2.8. Profile of Tools for Generating of Helical Surfaces by Kinematic Method

2.8.1. Profiling of the Disk Tool for the Generating of Helical Surfaces—Profiling Algorithm

The generating of helical surfaces with disc tools has a wide spread, both in mass production (processing of helical drill channels), large series production (in the case of milling of machine tool screws) [64–67] as well as in the production of unique [68].

Determining the form of the primary peripheral surface of the tool involves determining the axial section of the rotating surface—the most commonly used case in practice is when the clearance angle of the tool teeth is zero, therefore the axial section of the rotating surface is given by the cutting edge configuration of the tool teeth [2,69].

Thus, with the helical surface Σ being known, of axis $\vec{V}$ and parameter p, the rotation surface S can be determined, this being reciprocally enveloping with the surface Σ.

The reference systems are defined according to Figure 12.

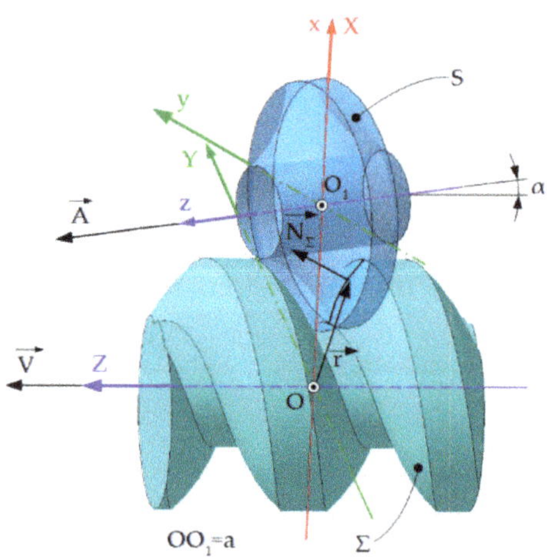

Figure 12. Disc tool-reference systems: xyz—fixed reference system, the z axis being superimposed on the axis of the rotation surface S; X—mobile reference system, integral with the Σ surface, the Z axis being superimposed on the axis $\vec{V}$ at the initial moment, with the plane ZO_1Y the mobile reference system is parallel to the plane zOy.

The axis of the S surface, the axis $\vec{A}$ is at a distance a from the axis of the helical surface $\vec{V}$ and, together with it, forms the angle of inclination, α. This angle is determined by the pitch of the helical surface being processed.

If the surfaces S and Σ are reciprocally enveloping, the surface S is considered to be the family envelope (Σ), in the absolute movement performed by the surface.

Although the surface Σ performs a compound movement: a rotational movement about the axis $\vec{A}$ and a helical movement about the axis $\vec{V}$ and parameter p, the characteristic of the Σ surface depends only on the first mentioned movement (rotational movement), because in the case of the helical movement, Σ self-generates.

Thus, the movement of the Σ surface, being joined with the mobile reference system, *XYZ*, will be known from the transformation of coordinates:

$$X = \alpha \cdot [\omega_3(\varphi) \cdot x - A], \tag{39}$$

where:

- α represents the transformation matrix between the versors of the *XYZ* and *xyz* systems and is given by the relation:

$$\alpha = \begin{Vmatrix} 1 & 0 & 0 \\ 0 & \cos\alpha & -\sin\alpha \\ 0 & \sin\alpha & \cos\alpha \end{Vmatrix}; \; A = \begin{Vmatrix} -a \\ 0 \\ 0 \end{Vmatrix}; \tag{40}$$

- *A*-the matrix formed by the coordinates of point O_1, compared to the fixed coordinate system.

Given that the parameter φ is a function of time, after development, the relationship

$$x = \omega_3^T(\varphi) \cdot \left[\alpha^T \cdot X + A\right] \tag{41}$$

represents the movement law of a point of space *XYZ* towards to the fixed reference system.

The characteristic curve is determined by calculating, in the first step, the velocity vector $\vec{R}_\varphi$, in movement of a point in the space *XYZ* towards to the fixed reference system, *xyz*:

$$R_\varphi = \frac{dx}{d\varphi}. \tag{42}$$

From Equation (41), it results:

$$R_\varphi = \dot{\omega}_3^T(\varphi) \cdot \alpha^T \cdot X + \dot{\omega}_3^T(\varphi) \cdot A \tag{43}$$

or

$$R_\varphi = \begin{Vmatrix} -\sin\varphi & -\cos\varphi & 0 \\ \cos\varphi & -\sin\varphi & 0 \\ 0 & 0 & 0 \end{Vmatrix} \cdot \begin{Vmatrix} 1 & 0 & 0 \\ 0 & \cos\alpha & \sin\alpha \\ 0 & -\sin\alpha & \cos\alpha \end{Vmatrix} \cdot \begin{Vmatrix} X \\ Y \\ Z \end{Vmatrix} + \begin{Vmatrix} -\sin\varphi & -\cos\varphi & 0 \\ \cos\varphi & -\sin\varphi & 0 \\ 0 & 0 & 0 \end{Vmatrix} \cdot \begin{Vmatrix} -a \\ 0 \\ 0 \end{Vmatrix}. \tag{44}$$

The result is the matrix:

$$R_\varphi = \begin{Vmatrix} -X \cdot \sin\varphi - Y \cdot \cos\varphi \cdot \cos\alpha - Z \cdot \cos\varphi \cdot \sin\alpha + a \cdot \sin\varphi \\ X \cdot \cos\varphi - Y \cdot \sin\varphi \cdot \cos\alpha - Z \cdot \sin\varphi \cdot \sin\alpha - a \cdot \cos\varphi \\ 0 \end{Vmatrix} \tag{45}$$

or, in vector form:

$$\begin{aligned} \vec{R}_\varphi = & [-X \cdot \sin\varphi - Y \cdot \cos\varphi \cdot \cos\alpha - Z \cdot \cos\varphi \cdot \sin\alpha + a \cdot \sin\varphi] \cdot \vec{i} + \\ & + [X \cdot \cos\varphi - Y \cdot \sin\varphi \cdot \cos\alpha - Z \cdot \sin\varphi \cdot \sin\alpha - a \cdot \cos\varphi] \cdot \vec{j}. \end{aligned} \tag{46}$$

In the mobile coordinate system, *XYZ*, the parametric equations of the helical surface Σ are known:

$$\Sigma \begin{cases} X = X(u,v); \\ Y = Y(u,v); \\ Z = Z(u,v), \end{cases} \tag{47}$$

with u and v independent variables.

The vector of the normal at the Σ surface, $\vec{N}_{\Sigma}$, is established:

$$\vec{N}_{\Sigma} = N_x \cdot \vec{i} + N_y \cdot \vec{j} + N_z \cdot \vec{k}. \tag{48}$$

The enwrapping condition is, according to Gohman's theorem, given by $\vec{N}_{\Sigma} \cdot \vec{R}_{\varphi} = 0$.

The condition requires that the vectors $\vec{N}_{\Sigma}$ and $\vec{R}_{\varphi}$ be defined in the same reference system.

Therefore, the vector $\vec{N}_{\Sigma}$ will be defined according to the fixed coordinate system, xyz, by the transformation:

$$\begin{Vmatrix} N_x \\ N_y \\ N_z \end{Vmatrix} = \begin{Vmatrix} 1 & 0 & 0 \\ 0 & \cos\alpha & \sin\alpha \\ 0 & -\sin\alpha & \cos\alpha \end{Vmatrix} \cdot \begin{Vmatrix} N_X \\ N_Y \\ N_Z \end{Vmatrix}. \tag{49}$$

The director parameters of the normal are:

$$\begin{aligned} &N_x = N_X; \\ &N_y = N_Y \cdot \cos\alpha + N_Z \cdot \sin\alpha; \\ &N_z = -N_Y \cdot \sin\alpha + N_Z \cdot \cos\alpha, \end{aligned} \tag{50}$$

where N_x, N_y and N_z represent the director parameters of normal at the Σ surface in the fixed coordinate system, xyz.

Thus, for a certain value of parameter φ, the expression of the condition for determining the characteristic curve is obtained, which allows the determination of a connection between the parameters u and v, of the form $v = v(u)$. Taking into account the Equations (26), (45) and (50), the condition is:

$$(X - a) \cdot (N_y \cdot \cos\alpha + N_z \cdot \sin\alpha) - (Y \cdot \cos\alpha + Z \cdot \sin\alpha) \cdot N_X = 0. \tag{51}$$

The condition (51) and the expression of the parametric equations of the helical surface Σ (47) determine the equations of the characteristic curve, C:

$$C: \begin{cases} X = X(u); \\ Y = Y(u); \\ Z = Z(u). \end{cases} \tag{52}$$

The axial section of the rotating surface, S, is, in fact, the primary peripheral surface of the tool, which, by means of the equations of the characteristic curve, C, at the fixed coordinate system, xyz, results in the transformation of coordinates:

$$C: \begin{cases} x = X - a; \\ y = Y \cdot \cos\alpha + Z \cdot \sin\alpha; \\ z = -Y \cdot \sin\alpha + Z \cdot \cos\alpha. \end{cases} \tag{53}$$

By replacing in Equation (52) the parameters X, Y and Z determine the parametric equations of the characteristic curve, C, in the coordinate system, xyz:

$$C: \begin{cases} x = x(u); \\ y = y(u); \\ z = z(u). \end{cases} \tag{54}$$

The characteristic curve has a rotational movement around the axis $\vec{A}$, thus generating the S surface, simplifying the determination of the axial section of the surface.

The points M and N on the characteristic curve, C and the generator, G, are on the same parallel circle, in the same plane perpendicular to the tool axis, Figure 13.

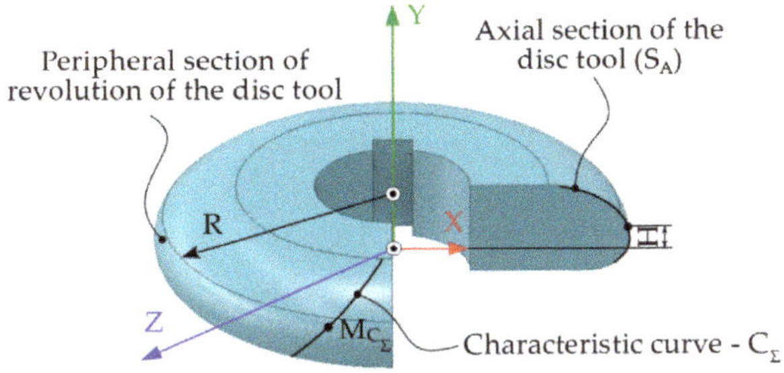

Figure 13. Axial section of the rotating surface [12].

The radius of the parallel circle is determined by the relation:

$$R = \sqrt{x^2(u) + y^2(u)}. \tag{55}$$

The equations of the axial section of the S_A surface are given by the relation:

$$S_A \quad \begin{cases} R = \sqrt{x^2(u) + y^2(u)}; \\ H = z(u). \end{cases} \tag{56}$$

The axis of the tool is defined by the axis $\vec{A}$ defined by the parameters a and α, and is perpendicular to the helix of minimum inclination of the helical surface—Figure 14.

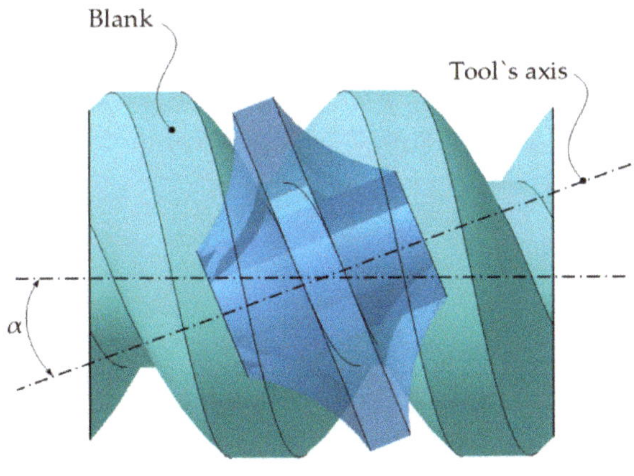

Figure 14. Axial section of the rotating surface.

Figure 15 shows the generating kinematics, namely the relative position of the tool (milling disc) and the blank (trapezoidal thread).

If D is the diameter of the cylinder, corresponding to the helical line, which belongs to the surface to be processed and P_E represents the pitch of the helical surface, the angle of inclination, α, can be determined as follows:

$$\alpha = arctg\frac{\pi \cdot D}{P_E}. \tag{57}$$

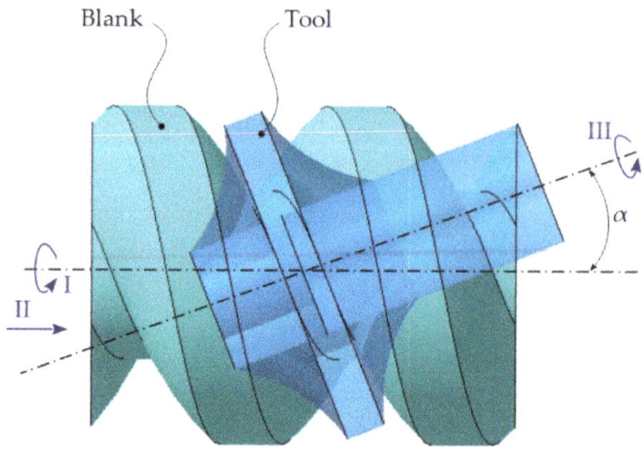

Figure 15. Generating kinematics with disc tool: movement *I*—rotational movement of the blank around its own axis, *II*—translational movement, along the same axis; *III*—cutting movement.

2.8.2. Profiling of Cylindrical-Front Tools (Finger Cutter Tools) for the Generating of Helical Surfaces—Profiling Algorithm

For technological or economic reasons, in some cases (for example zig-zag or V-teeth), the use of disc tools is not possible and it is necessary to use a tool with a smaller size for profiling helical surfaces [18,40,70].

Therefore, the disc tool is replaced with a finger cutter tool or a cylinder-front tool.

To determine the primary peripheral surface of the tool, a rotating surface is determined, perpendicular to the axis of the helical surface, Σ, already known and reciprocally enveloping with it [28,40].

The helical surface, Σ, performs a combined movement: a helical movement, of axis $\vec{V}$ and parameter p and a rotational movement around the axis $\vec{A}$. The surface envelope Σ, depends only on the rotational movement, because, in the helical movement, it self-generates.

Coordinate systems are defined according to Figure 16.

The movement of the helical surface is described by the transformation of coordinates:

$$x = \omega_1^T(\varphi) \cdot \alpha^T(X + \alpha \cdot a); \tag{58}$$

where:

$$\alpha = \begin{Vmatrix} \cos\alpha & -\sin\alpha & 0 \\ \sin\alpha & \cos\alpha & 0 \\ 0 & 0 & 1 \end{Vmatrix}; a = \begin{Vmatrix} 0 \\ 0 \\ -m \end{Vmatrix}. \tag{59}$$

The parameter m is a constant that depends on the relative position of the reference systems presented above. Usually, to simplify the calculations, it is consider $m = 0$.

The characteristic curve, C, of the Σ surface is determined by the equation: $\vec{N}_\Sigma \cdot \vec{R}_\varphi = 0$, for the parameter $\varphi = constant$.

The vector $\vec{R}_\varphi$ is calculated as follows:

$$R_\varphi = \frac{dx}{d\varphi} = \dot{\omega}_1^T(\varphi) \cdot \alpha^T \cdot (X + \alpha \cdot a), \tag{60}$$

which, after development, has the form:

$$R_\varphi = \begin{Vmatrix} -X \cdot \sin\varphi \cdot \sin\alpha - Y \cdot \sin\varphi \cdot \cos\alpha - (Z - m) \cdot \cos\varphi \\ -X \cdot \cos\varphi \cdot \sin\alpha + Y \cdot \cos\varphi \cdot \cos\alpha - (Z - m) \cdot \sin\varphi \end{Vmatrix}. \tag{61}$$

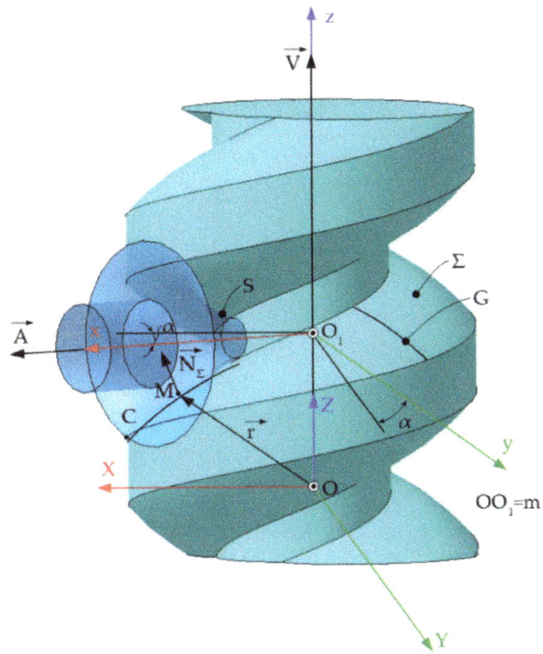

Figure 16. Generating of the helical surface with the finger cutter tool, reference systems: XYZ represents the mobile coordinate system, solid with the surface Σ, xyz—fixed reference system, having the x-axis superimposed on the tool axis, $\vec{A}$.

In the mobile coordinate system, XYZ, the parametric equations of the helical surface Σ are known by equations of type (47).

The normal $\vec{N_\Sigma}$ is defined with Equation (48), which, in the xyz coordinate system, has the director parameters:

$$\begin{aligned} n_x &= N_X \cdot \cos\alpha + N_Y \cdot \sin\alpha; \\ n_y &= -N_X \cdot \sin\alpha + N_Y \cdot \cos\alpha; \\ n_z &= N_Z. \end{aligned} \tag{62}$$

The condition for determining the characteristic curve, for $\varphi = 0$, is obtained from Equation (26):

$$[Y \cdot N_Z - (Z - m) \cdot N_Y] \cdot \cos\alpha - [X \cdot N_Z - (Z - m) \cdot N_X] \cdot \sin\alpha = 0, \tag{63}$$

which, associated with Equation (47), allows to obtain the parametric equations of the characteristic curve in form (52).

The Equation (63) is equivalent to the coplanarity condition of the vectors $\vec{N_\Sigma}$ and $\vec{A}$ is described by an equation of form, with X, Y and Z defined by (52):

$$\begin{vmatrix} N_X & N_Y & N_Z \\ \cos\alpha & \sin\alpha & 0 \\ X & Y & Z \end{vmatrix} = 0, \tag{64}$$

being known as "Nicolaev condition".

From a technological point of view, it is necessary to determine the axial section of the finger cutter tool, Figure 17.

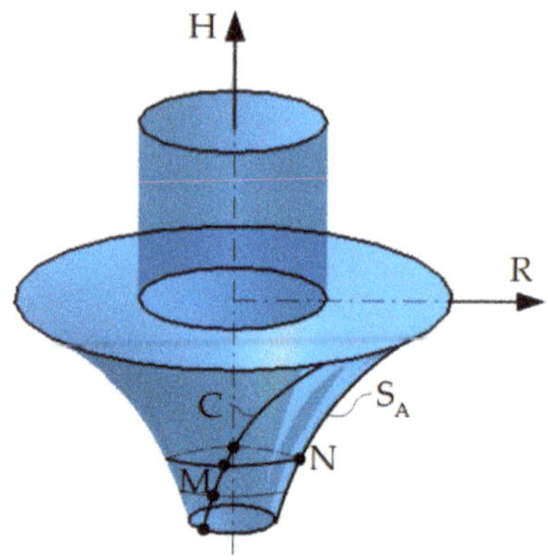

Figure 17. The axial section of the finger cutter tool.

The parametric equations of the characteristic curve are related to the coordinate system, *xyz*, by the transformation:

$$X = \alpha \cdot (x - a); \tag{65}$$

with C given by (54).

The parametric equations of the axial section are:

$$S_A \begin{cases} R = \sqrt{y^2(u) + z^2(u)}; \\ H = x(u). \end{cases} \tag{66}$$

Figure 18 shows the kinematic principle when processing helical surfaces with finger cutter tools [1,18].

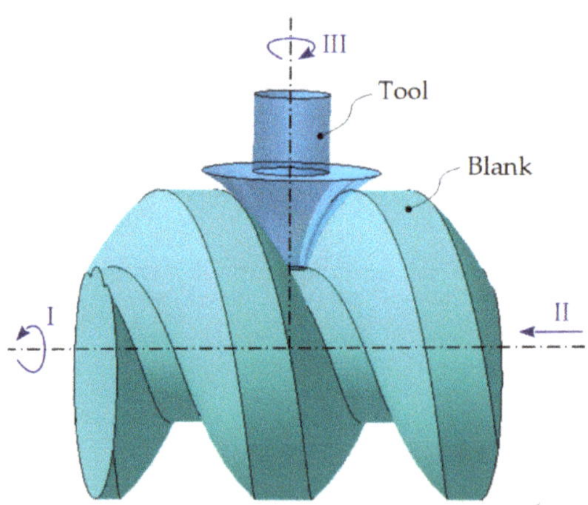

Figure 18. Kinematic principle.

Movement *I* represents the rotation of the blank around its own axis. This is correlated with the translational movement *II*, along the same axis, to ensure the helical movement.

Movement *III*, represents the rotational movement of the tool, being ensured by a special kinematic chain. The velocity with which the generator travels through the helical direction is conditioned by the possibilities of the technological process used.

2.8.3. Profiling of Cylindrical Generating Tools (Slotting Tools) for Generating of Helical Surfaces—Profiling Algorithm

Slotting is the process of generating helical surfaces, [7,53], most often used in the case of tools with cutting edges that belong to cylindrical surfaces or which, in relative movement to the blank, generates a cylindrical surface [1].

Although the processing productivity is relatively low, the operation of a slotting tool is simpler, due to the ease of its profiling process.

Figure 19 shows the relative position of the cylindrical (generating) and helical surfaces. The cylindrical surface is determined, reciprocally enveloping with the given surface.

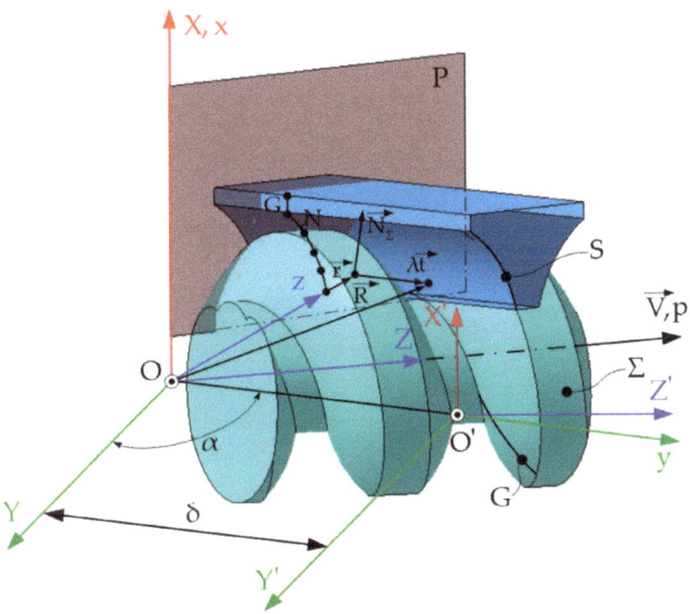

Figure 19. The relative position of the cylindrical (generating) and helical surfaces. Reference systems: *XYZ* represents the mobile reference system, joined with the helical surface; *xyz*—fixed coordinate system; the *y*-axis is parallel to the direction of the cylindrical surface generators; *X′Y′Z′*—mobile reference system resulting from the movement of the *XYZ* system with the variable parameter δ.

To generate the helical surface, *S* tool performs two movements:

- Translational movement, along the own generators, determining the main cutting movement;
- Helical movement of the axis $\vec{V}$ and helical parameter p, determining the generating movement of Σ surface.

As in the previous cases, the characteristic of the Σ surface does not depend on the helical movement, in which the surface self-generates.

Reference systems are defined according to Figure 19.

In the mobile reference system, the equations of the helical surface are known by equations of type (47), with u and v independent variables.

The movement of the *XYZ* system and of the Σ surface, along the generators of the cylindrical surface, is given by the relation:

$$x = \alpha^T \cdot X + a, \tag{67}$$

where:

$$\alpha = \begin{Vmatrix} 1 & 0 & 0 \\ 0 & \cos\alpha & -\sin\alpha \\ 0 & \sin\alpha & \cos\alpha \end{Vmatrix}; a = \begin{Vmatrix} 0 \\ \delta \\ 0 \end{Vmatrix}, \tag{68}$$

with δ variable parameter.

The director parameters of the normal, $\vec{N}_\Sigma$, are expressed as follows:

$$\vec{N}_\Sigma = N_Z \cdot \vec{i} + N_Y \cdot \vec{j} + N_Z \cdot \vec{k}. \tag{69}$$

In the xyz coordinate system, the director parameters of the normal are:

$$\begin{aligned} n_x &= N_X; \\ n_y &= N_Y \cdot \cos\alpha + N_Z \cdot \sin\alpha; \\ n_z &= -N_Y \cdot \sin\alpha + N_Z \cdot \cos\alpha. \end{aligned} \tag{70}$$

The vector $\vec{R}_\delta$ is calculated as follows:

$$R_\delta = \frac{dx}{d\delta} = \begin{Vmatrix} 0 \\ 1 \\ 0 \end{Vmatrix} \text{ or } \vec{R}_\delta = \vec{j}. \tag{71}$$

Thus, the condition for determining the characteristic curve, $\vec{N}_\Sigma \cdot \vec{R}_\varphi = 0$, is expressed as follows:

$$N_Y \cdot cos\alpha + N_Z \cdot sin\alpha = 0, \tag{72}$$

condition that allows the expression of the parametric equations of the characteristic curve of the form:

$$C \begin{cases} X = X(v); \\ Y = Y(v); \\ Z = Z(v). \end{cases} \tag{73}$$

In the xyz coordinate system, the equations of the characteristic curve are determined from:

$$x = \alpha^T \cdot X, \tag{74}$$

having the form:

$$C \begin{cases} x = X(v); \\ y = Y(v) \cdot \cos\alpha + Z(v) \cdot \sin\alpha; \\ z = -Y(v) \cdot \sin\alpha + Z(v) \cdot \cos\alpha. \end{cases} \tag{75}$$

Thus, the equation of the S surface will be:

$$\vec{R} = \vec{r} + \lambda \cdot \vec{t}, \tag{76}$$

where:

- $\vec{R}$ represents the vector of a current point on the cylindrical surface;
- $\vec{r}$—the vector of a current point on the characteristic curve;
- λ—variable parameter;
- $\vec{t} = \vec{j}$—the versor of the cylindrical surface generators.

Equation (76) is transcribed, in analytical form, as follows:

$$\begin{cases} x = X(v); \\ y = Y(v) \cdot \cos\alpha + Z(v) \cdot \sin\alpha + \lambda; \\ z = -Y(v) \cdot \sin\alpha + Z(v) \cdot \cos\alpha. \end{cases} \tag{77}$$

From (77), the condition results, representing the section $Y = 0$:

$$Y(v) \cdot cos\alpha + Z(v) \cdot sin\alpha + \lambda = 0, \quad (78)$$

that allows the determination of the cross section equations, ST:

$$S_T \begin{cases} y = y(\lambda); \\ z = z(\lambda). \end{cases} \quad (79)$$

Figure 20 shows the kinematic principle of generating in the processing of helical surfaces with slotting tools.

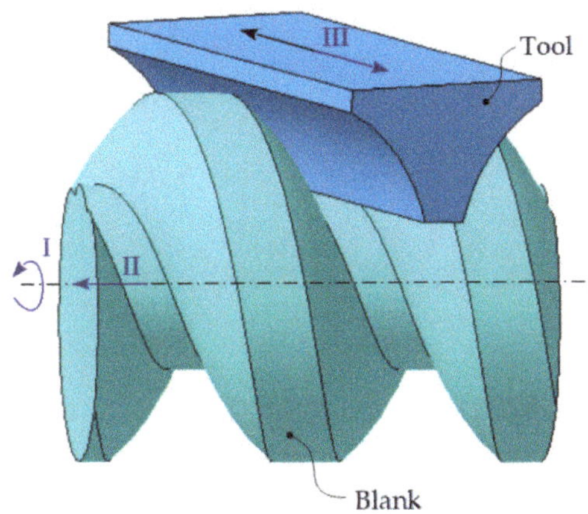

Figure 20. The kinematic principle of generating: I represents the rotational movement of the blank, II—the translational movement of the blank along its own axis, III—the cutting movement performed by the tool.

According to Figure 20, the movements are defined.

Movements *I* and *II* represent the helical movements of definition of the helical surface Σ.

2.9. Generating of Helical Surfaces by the Method of Decomposing the Helical Movement—Nikolaev Condition

2.9.1. Generating of Helical Surfaces with Disc Tools—Profiling Algorithm

The Nikolaev condition [2] is also known as the method of decomposing helical movement according to which: "the characteristic of a revolution surface, reciprocally enveloping a cylindrical and constant pitch helical surface, is the geometric locus of the points belonging to the helical surface for which the normal at this intersects the axis of the revolution surface" [2].

According to Figure 21, the problem involves determining the rotation surface S, the helical surface, Σ, being obtained by means of the helical movement of the generator, G.

The axis of the rotation surface—$\vec{A}$ forms, together with the axis of the Σ surface—$\vec{V}$ the angle of inclination α, the distance between the axes being determined by the scalar parameter a [2,12].

Reference systems are defined:

- xyz represents the fixed reference system, the z axis being superimposed on the surface axis Σ;
- $x_1y_1z_1$—fixed coordinate system with z_1 axis being superimposed on the axis $\vec{A}$;
- $x_2y_2z_2$—fixed coordinate system, z_2 axis being superimposed on the axis of rotation $\vec{B}$.

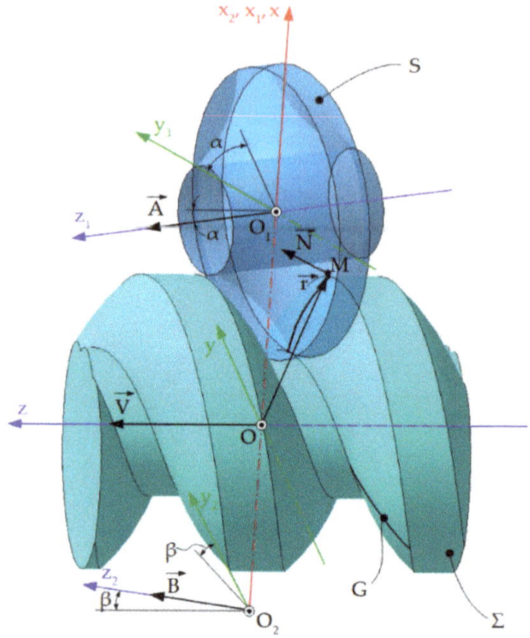

Figure 21. Generating with disc-tool.

The determination of the S surface results from the determination of its characteristic curve C, in the helical movement (V, ω, p).

In the helical movement of the characteristic curve C, the curve generates the Σ surface and, if rotational movement is applied around the axis $\vec{A}$, the S surface is generated.

The helical movement of the rotating surface is decomposed into rotational movements of axes $\vec{A}$ and $\vec{B}$, their sum being equivalent to the helical movement [2].

If the normal $\vec{N}$ is required to intersect the axis $\vec{A}$, therefore, according to the properties of the helical radii, the normal $\vec{N}$ also intersects the second conjugated axis.

The movement of the point M, corresponding to the S surface, represents a sum of rotations around the axes $\vec{A}$ and $\vec{B}$, the normal $\vec{N}$ being perpendicular to the velocity vector of this point. Therefore, the normal $\vec{N}$ determines the characteristic curve.

The parametric equations of the helical surface Σ are known:

$$\Sigma \left| \begin{array}{l} x = x(u,v); \\ y = y(u,v); \\ z = z(u,v). \end{array} \right. \tag{80}$$

The position vector of the point M is given by:

$$\vec{r} = x(u,\ v) \cdot \vec{i} + y(u,\ v) \cdot \vec{j} + z(u,\ v) \cdot \vec{k}. \tag{81}$$

The following are defined:

- Axis of rotation surface S:

$$\vec{A} = -\sin\alpha \cdot \vec{j} + \cos\alpha \cdot \vec{k}; \tag{82}$$

- The vector $\overline{O_1M}$, joining the origin of the fixed coordinate system, $x_1y_1z_1$ with the point M:

$$\overline{O_1M} = [x(u,\, v) - a] \cdot \vec{i} + y(u,\, v) \cdot \vec{j} + z(u,\, v) \cdot \vec{k}. \tag{83}$$

The characteristic curve, C, is determined from the coplanarity condition of the vectors $\vec{N}$, $\vec{A}$ and $\overline{O_1M}$, the condition being of the form:

$$\left(\overline{O_1M},\, \vec{N},\, \vec{A}\right) = 0, \tag{84}$$

or, taking into account the Equations (80)–(83):

$$\begin{vmatrix} x-a & y & z \\ N_x & N_y & N_z \\ 0 & -\sin\alpha & \cos\alpha \end{vmatrix} = 0, \tag{85}$$

N_x, N_y and N_z being the guiding parameters of the normal at the Σ surface.

By developing the Equation (85) the condition $\vec{N} \cdot \vec{V} = 0$ can be rewritten in the form:

$$[(x-a) \cdot N_z - z \cdot N_x] \cdot \sin\alpha + \left[(x-a) \cdot N_y - y \cdot N_x\right] \cdot \cos\alpha = 0. \tag{86}$$

Thus, the parametric equations of the characteristic curve, C, can be determined in form given by Equation (54), associating to the Σ surface equations the enwrapping condition (86).

To determine the axial section of the S surface, the characteristic curve is transposed into the fixed coordinate system, $x_1y_1z_1$, by transforming the coordinates:

$$\begin{Vmatrix} x_1 \\ y_1 \\ z_1 \end{Vmatrix} = \begin{Vmatrix} 1 & 0 & 0 \\ 0 & \cos\alpha & \sin\alpha \\ 0 & -\sin\alpha & \cos\alpha \end{Vmatrix} \cdot \left[\begin{Vmatrix} x \\ y \\ z \end{Vmatrix} - \begin{Vmatrix} a \\ 0 \\ 0 \end{Vmatrix}\right], \tag{87}$$

or, after development:

$$\left| \begin{array}{l} x_1 = x - a; \\ y_1 = y\cos\alpha + z\sin\alpha; \\ z_1 = -y\sin\alpha + z\cos\alpha. \end{array} \right. \tag{88}$$

If the parameters x, y and z from Equations (54) and (88) are replaced, in the fixed coordinate system, $x_1y_1z_1$, the equations of the characteristic curve C will be obtained:

$$C \left| \begin{array}{l} x_1 = x_1(u); \\ y_1 = y_2(u); \\ z_1 = z_3(u). \end{array} \right. \tag{89}$$

By the rotational movement of the characteristic curve C around the axis $\vec{A}$, the searched surface S can be generated.

Figure 22 shows the axial section of the disc tool.

The points M and N on the characteristic curve C and the generator G are on the same parallel circle. The radius of the circle is determined by the relation:

$$r = \sqrt{x_1^2(u) + y_1^2(u)}. \tag{90}$$

Therefore, the parametric equations of the generator G are determined:

$$G \left| \begin{array}{l} R = \sqrt{x_1^2(u) + y_1^2(u)}; \\ H = z_1(u). \end{array} \right. \tag{91}$$

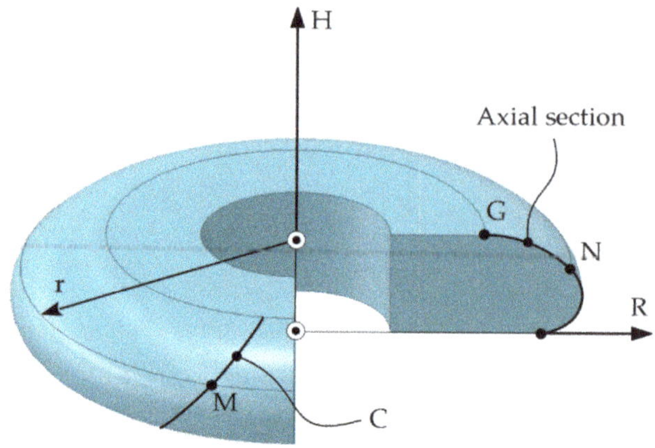

Figure 22. Axial section of the disc tool.

2.9.2. Generating of Helical Surfaces with Cylindrical-Front Tool—Profiling Algorithm

According to Figure 23, the helical surface Σ is known, this being determined in the helical movement (V, ω, p) of the generating curve, *G*, [2].

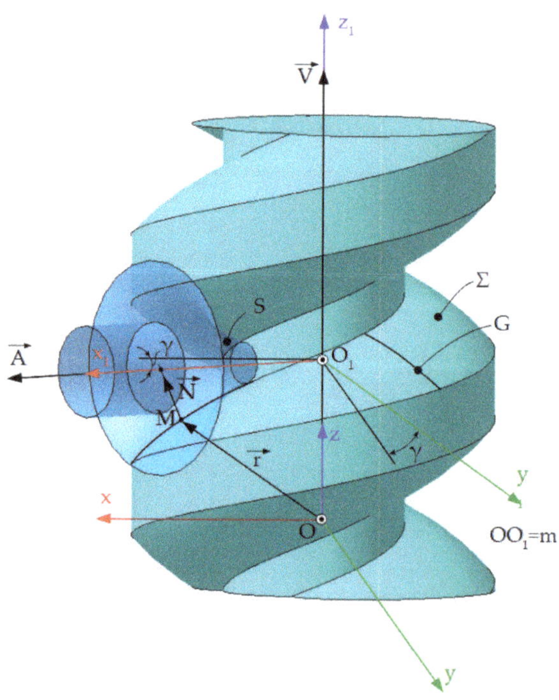

Figure 23. Generating with cylinder-front surface; reference systems: *z* represents the fixed reference system, with the *z* axis superimposed on the surface axis Σ; xz_1—fixed reference system, the x_1 axis being superimposed on the axis of the searched surface, *S*.

It is required to determine a rotation surface, *S*, with the axis perpendicular to the axis of the Σ surface. This surface is obtained from the condition that the Σ surface is the envelope of the *(S)* family in the helical movement and admits a common characteristic with the helical surface, Σ.

The *S* surface performs the following movements:

- Rotational movement around its own axis, constituting the cutting movement;
- Helical movement of the axis $\vec{V}$ and parameter p (the movement of generating the helical surface).

The normal to the Σ surface is perpendicular to the velocity vector and must be in the axial plane of the *S* surface.

Thus, for determination of the characteristic curve common to the reciprocally enveloping surfaces, the definition can be accepted: "the characteristic of a rotating surface, having the axis perpendicular to the axis of the helical surface with which it is reciprocally enveloping, represents the geometric locus of the points on the helical surface, where its normal intersects the axis of the rotating surface" [2].

According to Figure 23, the coordinate systems are defined.

In the *xyz* coordinate system, the parametric equations of the Σ surface are known in for given by Equation (80).

The position of the axis of the rotation surface of axis $\vec{A}$ is also known:

$$\vec{A} = \cos\gamma \cdot \vec{i} + \sin\gamma \cdot \vec{j}. \tag{92}$$

The vector of the current point on the Σ surface $\vec{r}$, is determined by the relation (81).

At the same point, the normal to the helical surface Σ is given by (69).

The vector $\overline{O_1M}$ is calculated as follows:

$$\vec{O_1M} = \vec{r} - \vec{OO_1} = \vec{r} - m \cdot \vec{k}, \tag{93}$$

where the m parameter is arbitrary and establishes a convenient position of the axis towards to the Σ surface.

The condition of coplanarity is given by (84), or, after replacement:

$$\begin{vmatrix} x & y & z-m \\ \cos\gamma & \sin\gamma & 0 \\ N_x & N_y & N_z \end{vmatrix} = 0. \tag{94}$$

By development, it results the condition for determining the characteristic curve of form:

$$[y \cdot Nz - (z-m) \cdot Ny] \cdot \cos\gamma - [x \cdot Nz - (z-m) \cdot Nx] \cdot \sin\gamma = 0. \tag{95}$$

Equations (54) and (95) define the equations of the characteristic curve C, common to the surfaces *S* and Σ.

If it is chosen that the origins *O* and O_1 of the reference systems coincide and that the axis $\vec{A}$ is parallel, for example, with the *x*-axis, then the relation (95) becomes:

$$x \cdot N_z - z \cdot N_x = 0. \tag{96}$$

If a connection is established between the parameters u and v, of form $u = u(v)$, then the equations of the characteristic curve will be in form (52):

$$C \left| \begin{array}{l} x = x(v); \\ y = y(v); \\ z = z(v). \end{array} \right. \tag{97}$$

By relating the equations from the relation (97) to the reference system $x_1y_1z_1$, joined with the S surface, the transformation of coordinates is obtained:

$$\begin{Vmatrix} x_1 \\ y_1 \\ z_1 \end{Vmatrix} = \begin{Vmatrix} \cos\gamma & \sin\gamma & 0 \\ -\sin\gamma & \cos\gamma & 0 \\ 0 & 0 & 1 \end{Vmatrix} \cdot \left[\begin{Vmatrix} x \\ y \\ z \end{Vmatrix} - \begin{Vmatrix} 0 \\ 0 \\ m \end{Vmatrix} \right]. \tag{98}$$

Thus, the parametric equations of the characteristic curve will be brought, in principle, to the form:

$$C \left| \begin{array}{l} x_1 = x_1(v); \\ y_1 = y_1(v); \\ z_1 = z_1(v). \end{array} \right. \tag{99}$$

The peripheral surface of the finger cutter tool is obtained by rotating the characteristic around the axis $\vec{A}$.

According to Figure 24, the points M and N on the characteristic curve and on the generator are at the same distance from the surface axis x_1 (with x_1 axis superimposed on the H axis).

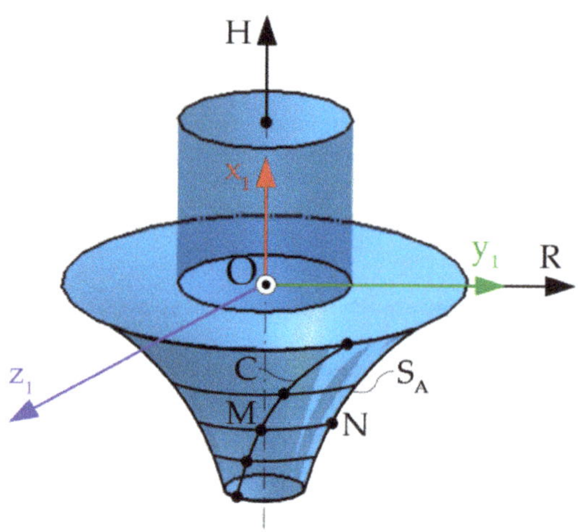

Figure 24. Axial section.

The radius of the parallel circle is determined by the equation:

$$R = \sqrt{y_1^2(v) + z_1^2(v)}, \tag{100}$$

where $y_1(v)$ and $z_1(v)$ have the meanings given by the Equation (99).

Therefore, the parametric equations of the axial section, S_A, are given by the relation:

$$S_A \left| \begin{array}{l} R = \sqrt{y_1^2(v) + z_1^2(v)}; \\ H = x_1(u). \end{array} \right. \tag{101}$$

2.9.3. Generating of Helical Surfaces with Cylindrical Tools—Profiling Algorithm

The problem involves knowing the helical surface Σ, obtained by the helical movement of the generating curve, G and determining a cylindrical surface, S, generated in the directional movement by the edge of a planing tool, Figure 25 [2,12].

The S surface is determined from the condition that the surface is the envelope of the (S) family, in the helical movement of axis $\vec{V}$ and parameter p. Therefore, the S surface admits a contact curve with the helical surface, this curve being the characteristic curve, C.

Therefore, the cylindrical tool must perform the following movements:

- Translational movement along the own generators, which constitutes the cutting movement;
- Helical movement, identical with the movement of generating the helical surface by the generator, G.

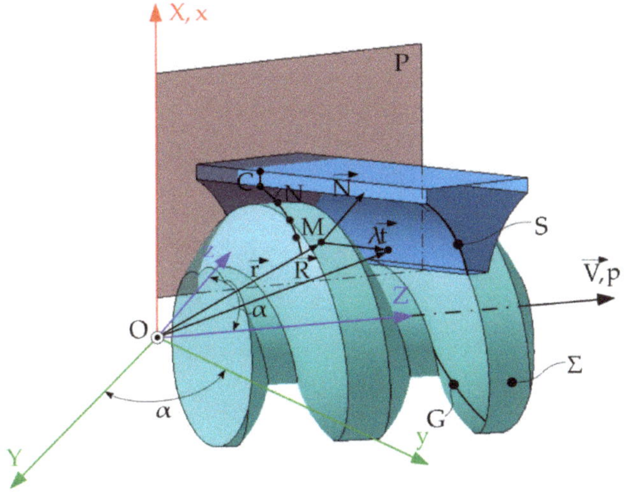

Figure 25. Generating with cylindrical tools.

The determination of the characteristic curve assumes that the normal to the Σ surface is perpendicular to the direction of velocity, in the translational movement along the generators of the S surface, therefore the condition for determining the characteristic curve of the reciprocally enveloping surfaces, Σ and S can be stated as follows: "the characteristic of the helically enwrapping surface with a cylindrical surface is the geometric locus of the points on the helical surface, where the normal to it is perpendicular to the direction of the cylindrical surface generators".

In the xyz coordinate system, the Σ surface is known in form (80), the direction versor of the generators of the cylindrical surface $\vec{t}$, is expressed in the form $\vec{t} = \sin\alpha \cdot \vec{j} + \cos\alpha \cdot \vec{k}$, where α represents the angle that the generators make with the axis $\vec{V}$.

If denoted by $\vec{N}$ the normal at the current point on the helical Σ surface, having the Equation (69), then, taking into account the Equations (80) and (81), the condition for determining the characteristic C, is expressed in the form:

$$\vec{N} \cdot \vec{t} = 0, \tag{102}$$

or, after replacing:

$$N_y \cdot \sin\alpha + N_z \cdot \cos\alpha = 0. \tag{103}$$

Equations (80) and (103) represent the parametric equations of the characteristic curve C on the helical surface, Σ.

The Equation (103) determines a connection between the parameters u and v of the form $u = u(v)$, so that the parametric equations of the characteristic C can be expressed as in (97).

Characteristic C from Figure 25 together with the versor $\vec{t}$ generates the S surface, which is in fact the searched surface.

The equation of the S surface, in vector form, is $\vec{R} = \vec{r} + \lambda \vec{t}$, where λ is a variable parameter and the vector $\vec{r}$ of a point on the characteristic curve.

The parametric equations of the cylindrical surface, S, are:

$$\left| \begin{array}{l} x = x(v); \\ y = y(v) + \lambda \cdot \sin\alpha\ ; \\ z = z(v) + \lambda \cdot \cos\alpha. \end{array} \right. \tag{104}$$

The cross section, S_T, of the S surface determines the form of the template with which the profile of the planing cutter is controlled.

Therefore, it is considered a plane containing the x-axis and is perpendicular to the direction of the versor $\vec{t}$ denoted by P.

The equation of the P plane is:

$$\vec{r}_1 \cdot \vec{t} = 0, \tag{105}$$

where $\vec{r}_1$ represents the position vector of a current point in the P plane calculated by the relation:

$$\vec{r}_1 = x \cdot \vec{i} + y \cdot \vec{j} + z \cdot \vec{k}. \tag{106}$$

After replacing the vectors, it results:

$$y \cdot \sin\alpha + z \cdot \cos\alpha = 0. \tag{107}$$

Equations (104) and (107) represent the parametric equations of the cross section of the cylindrical surface, S, the G curve:

$$G \left| \begin{array}{l} x = x(\lambda); \\ y = y(\lambda); \\ z = z(\lambda). \end{array} \right. \tag{108}$$

By coordinate transformation, changing the reference system, from xyz to $x_1y_1z_1$, the equations of the cross section take the form:

$$\left\| \begin{array}{c} x_1 \\ y_1 \\ z_1 \end{array} \right\| = \left\| \begin{array}{ccc} 1 & 0 & 0 \\ 0 & \cos\alpha & \sin\alpha \\ 0 & -\sin\alpha & \cos\alpha \end{array} \right\| \cdot \left\| \begin{array}{c} x \\ y \\ z \end{array} \right\|. \tag{109}$$

Taking into account the equation in relation (109), the parametric equations of the cross section, S are determined:

$$S \left| \begin{array}{l} x_1 = x(\lambda); \\ z_1 = -y(\lambda) \cdot \cos\alpha + z(\lambda) \cdot \sin\alpha. \end{array} \right. \tag{110}$$

Equation (110) is the profile of the planing cutter that performs a translational movement, representing the cutting movement, according to the direction of the versor $\vec{t}$.

3. Conclusions

The generating of surfaces by enwrapping presupposes the existence of specific methodologies for solving the problem of profiling generating tools.

The profiling of the tools for generating by enwrapping can be done using the fundamental theorems, applied in the conditions in which the surfaces are considered non-deformable.

These theorems have a high degree of generality and mathematical rigor, but can sometimes be difficult to apply. The way of expressing the enwrapping conditions leads, in many of the practical cases, to laborious analytical calculations and to complicated analytical forms of expression of the equations of the enwrapping surfaces.

This led to the development of complementary methods, characterized by a simpler expression of the enwrapping conditions.

Complementary methods are methods that use the direct way of expressing the enwrapping conditions and treat problems specific to the generating of reciprocally enwrapping surfaces.

These methods have a simpler form of expression, but a lower degree of generality.

The paper presented a review on study methods for reciprocally enwrapping surfaces, as well as methods for profiling tools for generating helical surfaces by the kinematic method and by the Nicolaev condition, forming a basis for future research papers and projects.

Author Contributions: Conceptualization, N.O., V.G.T. and N.B.; methodology, N.O. and V.G.T.; validation, N.B., V.G.T., N.O. and V.P.; formal analysis, N.O.; investigation, G.A.M.; resources, N.O., V.P., V.G.T. and N.B.; writing—original draft preparation, G.A.M.; writing—review and editing, G.A.M.; visualization, N.B. and V.G.T.; supervision, N.O.; project administration, V.G.T. and N.B.; funding acquisition, G.A.M. All authors have read and agreed to the published version of the manuscript.

Funding: Academic excellence and entrepreneurial values—scholarship system to ensure opportunities for training and development of entrepreneurial skills of doctoral and postdoctoral students—ANTREPRENORDOC, POCU project/380/6/13, contract number 36355/23.05.2019 HRD OP /380/6/13 SMIS Code: 123847 http://www.antreprenordoc.ugal.ro/index.php/ro/ (accessed on 1 December 2021).

Institutional Review Board Statement: Not applicable.

Informed Consent Statement: Not applicable.

Data Availability Statement: Not applicable.

Acknowledgments: This work is supported by the project *ANTREPRENORDOC*, in the framework of Human Resources Development Operational Programme 2014–2020, financed from the European Social Fund under the contract number 36355/23.05.2019 HRD OP /380/6/13–SMIS Code: 123847.

Conflicts of Interest: The authors declare no conflict of interest.

References

1. Radzevich, S.P. *Kinematics Geometry of Surface Machining*; CRC Press: London, UK, 2008.
2. Oancea, N. *Surfaces Generation through Winding, Vol. I–III*; Galati University Press: Galati, Romania, 2004.
3. Cioară, R. Kinematic Structures for Processing of Surfaces with a Circle Directrix and a Straight Line Generatrix (Part IV). *IOP Conf. Ser. Mater. Sci. Eng.* **2016**, *161*, 1–8.
4. Airy, G.B. On the Forms of the Teeth of Wheels. *Trans. Camb. Philos. Soc.* **1827**, *2*, 277–286.
5. Figliolini, G.; Stachel, H.; Angeles, J. Kinematic Properties of Planar and Spherical Logarithmic Spirals: Applications to the Synthesis of Involute Tooth Profiles. *Mech. Mach. Theory* **2019**, *136*, 14–26. [CrossRef]
6. He, F.; Shi, Z.; Yu, B. Effects of Tooth Surface Modification on Planar Double-Enveloping Hourglass Worm Gear Drives. *J. Adv. Mech. Des. Syst. Manuf.* **2018**, *12*, 1–13. [CrossRef]
7. Stanieck, R. Shaping of Face Toothing in Flat Spiroid Gears. *J. Mech. Eng.* **2011**, *57*, 47–54. [CrossRef]
8. Totolici, S.; Teodor, V.G.; Baroiu, N.; Oancea, N. A New Profile for the Worm Gear Drive of a Spiral Gear. *IOP Conf. Ser. Mater. Sci. Eng.* **2018**, *400*, 1–9. [CrossRef]
9. Carmo, M.P. *Differential Geometry of Curves and Surfaces*; Courier Dover Publications: New York, NY, USA, 2016.
10. Phillips, J. *General Spatial Involute Gearing*; Springer: Berlin/Heidelberg, Germany; New York, NY, USA, 2003.
11. Litvin, F.L. *Theory of Gearing, Reference Publication 1212*; NASA, Scientific and Technical Information Division: Washington, DC, USA, 1984.
12. Teodor, V.G. *Contributions to the Elaboration of a Method for Profiling Tools. Tools Which Generate by Enwrapping*; Lambert Academic Publishing: Saarbrücken, Germany, 2010.
13. Chan, C.L.; Ting, K.L. Extended Camus Theory and Higher Order Conjugated Curves. *J. Mech. Robot.* **2019**, *11*, 1–9. [CrossRef]

14. Dana-Picard, T.; Zehavi, N. Automated study of envelopes of one-parameter families of surfaces. *Springer Proc. Math. Stat.* **2017**, *198*, 29–44.
15. Dooner, D.B. On the Third Law of Gearing: A Study on Hypoid Gear Tooth Contact. *Mech. Mach. Theory* **2019**, *134*, 224–248. [CrossRef]
16. Guo, Q.; Qing, G.G.; Jiang, Y. An Analytic Method of Computing the Envelope Surface of General Cutter with Runout in 5-axis Machining for Manufacturing Systems. *J. Chin. Soc. Mech. Eng.* **2017**, *38*, 403–412.
17. Ding, H.; Zhou, Y.S.; Tang, J.Y.; Zhong, J.; Zhou, Z.; Wan, G. A Novel Operation Approach to Determine Initial Contact Point for Tooth Contact Analysis with Errors of Spiral Bevel and Hypoid Gears. *Mech. Mach. Theory* **2017**, *109*, 155–170. [CrossRef]
18. Albu, S.C. Simulation of Processing of a Helical Surface with the Aid of a Frontal-Cylindrical Milling Tool. *Procedia Manuf.* **2019**, *32*, 36–41. [CrossRef]
19. Huai, C.; Zhao, Y. Meshing Theory and Tooth Profile Geometry of Toroidal Surface Enveloping Conical Worm Drive. *Mech. Mach. Theory.* **2019**, *134*, 476–498. [CrossRef]
20. Jia, K.; Zheng, S.; Guo, J.; Hong, J. A Surface Enveloping-Assisted Approach on Cutting Edge Calculation and Machining Process Simulation for Skiving. *Int. J. Adv. Manuf. Technol.* **2019**, *100*, 1635–1645. [CrossRef]
21. Liu, Z.; Lu, H.; Yu, G.; Wang, S. A Novel CNC Machining Method for Enveloping Surface. *Int. J. Adv. Manuf. Technol.* **2016**, *85*, 779–790. [CrossRef]
22. Liu, Z.; Tang, Q.; Liu, N.; Song, J. A Profile Error Compensation Method in Precision Grinding of Screw Rotors. *Int. J. Adv. Manuf. Technol.* **2019**, *100*, 2557–2567. [CrossRef]
23. Meng, Q.; Zhao, Y.; Yang, Z. Meshing Limit Line of the Conical Surface Enveloping Conical Worm Pair. *Proc. Inst. Mech. Eng. Part C J. Mech. Eng. Sci.* **2020**, *234*, 693–703. [CrossRef]
24. Meng, Q.; Zhao, Y.; Yang, Z.; Cui, J. Meshing Theory and Error Sensitivity of Mismatched Conical Surface Enveloping Conical Worm Pair. *Mech. Mach. Theory* **2020**, *145*, 103681. [CrossRef]
25. Yang, J.; Li, H.; Rui, C.; Wei, W.; Dong, X. A Method to Generate the Spiral Flutes of an Hourglass Worm Gear Hob. *J. Mech. Des. Trans. ASME* **2018**, *140*, 1–12. [CrossRef]
26. Zhao, Y.; Kong, X. Meshing Principle of Conical Surface Enveloping Spiroid Drive. *Mech. Mach. Theory* **2018**, *123*, 1–26. [CrossRef]
27. Zhou, Y.; Wu, Y.; Wang, L.; Tang, J.; Ouyang, H. A New Closed-Form Calculation of Envelope Surface for Modeling Face Gears. *Mech. Mach. Theory* **2019**, *137*, 211–226. [CrossRef]
28. Berbinschi, S.; Teodor, V.G.; Baroiu, N.; Oancea, N. The Substitutive Circles Family Method-Graphical Aproach in CATIA Design Environment. *Ann. Dunarea Jos Univ. Galati Fascicle V Technol. Mach. Build.* **2013**, *1*, 53–66.
29. Baroiu, N.; Berbinschi, S.; Teodor, V.G.; Oancea, N. Comparative Study of Drill's Flank Geometry Developed with the CATIA Software. *Ann. Dunarea Jos Univ. Galati Fascicle V Technol. Mach. Build.* **2012**, *1*, 27–32.
30. Baroiu, N.; Berbinschi, S.; Teodor, V.G.; Susac, F.; Oancea, N. The Complementary Graphical Method used for Profiling Side Mill for Generation of Helical Surface. In Proceedings of the IOP Conf. Series: Materials Science and Engineering, 2017, 227, ModTech International Conference Modern Technologies in Industrial Engineering, Sibiu, Romania, 14–17 June 2017.
31. Baroiu, N.; Baroiu, L.; Teodor, V.G.; Ciocan, T.L. Graphical Method for Profiling the Side Mill which Generate Helical Flute of Tungsten Carbide Dental Cross Cut Bur. *Rom. J. Mater.* **2018**, *48*, 131–139.
32. Berbinschi, S.; Teodor, V.G.; Oancea, N. 3D Graphical Method for Profiling Gear Hob Tools. *Int. J. Adv. Manuf. Technol.* **2013**, *64*, 291–304. [CrossRef]
33. Berbinschi, S.; Teodor, V.G.; Baroiu, N.; Oancea, N. Enwrapping Surfaces with Point Contact-Comparison between CATIA Method and Analytical One. *Ann. Dunarea Jos Univ. Galati Fascicle V Technol. Mach. Build.* **2011**, *2*, 117–122.
34. Berbinschi, S.; Teodor, V.G.; Frumuşanu, G.R.; Oancea, N. Graphical Method for Profiling the Tools which Generate Internal Surfaces by Rolling. *Acad. J. Manuf. Eng.* **2014**, *12*, 12–17.
35. Berbinschi, S.; Teodor, V.G.; Oancea, N. A Graphical Method Developed in CATIA Design Environment for the Modeling of Generation by Enveloping. *Ann. Dunarea Jos Univ. Galati Fascicle V Technol. Mach. Build.* **2011**, *1*, 25–30.
36. Berbinschi, S.; Baroiu, N.; Teodor, V.G.; Oancea, N. Profiling Method of Side Mill for Threading Screw for Dental Implants. *Adv. Mater. Res.* **2014**, *837*, 22–27. [CrossRef]
37. Teodor, V.G.; Popa, I.; Oancea, N. The profiling of End Mill and Planing Tools to Generate Helical Surfaces Known by Sampled Points. *Int. J. Adv. Manuf. Technol.* **2010**, *51*, 439–452. [CrossRef]
38. Teodor, V.G.; Popa, I.; Popa, C.; Dura, G.; Oancea, N. Algorithm for Gear Hub Tool Profiling by Bezier Polynomial Approximation. The Rotary Helical Screw Compressor Case. In Proceedings of the 8th WSEAS International Conference on System Science and Simulation in Engineering (ICOSSSE'09), Genova, Italy, 17–19 October 2009.
39. Baroiu, N.; Teodor, V.G.; Susac, F.; Oancea, N. Hob mill for trilobed rotor-Graphical method in CATIA. *IOP Conf. Ser. Mater. Sci. Eng.* **2018**, *448*, 1–17. [CrossRef]
40. Teodor, V.G.; Baroiu, N.; Berbinschi, S.; Susac, F.; Oancea, N. A Graphical Solution in CATIA for Profiling End Mill Tool which Generates a Helical Surface. *IOP Conf. Ser. Mater. Sci. Eng.* **2017**, *227*, 1–15. [CrossRef]
41. Teodor, V.G.; Baroiu, N.; Susac, F. *The Synthesis of New Algorithms for CAD Profiling of Cutting Tools*; Lambert Academic Publishing: Saarbrücken, Germany, 2018.
42. Baroiu, N.; Teodor, V.G.; Susac, F.; Oancea, N. Hob Mill Profiling Method for Generation of Timing Belt Pulley. In Proceedings of the 5th International Conference on Advanced Manufacturing and Technologies (NewTech), Belgrade, Serbia, 6–9 June 2017; pp. 13–26.

43. Hong, X.; Hong, R.; Lin, X. Tool Orientations' Generation and Nonlinear Error Control based on Complex Surface Meshing. *Int. J. Adv. Manuf. Technol.* **2019**, *105*, 4279–4288. [CrossRef]
44. Wang, Q.; Feng, Y.; Gao, Y.; Li, Z.; Tan, J. Smooth Fillet-End Cutter Tool Path Generation Method on Triangular-Mesh Surface Based on Modified Butterfly Subdivision. *Int. J. Adv. Manuf. Technol.* **2018**, *98*, 2831–2847. [CrossRef]
45. Cera, M.; Pennestrì, E. The Mechanical Generation of Planar Curves by Means of Point Trajectories, Line and Circle Envelopes: A Unified Treatment of the Classic and Generalized Burmester Problem. *Mech. Mach. Theory* **2019**, *142*, 103580. [CrossRef]
46. Cera, M.; Pennestrì, E. Higher-order curvature analysis of planar curves enveloped by straight-lines. *Mech. Mach. Theory* **2019**, *134*, 213–223. [CrossRef]
47. Gouchao, L.; Sun, J.; Jianfeng, L. Process Modeling of End Mill Groove Machining Based on Boolean Method. *Int. J. Adv. Manuf. Technol.* **2014**, *57*, 959–966.
48. Guo, Q.; Sun, Y.; Jiang, Y.; Yan, Y.; Zhao, B.; Ming, P. Tool Path Optimization for Five-Axis Flank Milling with Cutter Runout Effect Using the Theory of Envelope Surface Based on CL Data for General Tools. *J. Manuf. Syst.* **2016**, *38*, 87–97. [CrossRef]
49. Liu, G.; Wei, W.; Dong, X.; Rui, C.; Liu, P.; Li, H. Relief Grinding of Planar Double-Enveloping Worm Gear Hob Using a Four-Axis CNC Grinding Machine. *Int. J. Adv. Manuf. Technol.* **2016**, *89*, 9–12. [CrossRef]
50. Nieszporek, T.; Boral, P.; Gołębski, R. Particular Solution of Cutting Tool Path Applied on Helical Surface with Circular Profile. *Teh. Vjesn.* **2019**, *26*, 22–28.
51. Oancea, N. Méthode numérique pour l'étude des surfaces enveloppées. *Mech. Mach. Theory* **1996**, *31*, 957–972.
52. Dooner, D.B.; Griffis, M.W. On spatial Euler-Savary Equations for Envelopes. *J. Mech. Des. Trans. ASME* **2007**, *129*, 865–875. [CrossRef]
53. Berbinschi, S.; Teodor, V.G.; Oancea, N. A 3D Method for Profiling the Shaping Tool for the Generation of Helical Surfaces. *Ann. Dunarea Jos Univ. Galati Fascicle V Technol. Mach. Build.* **2011**, *1*, 19–24.
54. Baroiu, N.; Teodor, V.G.; Oancea, N. A New Form of Plane Trajectories Theorem. Generation with Rotary Cutter. *Bull. Polytech. Inst. Iasi* **2015**, *3*, 27–36.
55. Baroiu, N.; Teodor, V.G.; Popa, C.; Oancea, N. Gear Shaped Cutter–A Profiling Method Developed in Graphical Form. *Ann. Dunarea Jos Univ. Galati Fascicle V Technol. Mach. Build.* **2015**, *33*, 9–16.
56. Teodor, V.G.; Baroiu, N.; Susac, F.; Oancea, N. The Modelling of Involute Teeth Generation with the Relative Generating Trajectories Method. *Acad. J. Manuf. Eng.* **2016**, *14*, 126–131.
57. Teodor, V.G.; Baroiu, N.; Susac, F.; Oancea, N. The Rack-Gear Tool Generation Modelling. Non-Analytical Method Developed in CATIA, Using the Relative Generating Trajectories Method. *IOP Conf. Ser. Mater. Sci. Eng.* **2016**, *161*, 1–6. [CrossRef]
58. Teodor, V.G.; Păunoiu, V.; Berbinschi, S.; Baroiu, N.; Oancea, N. The Method of "In-Plane Generating Trajectories" for Tools Which Generate By Enveloping-Application in CATIA. *J. Mach. Eng.* **2015**, *15*, 69–80.
59. Berbinschi, S.; Frumuşanu, G.; Teodor, V.G.; Oancea, N. The method of substitutive circles family. Application in CATIA design environment for gear shaped tool profiling. *Adv. Mater. Res.* **2014**, *1036*, 370–375. [CrossRef]
60. Berbinschi, S.; Teodor, V.G.; Frumuşanu, G.R.; Oancea, N. The pinion cutter for profiled holes-Graphical method in CATIA. *Appl. Mech. Mater.* **2014**, *657*, 720–724. [CrossRef]
61. Teodor, V.G.; Berbinschi, S.; Baroiu, N.; Oancea, N. Study of the Enwrapping Profiles Associated with Rolling Centrodes by the Minimum Distance Method. Graphical Solution Developed in the CATIA Design Environment. *Appl. Mech. Mater.* **2014**, *656*, 181–191. [CrossRef]
62. Popa, L.C.; Popa, I.; Teodor, V.G.; Baroiu, N. Profiling Tool Generating Method using the Profile Measurement of Rotor of Screw Compressor Components. *Ann. Dunarea Jos Univ. Galati Fascicle V Technol. Mach. Build.* **2011**, *2*, 123–128.
63. Baroiu, N.; Moroşanu, G.A.; Teodor, V.G.; Oancea, N. Roller Profiling for Generating the Screw of a Pump with Progressive Cavities. *Inventions* **2021**, *6*, 34. [CrossRef]
64. Bo, P.; Bartoň, M.; Plakhotnik, D.; Pottmann, H. Towards efficient 5-axis flank CNC machining of free-form surfaces via fitting envelopes of surfaces of revolution. *Comput. -Aided Des.* **2016**, *79*, 1–11. [CrossRef]
65. Wang, W.; Wang, D. Curvature Theory of Envelope Curve in Two-Dimensional Motion and Envelope Surface in Three-Dimensional Motion. *J. Mech. Robot.* **2015**, *7*, 1–9. [CrossRef]
66. Zanger, F.; Sellmeier, V.; Klose, J.; Bartkowiak, M.; Schulze, V. Comparison of Modeling Methods to Determine Cutting Tool Profile for Conventional and Synchronized Whirling. *Procedia CIRP* **2017**, *58*, 222–227. [CrossRef]
67. Hu, P.; Chen, L.; Tang, K. Efficiency-Optimal Iso-Planar Tool Path Generation for Five-Axis Finishing Machining of Freeform Surfaces. *Comput. -Aided Des.* **2018**, *83*, 33–50. [CrossRef]
68. Berbinschi, S.; Teodor, V.G.; Baroiu, N.; Oancea, N. Profiling Methodology for Side Mill Tools for Generation of Helical Compressor Rotor Using Reverse Engineering. *Ann. Dunarea Jos Univ. Galati Fascicle V Technol. Mach. Build.* **2011**, *2*, 111–116.
69. Berbinschi, S.; Teodor, V.G.; Oancea, N. 3D Graphical Method for Profiling Tools that Generate Helical Surfaces. *Int. J. Adv. Manuf. Technol.* **2012**, *60*, 505–512. [CrossRef]
70. Berbinschi, S.; Baroiu, N.; Teodor, V.G.; Oancea, N. End Mill Tool Profiling by CAD Method. *Indian J. Eng. Mater. Sci.* **2014**, *21*, 296–302.

inventions

MDPI

Review

An Overview of the Ship Ventilation Systems and Measures to Avoid the Spread of Diseases

Victor Mihai * and Liliana Rusu

Faculty of Engineering, University "Dunarea de Jos" of Galati, 47 Domnească Street, 800008 Galati, Romania; liliana.rusu@ugal.ro
* Correspondence: victor.mihai@ugal.ro; Tel.: +40-749065510

Abstract: This article presents a review of the main aspects regarding the current rules of classification societies, standards, and practice regarding the design and construction of ventilation and air conditioning systems for different compartments in different types of ships. In the context of the COVID-19 pandemic, this paper also presents the usual practice of the actual heating ventilation and air conditioning (HVAC) systems used on large ships, which recirculate the air between living compartments, in comparison with the new requirements to avoid the risk of spreading diseases. According to the rules, the technical compartments are provided with independent ventilation systems that ensure high air flow rates; therefore, the spread of diseases through this system is not an issue. The living spaces are provided with common ventilation and air conditioning systems that recirculate the air in all compartments served. The current practice of air recirculation in various living rooms leads to the spread of diseases, which should therefore be analyzed and improved by adding high-efficiency particulate air (HEPA) filters and UV disinfection or be replaced with individual systems that provide local heating or cooling without air recirculation between different rooms and fresh air supply with complete evacuation. For existing ships, different solutions should be analyzed such as reducing or cancelling recirculation and increasing filtration.

Keywords: ventilation on ships; HVAC requirements; COVID-19; spread the disease; airborne transmission; HEPA filtration; UV disinfection; heat recovery

Citation: Mihai, V.; Rusu, L. An Overview of the Ship Ventilation Systems and Measures to Avoid the Spread of Diseases. *Inventions* **2021**, *6*, 55. https://doi.org/10.3390/inventions6030055

Academic Editors: Amir H. Alavi and Kambiz Vafai

Received: 7 June 2021
Accepted: 22 July 2021
Published: 27 July 2021

1. Introduction

Ship ventilation systems have different requirements depending on the destination of the compartments and the type of ship. Thus, there are compartments in which natural ventilation is sufficient while other compartments require forced ventilation using extraction and/or the introduction fans, and for other compartments, the ventilation is provided through an air conditioning system.

Therefore, according to classification society rules, the special compartments in which flammable or toxic gases can be released should be provided with exhaust mechanical ventilation. The air change rate required is different, as it is necessary to reduce the risk of flammable or toxic gases being allowed to escape and accumulate.

Other compartments in which installed equipment needs air to run should be provided with an independent mechanical supply or both supply and exhaust ventilation, which should ensure the required air for the running and cooling of equipment. Compartments in which there are no special requirements from equipment makers or classification societies are provided with natural ventilation only.

The living rooms, in general, are provided with common ventilation and air conditioning systems to ensure a proper climate for crew and passengers. The minimum requested fresh air flow is related to the number of persons but is not less than 40% of the total air flow [1]. In addition, there are rooms such as the galley, toilets, mess rooms, and laundry where recirculation is not allowed. The common ventilation and air conditioning systems, which are provided with a common air handling unit (AHU), have the advantage that the noise inside the rooms served is low and the maintenance of the equipment is quite easy because the main equipment is installed in a dedicated room.

Nevertheless, it should be mentioned that, in the case of common ventilation and air conditioning systems, which mix the air from rooms with fresh air and recirculate it back to the rooms, the spread of diseases through the air conditioning system could be high. The spread of diseases through the air conditioning system can be reduced by adding HEPA filters [2], but in this case, the protection is limited and the fans should have increased pressure. If the HEPA filters are installed on the return ventilation duct, and if the air handling unit (AHU) is not provided with a recirculation fan, the pressure inside the rooms will increase by 150–900 Pa due to the pressure drop across the new filter. Therefore, the system should be provided with an additional exhaust fan.

The spread of diseases through air conditioning systems can also be reduced by adding ultraviolet (UV) disinfection and "more than 90% of the SARS-CoV-2 virus, more than 97% of Influenza A virus, and 100% of Legionella pneumophila can be inactivated" [3]. However according to the IES [4], if the UV exposure time is increased by 50%, the efficiency of disinfection will increase from 90% to 99%.

Many studies have confirmed the high level of aerosol and droplet transmission of the SARS-CoV-2 virus. Hwang et al. [5] reported that the virus spreads to all floors through the ventilation duct and suggested aerosol transmission. Priyanka et al. [6] reported the aerosol transmission of SARS-CoV-2, especially in enclosed spaces, and Li et al. [7] concluded that the aerosol transmission of SARS-CoV-2 due to poor ventilation may explain the community spread of the disease.

M. Guo et al. [8] mention that respiratory droplets will evaporate and become smaller particles in only 2 s. These small-diameter droplets (<1 μm) can remain suspended inside the HVAC system for a long time and will have a high probability of transmission. However, in general, most viruses, including SARS-CoV-2, are combined with water and other components as droplets, as well as aerosol particles that have a size of 0.25–0.5 μm. In this case, the HEPA filters have a good efficiency of filtration.

According to class recommendations, recirculation should be avoided, but this increases the energy consumption a great deal. In this case, but also for other alternatives, heat recovery devices should be added to transfer the energy from the exhausted air to the freshly supplied air. The main challenge is to find a heat recovery system with a low initial cost, low air pressure drop, and high efficiency, and that does not suffer from contamination between the air supply and exhaust. This article analyzes the strengths and weaknesses of different possible alternatives to improve HVAC systems to avoid the spread of diseases in an actual pandemic but also in case of a future, unexpected epidemic. Other aspects related to the general design of the vessel, epidemy inventory, and emergency plans needed are highlighted.

Regarding the technical spaces of ships, the risk of viral transmission of COVID-19 or other viruses through the ventilation system is low because these rooms are provided with independent ventilation with a high air-change rate.

2. Main Types of Ventilation and Air Conditioning Systems

The following ventilation and air conditioning systems are used to achieve ventilation for the different compartments of a ship:

2.1. Independent Natural Ventilation

In general, technical compartments with no specific requirements for the operation of equipment or the safety of the ship and passengers are provided with independent natural ventilation. To improve the natural circulation, inlet grilles are placed at the bottom of the compartment and exhaust grilles are placed at the top. In Figure 1, a typical natural ventilation system for a room located below the weather deck is presented. The ventilation openings are located above the weather deck at a minimum height of 900 mm in position 1 and a minimum of 760 mm in position 2 according to International Maritime Organization (IMO) load line requirements [9].

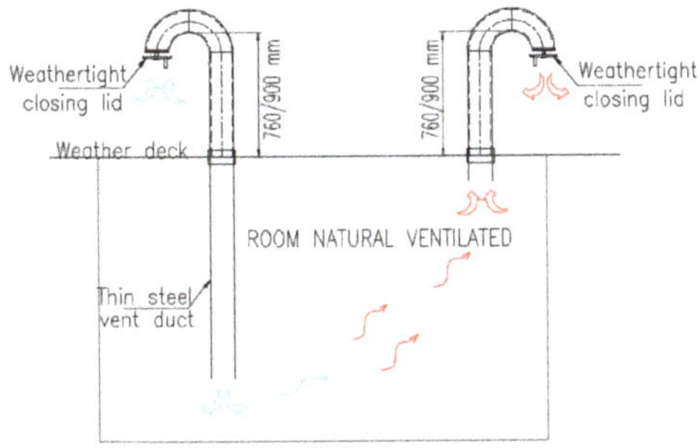

Figure 1. General arrangement. Independent natural ventilation for a room located below the weather deck.

2.2. Independent Mechanical Ventilation

According to the requirements of the Classification Society [10–13], the technical and cargo compartments of the ships are provided with independent mechanical ventilation systems equipped with supply or exhaust fans. The ventilation system provided for these compartments must ensure the cooling of the room, the necessary air for operation of the equipment, and remove the flammable gases that can accumulate inside the room. In certain situations, the ventilation systems are supplemented with local cooling systems provided with fan coils with 100% recirculation.

Sometimes, both exhaust and air supply fans are provided to ensure the desired pressure inside the compartment. In Figure 2, a typical mechanical ventilation system for a room located below the weather deck (with supply fan, exhaust fan, and both supply and exhaust fans) is presented.

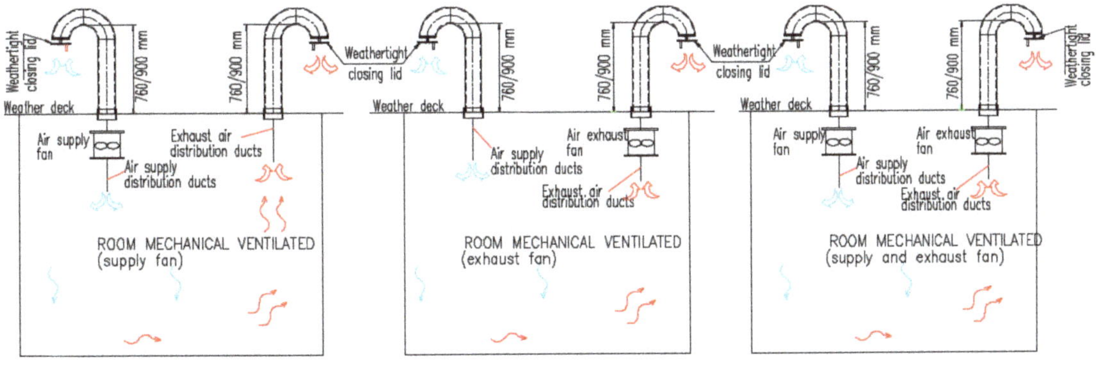

Figure 2. Typical arrangement. Independent mechanical ventilation for a room located below the weather deck.

In the case of the engine room, the openings on the weather deck should be kept without a closing device. In Figure 3, a typical example of ventilation for the engine room is presented. In this case, according to the IMO load line requirements [9], the openings should be located at a height of a minimum of 4.5 m above deck in position 1 and a minimum of 2.3 m in position 2.

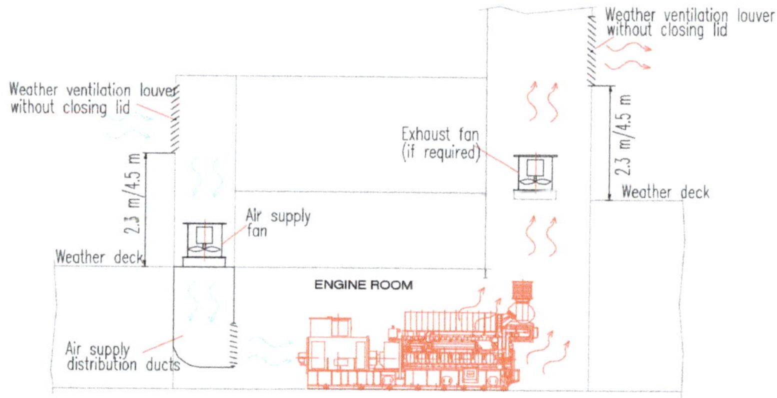

Figure 3. Typical arrangement. Independent mechanical ventilation for the engine room.

2.3. Common Mechanical Ventilation for Several Rooms, Combined with Air Conditioning System

In the case of living spaces, in general, common ventilation and air conditioning systems are used to ensure the cooling or heating of the compartments as well as the requested fresh air. The air is recirculated through the air handling unit (AHU) back to rooms for cooling/heating purposes. Figure 4 shows a common usual ventilation and air conditioning system where the fresh air is mixed inside the AHU unit with air recirculated from the room served.

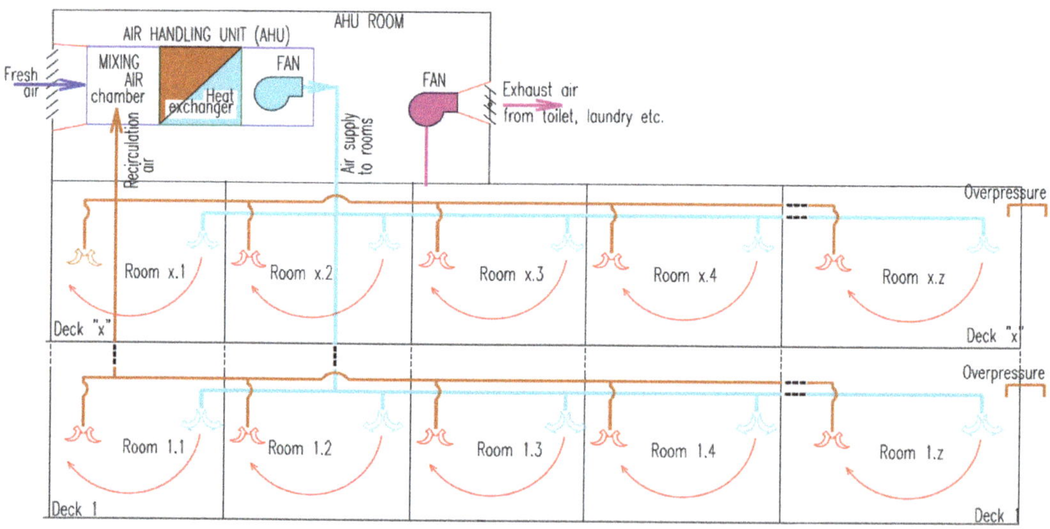

Figure 4. Typical arrangement. Common ventilation and air conditioning system for large ships, with recirculation and air mixing.

To increase the efficiency of the system, some energy can be recovered from the exhaust air but the possibility to recover the energy is not always profitable, as concluded by Zender–Swiercz [14].

In the case of small ships and boats, the common air handling unit (AHU) is replaced by local small fan coils and local fresh air connections, without air recirculation from different rooms (see Figure 5 for reference).

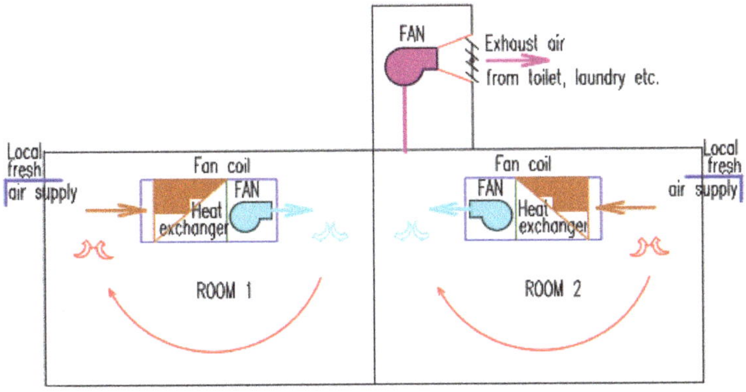

Figure 5. Typical arrangement. Independent ventilation and air conditioning system for small ships and boats, without air recirculation/mixing from/to different rooms.

3. Review of Main Rules and Requirements for Ventilation of Different Compartments

The ventilation of ship compartments is carried out based on the requirements of the classification society rules, IMO rules, and international standards.

3.1. General Requirements

In general, the Classification Societies require the following information to check and approve the ventilation systems installed on the ships [10]:

- Machinery spaces and technical spaces—ventilation system
- Accommodation and service spaces (galleys, pantries, lockers, etc.)—ventilation and air condition diagrams
- Cargo spaces—ventilation system including hazardous areas if applicable
- Details of fire dampers and weather-tight dampers and approvals
- Details for duct penetrations and approvals

According to Classification Societies rules [10], which follow IMO rules [15], the ventilation systems for machinery spaces of category A, vehicle spaces, ro-ro spaces, galleys, special category spaces, and cargo spaces should be separate from each other's and should be separate from the ventilation systems serving other spaces. There is an exemption for the ventilation system of the galley in the case of cargo ships with a gross tonnage below 4000 and for small passenger vessels carrying a maximum of 36 passengers.

The penetration through watertight and fire divisions should be kept at a minimum and the watertight/fire integrity should be kept unaffected.

According to IMO—SOLAS rules [15], in case a thinly plated ventilation duct penetrates a fire deck/bulkhead, the penetration sleeve should have the length and minimum thickness outlined in Figures 6–8.

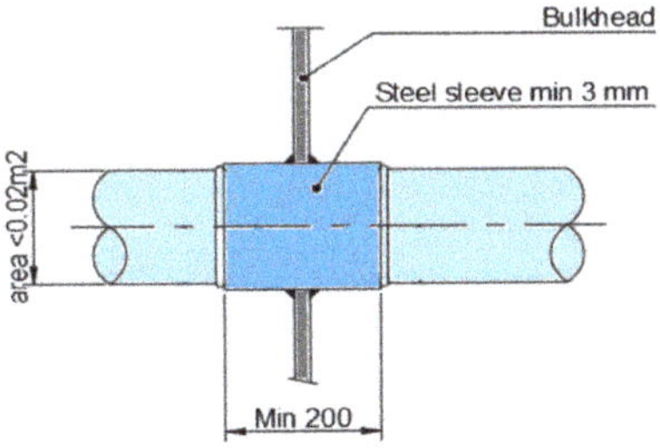

Figure 6. A class fire division—penetration of thin plate ventilation duct, area ≤ 0.02 m^2.

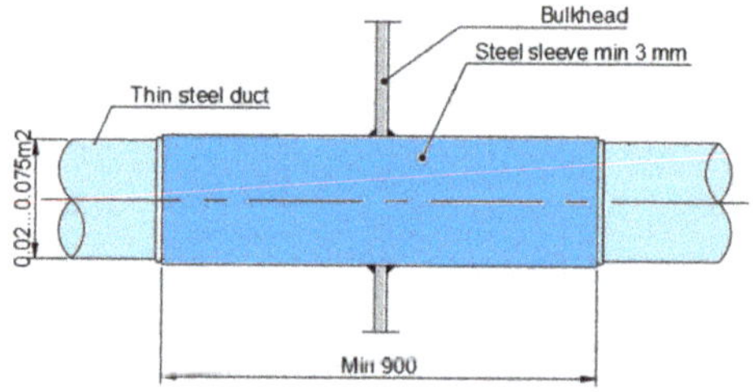

Figure 7. A class fire division—penetration of thin plate ventilation duct, area 0.02–0.075 m^2.

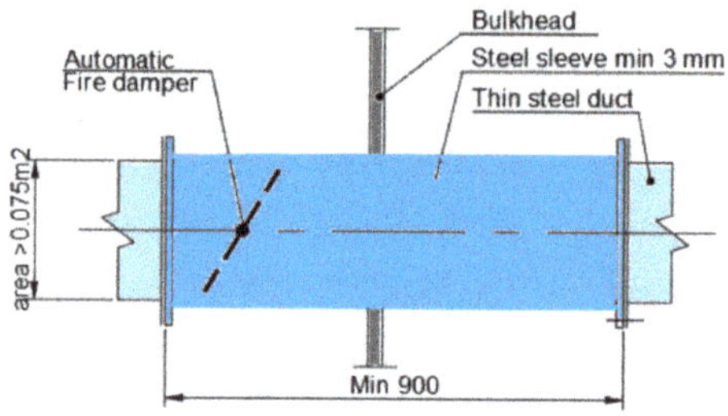

Figure 8. A class fire division—penetration of thin plate ventilation duct, area $\geq$ 0.075 m^2.

The requirements regarding avoiding flooding through the ventilation system are indicated in Figure 9 [9,16–18].

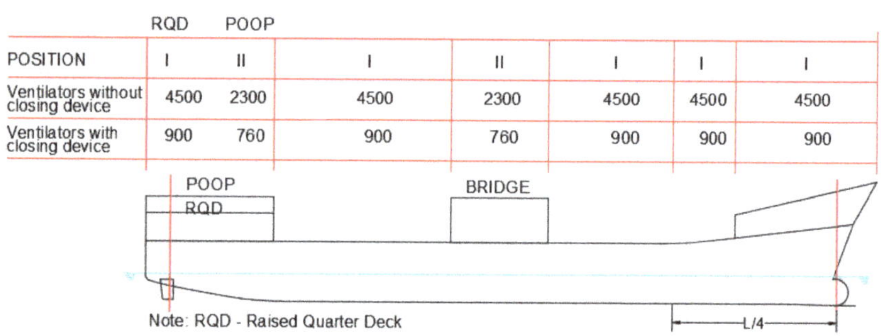

Figure 9. Requested height of ventilation openings above weather deck (data processed from [9,10]).

Strength requirements for ventilation ducts located on the weather deck.

In the case of vessels with a length of 80 m and above, the ventilation ducts located on the weather deck in the forequarter of the vessel must respect the strength requirements from the International Association of Classification Societies (IACS) Unified Requirements (UR) S27 [19].

Avoiding explosions in hazardous areas.

In hazardous areas, fans with anti-explosion protection are installed and the ventilation openings should be provided with a protection screen of a minimum of 13 mm square mesh to prevent objects from entering the fan [10,11,13,20]. The ventilation openings for other compartments should be located far from hazardous areas.

3.2. Main Ventilation Requirements for Different Compartments

The ventilation system for machinery spaces is provided with supply and exhaust fans with high airflow, which should ensure a maximum of 45 °C RH60% inside the room [21,22].

For ships not certificated for unrestricted services, the Classification Society can approve an increased temperature inside the engine room [23].

The airflow needed for the ventilation of the engine room should be calculated according to ISO standard 8861 [21]. The equipment makers also have some requirements and recommendations; however, to obtain a good efficiency of the ventilation, detailed computational fluid dynamics (CFD) analyses could be necessary [24].

In addition according to Classification Societies rules [13,25] the gas tank connection space and fuel preparation room should be provided with extraction type ventilation with a rate of at least 30 changes per hour.

In general, the engine control rooms are provided with dedicated ventilation and air conditioning systems that must keep the temperature below +27 °C, considering the heat dissipation from equipment installed inside and moderate fresh air exchange rate [26].

The compartments wherein flammable liquids are stored must be provided with an independent ventilation system that ensures at least 10 air changes per hour [10].

Figure 10 presents a typical example of ventilation for flammable store compartments with extended hazardous areas. The ventilation openings for other systems should not be installed in the extended hazardous area indicated in Figure 10.

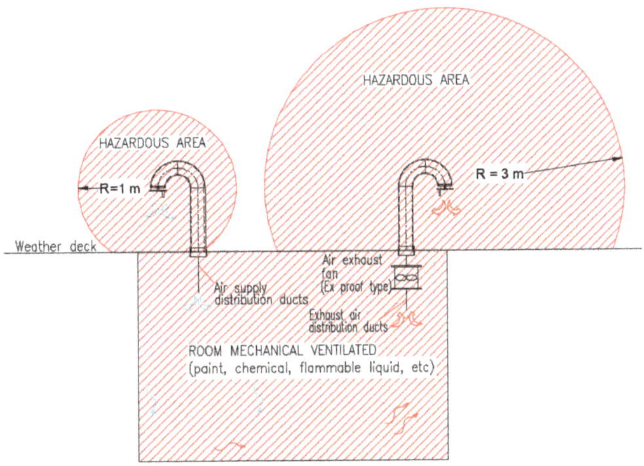

Figure 10. Typical arrangement. Ventilation of pain store, chemical store, another flammable liquid store, and oxygen-acetylene storage rooms, with extended hazardous areas.

For oxygen-acetylene storage rooms and CO_2 rooms, the independent ventilation system ensures a minimum of six air changes per hour and for refrigerating machinery and at least 30 air changes per hour for the battery rooms [10,11].

In case the fan is installed at a distance less than 3 m from the exhaust outlet, it should be a safe type [27].

The galley is provided with a mechanical ventilation system with the fresh air supply and exhaust fan calculated according to ISO standards [28].

The wheelhouse should be provided with a ventilation and air conditioning system calculated according to ISO standard 8864 [29].

In general, the ventilation system should ensure a small positive pressure inside the room [29].

3.3. *Additional Requirements for Dedicated Vessels*

All enclosed spaces containing livestock should be provided with mechanical ventilation with a capacity of not less than 20 changes per hour.

The cargo pump rooms for oil tankers and FLS tankers are provided with an exhaust ventilation system that ensures a minimum of 20 air changes per hour, but in the case of oil recovery ships, only eight air changes per hour are required [30].

In the case of liquefied gas carriers, the exhaust ventilation openings from a hazardous area should be located at a distance of at least 10 m in the horizontal direction against other openings from non-hazardous spaces.

The public atriums spaces in passenger's vessels should be provided with an extraction ventilation system with the capacity to remove the smoke in a maximum of 10 min, according to Maritime Safety Committee (MSC)/Circ 1034 [31].

In the case of inland navigation passenger vessels, at least two separated gas-tight engine rooms with dedicated independent ventilation should be provided [32].

In the case of ro-ro vessels, the ventilation of the cargo spaces, where vehicles provided with internal combustion engines can be driven, is calculated and arranged according to ISO standard 9785 [33].

The ventilation system should ensure a minimum of 10 air changes per hour, but during loading and unloading the airflow should be increased to 20 air changes per hour, according to the International Convention for the Safety of Life at Sea (SOLAS) [15] requirements and Det Norske Veritas—Germanischer Lloyd (DNV-GL) interpretation [20].

In the case of ships operating in polar water, the air intake ventilation openings should be located on both sides of the ship [34].

4. Main IMO and Classification Societies Requirements and Guidelines for the Management of COVID-19 and Infectious Diseases

The International Maritime Organisation (IMO) and its member states are analyzing the impact of pandemic COVID-19 to find a solution for the prevention and control of infectious diseases, but these measures are related in general to seafarer movement across the border, protection of personnel, etc., and there are no rules or solutions for the construction and updating of the vessel's systems, which can spread the disease.

The Classification Societies have also provided guidelines for the management of COVID-19 and infection diseases on board the ship. Heating, ventilation, and air conditioning systems (HVAC) are considered by Classification Societies [35] as "an extremely complex system in an infectious protection environment . . . It is likely to be necessary to have a separate HVAC strategy".

According to class requirements, the owner or others responsible for vessel operation must analyze and understand the existing HVAC systems installed on the vessel "in line with future developments" [35] and take the appropriate measures to reduce the risk of infection. The Classification Society mentions that "it is not necessarily a requirement for ongoing expensive upgrade and replacement, but rather an understanding of how future designs seek to reduce risk and how such concepts may influence existing designs" [35].

Related to existing ventilation systems, Classification Societies recommend taking all reasonable measures to avoid propagation along with the system. The reasonable measures considered include the "introduction of clean air, reducing or cancelling recirculation and increasing filtration at reasonable cost" [35].

In addition to IMO and Classification Societies requirements and guidelines for the management of COVID-19 and infectious diseases on board the ships, there are also other authorities, agencies, and professional associations that provide standards and guidelines for the design and operation of the HVAC system in general. Below is a selection of the main institutions that have analyzed the HVAC operation issues and provided guidelines during the COVID-19 pandemic:

- American Society of Heating, Refrigerating and Air-Conditioning Engineers (ASHRAE),
- Federation of European Heating, Ventilation and Air Conditioning Associations (REHVA),
- European Centre for Disease Prevention and Control (ECDC),
- Canadian Committee on Indoor Air Quality (CCIAQ),
- Architectural Society of China (ASC),

- Society of Heating, Air-Conditioning and Sanitary Engineers in Japan (SHASE),
- Indian Society of Heating, Refrigerating and Air Conditioning Engineers (ISHRAE)
- Public Health Ontario (PHO),
- Chinese Association of Refrigeration (CAR),
- National Health Commission of China (NHC).

5. Ventilation and Air Conditioning Systems for Living Spaces

5.1. Environmental Conditions According to ISO Standard

The ventilation and air conditioning systems for accommodation are calculated and designed according to ISO standard 7547 [1]. The ventilation and air conditioning system should be designed to ensure the following indoor conditions based on the indicated outdoor air conditions and minimum outdoor air supply of "40% of the total air supplied to the spaces concerned" [1].

- Summer—temperatures and relative humidity:

Outdoor air	+35 °C and 70%RH
Indoor air	+27 °C and 50%RH

- Winter—temperatures

Outdoor air	−20 °C
Indoor air	+22 °C

5.2. Influence of Outdoor Air, Heat Gain, and Heat Loss on the Capacity of Ventilation and Air Conditioning System

The capacity of the ventilation and air conditioning system is calculated according to ISO standards considering the requested outdoor fresh air and the heat gain loss, as follows:

5.2.1. Heat Gain and Loses

According to ISO Standard:

- Heat transfer through the ceiling, walls, and floors have a low impact on total heating/cooling power, considering that all these areas are provided with heat insulation with an overall heat transfer coefficient of 0.4 to 0.9 W/m^2K.
- Heat transfer through windows. The overall heat transfer coefficient is 2.5–6.5 W/m^2K but the impact on the total heating/cooling power is also reduced because the area of windows on the ship is reduced.
- Solar heat gain. It is calculated for all exposed bulkheads and decks where the temperature is increasing above the air temperature with 12 °C for vertical light surfaces, 29 °C for vertical dark surfaces, 16 °C for horizontal light surfaces, and 32 °C for horizontal dark surfaces. However, this increased temperature has a low impact on the total cooling power because the bulkhead and decks are thermally insulated.
- Solar heat gain by solar radiation through windows. The total heat gain is calculated based on the total area of windows and solar radiation coefficient which is considered 350 W/m^2 in the case of clear glasses and 240 W/m^2 if the glasses are provided with interior shading. The heat radiation has a big impact on the cooling power for the rooms provided with large windows such as wheelhouses and other special spaces in passenger's vessels. To reduce the impact of the sun radiation, the windows can be provided with glasses with a high reflection capacity for infrared radiation. Additionally, different solutions for exterior and interior shading are used.
- Heat gain from persons and equipment. The heat gain from persons, in general, has a low impact on the cooling system, except the public spaces designed for a high number of persons.
- Heat gain from equipment. The heat gain from equipment is insignificant in living spaces.

5.2.2. Outdoor Air Supply

The outdoor air supply is requested to permit the air extraction from toilets, laundry, the galley, changing rooms, hospital, etc., and to ensure the requested fresh air of a minimum of 28.8 m^3/person. In addition, according to ISO standards, the fresh air rate should be increased to over 40% from total airflow. In general, this requirement increases the total cooling and heating power for the air handling unit.

However, it can be concluded that the cooling/heating capacity of the air conditioning system is mainly determined by the quantity of fresh air that should be cooled or heated and humidified or dehumidified. Therefore, from an energy point of view, it is very important to recover the energy from the exhaust air. This issue is even more important in the conditions of cancelling the air recirculation.

In general, the following heat recovery devices are used for HVAC systems:

(a) Rotary wheel heat recovery exemplified in Figure 11 is the most used solution for air handling units, which can have an efficiency of up to 85%. There is also a small energy consumption for rotating the wheel, but it has little effect on the recovery device efficiency. The pressure drop across the wheel heat recovery is about 200 Pa but it can be a higher or lower function of the size of the wheel and a function of the air velocity. The main disadvantage of this device, especially related to the spread of diseases, is that the two sides are not gas-tight; therefore, leakage of exhaust air to fresh air is expected.

(b) Fixed plate heat recovery, exemplified in Figure 12, has a good efficiency, which can reach up to 90%; however, the main disadvantage is that the pressure drop is high and can increase in cold weather when the frost limit is exceeded and the condensate inside the heat exchanger freezes. Regarding the leakage of exhaust air to fresh air, these heat recovery devices are very leak tight. According to Hoval's handbook [37] the leakage between supply and exhaust sides is below 0.1%, for a pressure difference of 250 Pa; therefore, it can be considered that there is no cross-contamination.

(c) Round around coil-type heat recovery systems, exemplified in Figure 13, have a lower efficiency than those presented above, increased cost, and need a separate system with a pump and control system. Its advantages include the low pressure drop and no mixing between fresh air and exhaust air.

(d) Heat pipe-type heat recovery systems are similar to coil-type systems but need a refrigerant system instead of water. Its advantages include the low pressure drop and no mixing between fresh air and exhaust air.

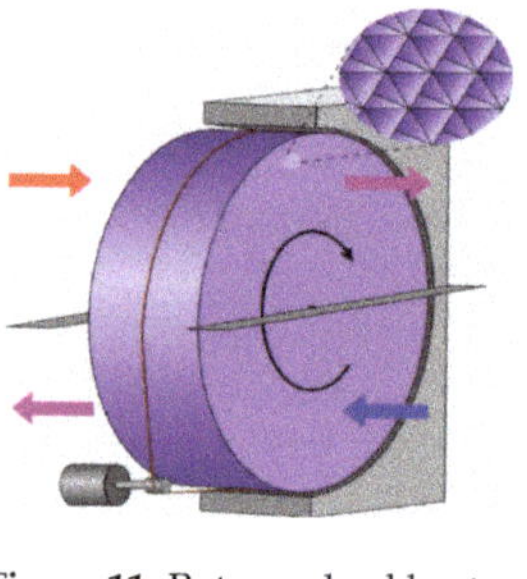

Figure 11. Rotary wheel heat recovery (data processed from Wikipedia [36]).

In their study, Zheng et al., 2021 [40] summarized the recommendation of different institutions regarding heat recovery devices:

- ASHRAE considers that the fact that "heat recovery devices can be utilized for leakage is acceptable".
- REHVA, ECDC, and SHASE consider that "heat recovery devices can be utilized for leakage [...] below 5%".

- CCIAQ and PHO consider that "cross-contamination between outdoor air and exhaust air should be avoided with the application of heat recovery devices".
- ASC, NHC, and ISHRAE consider that "Rotary heat exchangers should not be applied."

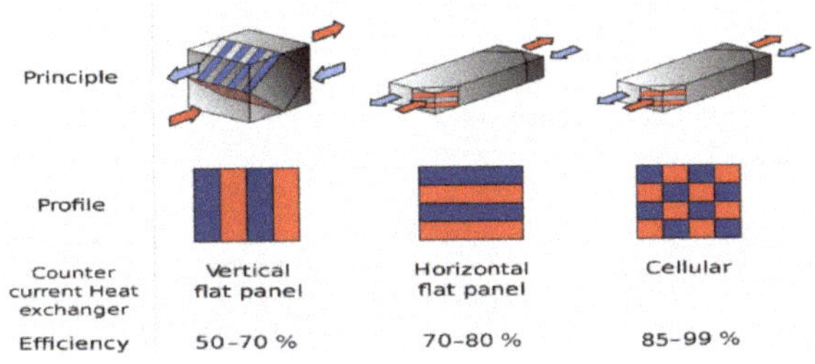

Figure 12. Plate heat recovery (data processed from Wikipedia [38]).

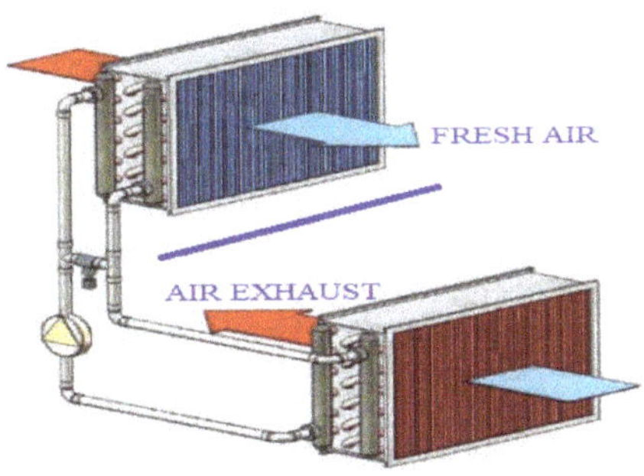

Figure 13. Round around coil-type heat recovery (data processed from reference [39]).

In general, the HVAC system for living spaces is provided with a common air conditioning system that ensures the necessary fresh air and heating or cooling of all living spaces. In the case of big ships, such as passenger vessels or special vessels with a large number of people on board, two or more independent air handling units (AHU) are provided for each main fire vertical zone. Each air handling works with 40–60% fresh air and 60–40% recirculation air from the served rooms.

5.3. Spread of Diseases on the Ships

By air recirculation from all rooms and mixing inside the AC unit, the viruses from one space are distributed in all living spaces.

In the case of the Diamond Princess Cruise Ship, a lot of persons were infected with COVID-19 in a short time, even if the vessel and the people on board quarantined from the first 10 cases confirmed.

Eisuke et al. [41] chronologically analyzed the evolution of the COVID-19 pandemic on the Diamond Princess Cruise Ship and indicated that over 600 persons (over 15% from the total people on board) were infected in the first two weeks, from 5 February to 19 February, even though it was decided to start the quarantine on 5 February.

Gupta A et al. [42] concluded that "it is just impossible to segregate infected from the non-infective in a space-constrained, air-conditioned setting" and mentioned that the quarantine on board the ship is not efficient.

Parham et al. [43] concluded in their study that "aerosol inhalation was likely the dominant contributor to COVID-19 transmission among passengers aboard the Diamond Princess Cruise Ship even with a very high ventilation rate (9–12 air changes per hour) and no recirculated air".

Hwang et al. [5] analyzed "Possible aerosol transmission of COVID-19 associated with an outbreak in an apartment in Seoul, South Korea, 2020" and concluded that the virus spread upstairs and downstairs through the ventilation duct, indicating aerosol transmission. Priyanka et al. [6] reported aerosol transmission of SARS-CoV-2, especially in enclosed spaces, due to that aerosols being suspended for a long time and travelling via airflow. Kulkarni et al. [44] have analyzed the spread of diseases with respiratory syncytial virus (RSV) and concluded that aerosol particles of RSV remained airborne for a long time and led to transmission of the infection. In his study, Melikov [45] highlighted that aerosol transmission should be considered; otherwise, it is difficult to control the spread of diseases. Corea et al. [46] consider the HVAC system "as a major source for indoor and environmental contamination".

Therefore, we can conclude that in the case of common air conditioning systems provided with recirculation, the closing of air recirculation after discovering a new infection will not solve the problem while the entire system is already contaminated and cannot be easily cleaned and disinfected.

5.4. Measures to Reduce the Risk of Disease Transmission through the HVAC System

It should be concluded that to avoid the spread of diseases through the ventilation system onboard the ship, the following solution can be adopted:

1. Common ventilation system with HEPA filters or UV disinfection

The common air conditioning and ventilation systems for living spaces can be provided with mechanical filtration. Unfortunately, the HEPA filter can remove 90–99.9% of the particle at 0.3 μm in size [47] but "most viruses, including CoVs range from 0.004 to 1 μm" which can be interpreted that HEPA filtration will not offer complete protection [2]. However, most viruses, including SARS-CoV-2, are combined with water and other components as droplets and aerosols particles with size of 0.25–0.5 μm [2]. In this case, the HEPA filters "play an important role in virus exclusion and hence lowering the transmission potential of SARS-CoV-2 within the pharmaceutical and healthcare setting" [2]. The spread of diseases through the air conditioning system can be reduced by adding HEPA filters, but in this case, the protection is limited and the fans should have increased pressure. In general, the air handling units are provided with filters that have a "minimum efficiency reporting value (MERV)" of 7–8 and ASHARE recommends filters rated MERV-13 or higher [2,48]. The pressure drop across these filters is about 30 Pa when the filters are clean and can increase to 300 Pa when the filters are dirty [49]. The existing filters can be replaced with HEPA filters or the existing filters can be kept as pre-filters to avoid rapid clogging of the HEPA filter. The pressure drop across the HEPA filters is about 150–300 Pa [49,50] when the filters are clean. According to the manufacturer's recommendation [49], the final pressure drop could be two to three times the initial pressure drop. However, although the manufacturer accepts a maximum pressure drop of 450–900 Pa across the HEPA filters, the system should be designed so that the total pressure drop is kept as low as possible to avoid high energy consumption and to keep the total pressure drop in line with available standard fans, which in general is 1200–2200 Pa [51].

According to the United States Environmental Protection Agency [52], the pre-filters should be replaced every 60 to 90 days, but the manufacturer's recommendation must be checked and followed. In general, HEPA filters cannot be cleaned and should be changed after one month up to one year [53]. In any case, the filter should be checked and cleaned/changed every time it becomes clogged; therefore, it is recommended to be provided with pressure drop measuring devices.

Figure 14 presents a typical example of a ventilation and air conditioning system, with recirculation and air mixing, provided with HEPA filters on the main supply duct and recirculation duct. The solution with HEPA filters installed on both the supply and the recirculation ducts could increase the protection while there are two stages of filtration. If the air recirculation is contaminated, the viruses are filtered on the main recirculation line; therefore, the fresh air will be mixed with almost clean air. Finally, the HEPA filter installed on the supply main distribution duct will clean the viruses or bacteria that escaped from the first stage of filtration or if outside fresh air is contaminated. Although the two HEPA filters increase the protection, the increase of the total pressure drop due to the second filter should be considered. If the total pressure drop increases, the final pressure drop of the filter after clogging should be reduced; therefore, the filters will need to be changed more often. In addition, if the recirculation line is provided with a HEPA filter, the pressure inside the rooms will increase with the backpressure of that filter. Taking into consideration a back pressure of 150–900 Pa across the HEPA filter, it is mandatory that the system be provided with a recirculation (exhaust) fan so that the pressure inside the room is kept in accordance with the requirements (e.g., −50 to +50 Pa). Regarding the pressure inside the accommodation area, it is necessary that the system be designed/balanced to keep the same pressure in all rooms, corridors, and common spaces so that natural circulation between different rooms/corridors is reduced to a minimum.

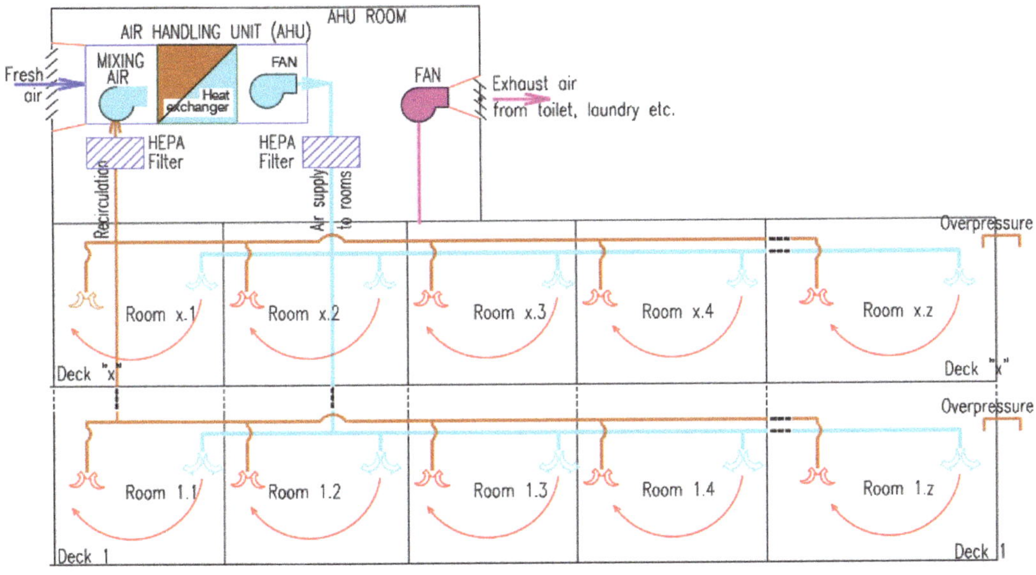

Figure 14. Example of typical arrangement. Common ventilation and air conditioning system for large ships, with recirculation and air mixing, provided with HEPA filters.

The solution indicated above can be simplified by cancelling the HEPA filter installed on the recirculation line, but in this case, the contaminated air is mixed with the fresh air and introduced into the AHU unit. The filtration occurs on the main supply ventilation duct only. The solution will be simpler when the recirculation (exhaust) fan is no longer mandatory.

Another solution is to keep the recirculation HEPA filter and to cancel the main air supply HEPA filter. In this case, the contaminated air is not mixed with fresh air and does not enter the AHU unit anymore, but the recirculation (exhaust) fan is mandatory. It should also be mentioned that the recirculation filter is 40–60% smaller than the main supply filter because the recirculation airflow is in general 40–60% lower than the supply airflow.

In case the recirculation and/or main supply HEPA filters cannot be installed, the filtration can be carried out locally by adding small HEPA filters in the supply line/diffuser of each room. Regarding installation and maintenance, this solution has the disadvantage that all rooms are affected when changing the filters and the time, costs, and discomfort will increase during filter replacing. Regarding the functionality, the system becomes unbalanced more easily due to the big pressure drop across the HEPA filters, which can easily change the airflow rate if the clogging rate of the filters is different between different rooms. The filter clogging rate is different due to different initial airflow rates or if the flow rate can be adjusted inside each room.

To reduce the risk of disease transmission through the HVAC system, Jones et al. [54] highlight the ultraviolet lamp's "(UV-C's) potential as an effective air and surface disinfectant". The ultraviolet lights can be used as a separate system or together with HEPA filters to reduce "and even prevent the growth of dangerous microbes in HVAC systems" [54].

In their article, Vranay et al. [3] concluded that by using a UV-C disinfection system, "more than 90% of the SARS-CoV-2 virus produced by humans in the internal environment can be inactivated". Additionally, according to their analysis, the energy consumption for disinfection with UV-C lamps was about 2.3% of the total energy consumed by the HVAC system.

In their article, Kowalski et al. 2020 [55], based on hundreds of laboratory studies, demonstrated that ultraviolet light is capable of inactivating viruses, bacteria, and fungi. The author has summarized the result of the exposure of coronaviruses to ultraviolet light and concluded that 90% (D90) virus inactivation is obtained for a radiation dose of 7–241 J/m^2 with an average of 67 J/m^2. Table 1 presents a summary of ultraviolet studies on coronaviruses according to Kowalski [55].

Table 1. Summary of ultraviolet studies on coronaviruses (data processed from [55]).

Microbe	D90 Dose J/m^2	UV k m^2/J
Coronavirus	6.6	0.35120
Berne virus (Coronaviridae)	7.2	0.32100
SARS-CoV-2 (Italy-INMI1)	12.3	0.18670
Murine Coronavirus (MHV)	15.0	0.15351
SARS Coronavirus (Frankfurt 1)	16.4	0.14040
Canine Coronavirus (CCV)	28.5	0.08079
Murine Coronavirus (MHV)	28.5	0.08079
SARS Coronavirus (CoV-P9)	40.0	0.05750
SARS-CoV-2 (SARS-CoV-2/Hu/DP/Kng/19-027)	41.7	0.05524
Murine Coronavirus (MHV)	103.0	0.02240
SARS Coronavirus (Hanoi)	133.9	0.01720
SARS Coronavirus (Urbani)	2410	0.00096
Average	237	0.00972
Average for SARS-CoV-2	27	0.08528

According to the Illuminating Engineering Society, 2020, [4] ultraviolet light has a spectrum of 100–400 nanometers (nm), and the most effective for disinfection is "UV-C" with a wavelength from 200 to 280 nm. According to ASHRAE [56], the maximum efficiency of the UV disinfection is for a UV-C wavelength of 265 nm. It should be noted that "a by-product of UV-C radiation with a wavelength of 185 nm is ozone" [3], which is a strong oxidant and harmful to health in high concentrations, but this emission is absorbed by glass, which is used for this kind of application. In practice, UV lamps generate about 90% of energy close to 254 nm [57]. Regarding UV-B lamps (280–315 nm), it should be mentioned that they can also be effective for disinfection, but this UV light penetrates the skin more deeply; therefore, in the case of accidental exposure, it is more dangerous. The UV-A (315–400 nm) is not capable to inactivate the viruses.

The efficacity of UV disinfection is proportional to the UV exposure dose (J/m^2), which represents the product of the irradiance rate (W/m^2) and time (s). It should be mentioned that there is a nonlinear relationship between UV exposure and germicidal efficacy. Therefore, if the exposure time is increased by 50%, the efficiency of disinfection will increase from 90% to 99% [4]. Another aspect that should be considered when designing the UV system is the humidity, which can reduce the efficiency of UV disinfection.

In in-duct systems, the exposure time is low (seconds or fraction of seconds); therefore, the UV irradiance rate should be high enough to produce the requested dose needed to inactivate the viruses in seconds or fraction of seconds, according to air velocity. To increase the effective UV dose, the inside of the air handler is provided with high reflective materials such as aluminum and other highly reflective materials, which will reflect the UV energy into the irradiated area. According to ASHRAE [56], in the general in-duct system, the requested average irradiance is 1000–10.000 μW/cm^2. Regarding the location of the UV lamps, they can be installed anywhere inside the HVAC system, but some locations can improve the UV disinfection system. In general, the most suitable place is inside the air handling unit (AHU) due to the low air velocity, at the same time also ensuring the irradiation of the cooling coils. To avoid increasing the system cost and power consumption, ASHRAE [56] recommends that the in-duct system be located so that the UV air exposure is a minimum of 0.25 s. The location of the UV system should be decided also keeping in mind that the UV system will need periodic inspection, maintenance, and lamp replacement. In general, the UV lamps have a rated life of 9000 h of operation; therefore, the lamps should be changed annually if they run continuously. However, in function of the ballast type used, if the lamps are often switched on/off, the lifetime of the lamps can be reduced. Nevertheless, the UV lamps can operate after the lifetime indicated by the manufacturer, albeit with reduced performance.

Morawska et al. [58] highlighted that air recirculation is a measure for saving energy but it increases the risk of airborne infection; therefore, the recommendation is to cancel air recirculation. Air cleaning and disinfection by ultraviolet irradiation may offer benefits and it is recommended where it is difficult to improve the HVAC system.

Federation of European Heating Ventilation and Air Conditioning Associations (REHVA) [59] and the European Centre for Disease Prevention and Control recommends cancelling central air recirculation to avoid the spread of COVID-19 aerosols through the HVAC system.

In case the air recirculation cannot be cancelled, it is recommended to increase the fresh air supply as much as possible in addition to HEPA filter or ultraviolet disinfection, devices to be provided in the main air duct. However, the solution with a high-class filter (e.g., HEPA filter) is recommended, if this is technically possible [59]. If the HVAC system cannot be improved by cancelling air recirculation, the Canadian government [60] recommends HEPA filtration.

2. The common fresh air supply system and local individual cooling and heating

The common AHU unit is replaced with a smaller one, which will ensure the requested fresh air only. The heating and cooling of the room is carried out locally by using local small fan coils or other local air conditioning equipment. In this case, the recirculation of air from one room to another room, and the spread of diseases through the HVAC system, will be cancelled. This system also has other advantages, such as smaller ventilation pipe sizes, the big AHU unit being replaced with a smaller one, and HEPA filters not being necessary; therefore, the pressure of the fans should not be increased, the control of temperature inside the room is better, etc. The main disadvantage is that each room should be provided with a small local fan coil to ensure the necessary cooling and heating of the room and these can generate additional noise. In addition, the total price of the system could be higher. Figure 15 proposes a typical example of an individual ventilation and air conditioning system, without recirculation and air mixing and HEPA filters. From an energy point of view, the solution can be improved with heat recovery devices without leakage between the supply and exhaust air. The heat recovery system increases the initial cost, but it is amortized over time and the energy consumption and pollution is reduced.

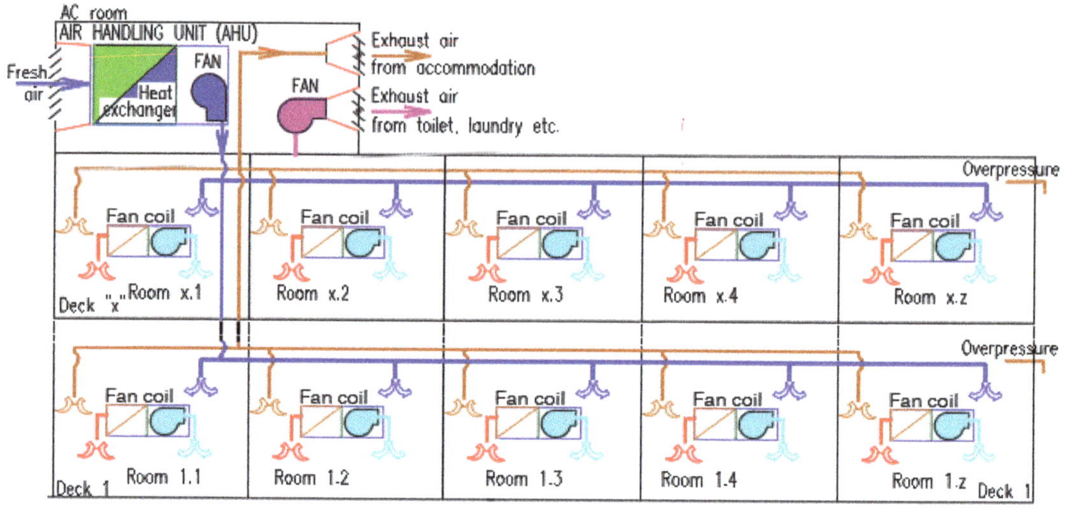

Figure 15. Typical arrangement. Ventilation and air conditioning system improved from a small ship as a proposal for large ships. The system without mixing and air recirculation between rooms, without HEPA filters.

Another solution that should be analyzed for the new vessels with many passengers is to provide different gas-tight zones with dedicated air handling units. These zones can be easily split by main fire vertical zones; however, in addition, some small separate gas-tight areas should be provided in hospital and sickbay areas for quarantine purposes. The intake and especially exhaust air for each gastight area should be located so that air recirculation between different HVAC systems is not possible. In addition, the intake and exhaust openings should also be located far from the other vessel hazardous areas.

According to ASHRAE [56] another effective and low-cost solution is to provide upper-air UV-C devices inside the rooms. As indicated in Figure 16, this solution is applicable in rooms with high ceilings to keep the UV disinfection zone above the minimum safe area of 7 ft [56]. This disinfection solution will be very useful in the quarantine area and large common spaces on the vessel; therefore, these spaces should be designed with high ceiling elevation.

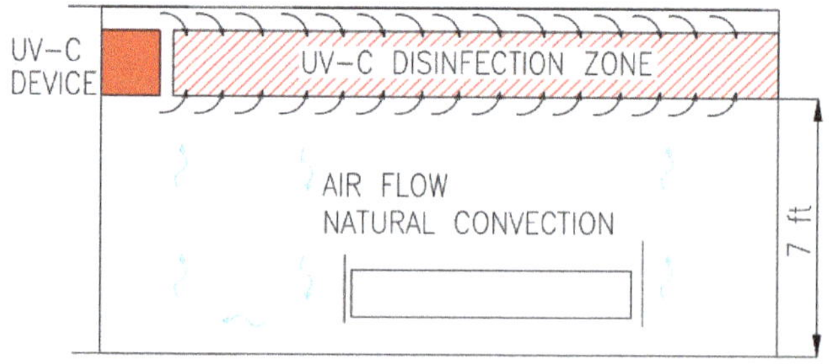

Figure 16. Example, elevation view of upper-air UV-C applied in hospital (data processed from ASHRAE [56]).

In case of an unwanted epidemic on board a ship, an emergency plan is necessary to avoid the spread of the infection, as are devices and tools to control the disease. Therefore, in addition to the new design for the HVAC system, other devices and tools such as disinfection tools, protective suits, masks, gloves, etc. should be provided in the inventory of each ship, sized according to the number of people on board.

There is different mobile UV disinfection equipment on the market in general for shore applications, but they can be certificated also for marine applications. According to American Ultraviolet [61] the mobile UV unit type MRS45-12, which weighs 43 kg, can inactivate coronavirus at 99.9%, at a distance of 3 m in less than 1 min. In the same way, the mobile UV unit type MRS33-8 [62], which weighs 36 kg, can inactivate coronavirus at 99.9%, at a distance of 3 m in less than 2 min. Smaller units can also be used if the weight should be reduced to be easier to transport, but the time of disinfection will increase.

6. Discussion

The existing vessels all have compartments provided with natural and mechanical ventilation. In general, the technical spaces are provided with independent ventilation and a high rate of air exchange; therefore, the spreading of viruses in these compartments is not an issue.

Unfortunately, it is different for living spaces, especially in large ships, which are provided in general with a common air conditioning system with air recirculation. Taking into consideration that HVAC systems are not provided with efficient filtration against viruses, the spreading of diseases in all living rooms is becoming a big issue for actual air conditioning and ventilation systems on the vessels.

COVID-19 has shown us that the effects of a new virus can be quite harsh for people's health and freedom. Therefore, the article intends to point out the need for a new approach to the solution for ventilation and air conditioning systems for living spaces. The standards, Classification Societies rules, IMO rules, and administration requirements regarding ventilation and air conditioning systems onboard the ships should be analyzed and revised so that the possibility of spreading viruses through these installations is reduced as much as possible.

IMO and Classification Societies are concerned about the COVID-19 pandemic and the spread of diseases aboard the ships. However, even if the Classification Societies consider that the HVAC system is "an extremely complex system in an infectious protection environment" and "it is likely to be necessary to have a separate HVAC strategy" [35], there are no clear rules and requirements regarding the necessary changes. The Classification Societies shifted the responsibility for finding the necessary solutions to the owner and other vessel operators. They must analyze and understand the existing HVAC systems installed on the vessel "in line with future developments" [35] and take measures to reduce the risk of infection.

Regarding the existing ventilation system, Classification Societies recommend "reducing or cancelling recirculation and increasing filtration at reasonable cost" [35]. For actual common HVAC systems, in general, the recirculation cannot be cancelled because the system will no longer be able to ensure the efficient cooling or heating of the rooms, while they are designed for recirculation of 40–60% from total airflow. However, if the heat exchanger has enough spare power to ensure the cooling/heating of the room with fresh air only, the energy consumption will increase significantly. Zheng et al. [40] highlighted in their study that "during the pandemic the energy consumption of HVAC systems increases by 128%" according to the study in China. This article presents two solutions to reduce the risk of spreading viruses between different rooms through the HVAC system, solutions which are in line with Classification Societies recommendation.

The first solution, shown in Figure 11, can be used for existing ships by providing HEPA filters in main recirculation and/or main supply ducts. In this case, the fans should be checked, and if the pressure available is not enough, the fans should be replaced. The HEPA filters can have a final back pressure of 450–900 Pa and in addition pre-filters (with a final pressure drop of abt. 300 Pa) are recommended to avoid rapid clogging of HEPA filters. The total pressure drop will then increase quite a lot; therefore, the solution with two stages of HEPA filtration becomes difficult to implement. It should be noted that the pressure drop inside the rooms will increase substantially if the HEPA filter is installed on the recirculation duct; therefore, an exhaust fan must be provided. The HEPA filters can be added to the supply line in each room, but in this situation, it is more difficult to keep the system balanced due to the high-pressure drop of the HEPA filter when clogging.

In addition, UV lamps can be added to improve the disinfection of the air inside the HVAC system. The UV disinfection can have good efficiency, but the main challenge is to ensure the exposure time of a minimum of 0.25 s, which means that a large space is needed to keep the air velocity low. Otherwise, the UV lamps can be installed anywhere inside the HVAC system.

Another effective and low-cost solution for large rooms with high ceilings, in addition to the HVAC UV disinfection system, is to provide inside the rooms upper-air UV-C devices.

The second solution, shown in Figure 12, is a proposal based on individual HVAC systems used on small vessels and boats, which was improved with a fresh air system. In this case, the spread of disease through the HVAC system between different rooms could be eliminated. The solution can be analyzed for new vessels or in case of large conversion because it will need changes in all rooms. It can be feasible also in case the existing vessel is not working in areas with high temperatures and the cooling can be carried out with fresh air only. In this case, the local fan coils can be replaced with electrical heaters.

A solution to improve the system from the energy point of view is to recover the energy from the exhausted air and to transfer it to the fresh air but without contamination of the fresh air. There are different solutions for heat recovery, some of them without cross-contamination and low-pressure drop (e.g., heat pipes and run-around), but these are more expensive to implement and have some particular requirements for air temperatures. Another solution (fixed plates) has a high-pressure drop which can increase more in case of cold air. One of the most used solutions (rotary wheel) has quite a large leakage and the fresh air can be contaminated by exhaust air; therefore, it is not recommended in the case of epidemy.

To improve protection in case of epidemy, especially for ships with a large number of passengers on board, the vessel should be split into more gastight zones, which can be quite easily executed by existing fire vertical zones. In addition to these, an emergency plan to avoid the spread of diseases onboard and a separated gas-tight space which can be used for quarantine purposes in case of epidemy should be provided. Additionally, mobile UV disinfection units should be added to the inventory of each ship to be used for disinfection of the rooms previously occupied by sick people. In future works, the aim is to make a budget comparison between different alternative solutions in the case of existing vessels and for a new ship.

Author Contributions: Conceptualization, investigation, visualization, methodology and writing—original draft preparation, V.M.; supervision, formal analysis, writing—review and editing, L.R. All authors have read and agreed to the published version of the manuscript.

Funding: The work was carried out in the framework of the research project DREAM (Dynamics of the REsources and technological Advance in harvesting Marine renewable energy), supported by the Romanian Executive Agency for Higher Education, Research, Development and Innovation Funding—UEFISCDI, grant number PN-III-P4-ID-PCE-2020-0008.

Acknowledgments: This work will also be presented to the 9th edition of the Scientific Conference organized by the Doctoral Schools of "Dunărea de Jos" University of Galati (SCDS-UDJG) http://www.cssd-udjg.ugal.ro/ (accessed on 10 July 2021), which will be held on 10–11 June 2021, in Galati, Romania.

Conflicts of Interest: The authors declare no conflict of interest.

References

1. Romanian Standardization Association. *SR ISO 7547, Shipbuilding—Air-Conditioning and Ventilation of Accommodation Spaces on Board Ships—Design Conditions and Basis of Calculations*; Romanian Standardization Association: Bucharest, Romania, 2005.
2. Sandle, T. Review of the efficacy of HEPA filtered air to control coronavirus risks in cleanrooms. *EJPPS Eur. J. Parenter. Pharm. Sci.* **2020**, 25. [CrossRef]
3. Vranay, F.; Pirsel, L.; Kacik, R.; Vranayova, Z. Adaptation of hvac systems to reduce the spread of covid-19 in buildings. *Sustainability* **2020**, *12*, 9992. [CrossRef]
4. IES. IES Committee Report: Germicidal Ultraviolet (GUV)—Frequently Asked Questions. *Illum. Eng. Soc.* **2020**, 1–24. Available online: https://www.ies.org/standards/committee-reports/ (accessed on 3 July 2021).
5. Hwang, S.E.; Chang, J.H.; Oh, B.; Heo, J. Possible aerosol transmission of COVID-19 associated with an outbreak in an apartment in Seoul, South Korea, 2020. *Int. J. Infect. Dis.* **2021**, *104*, 73–76. [CrossRef] [PubMed]
6. Priyanka, O.P.C.; Singh, I.; Patra, G. Aerosol transmission of SARS-CoV-2: The unresolved paradox. *Travel Med. Infect. Dis.* **2020**, *37*, 101869. [CrossRef] [PubMed]
7. Li, Y.; Qian, H.; Hang, J.; Chen, X.; Hong, L.; Liang, P.; Li, J.; Xiao, S.; Wei, J.; Liu, L.; et al. Evidence for probable aerosol transmission of SARS-CoV-2 in a poorly ventilated restaurant. *Build. Environ.* **2020**. [CrossRef]
8. Guo, M.; Xu, P.; Xiao, T.; He, R.; Dai, M.; Miller, S.L. Review and comparison of HVAC operation guidelines in different countries during the COVID-19 pandemic. *Build. Environ.* **2021**, *187*, 107368. [CrossRef] [PubMed]
9. International Maritime Organization—IMO. *Load Lines, 1966/1988—International Convention on Load Lines, 1966, as Amended by the Protocol of 1988*; Lloyd's Register Rulefinder 2020—Version 9.33—Fix; International Maritime Organization: London, UK, 2014.
10. Det Norske Veritas—Germanischer Lloyd. *Seagoing Ships—21 Ventilation*; 2014. Available online: https://rules.dnv.com/docs/pdf/gl/maritimerules/gl_i-1-21_e.pdf (accessed on 10 July 2021).
11. BV Marine & Offshore Division. *NR 467.C1 DT R12 E—Part C—Machinery, Electricity, Automation*; BV Marine & Offshore Division: Neuilly-sur-Seine, France, 2020.
12. BV Marine & Offshore Division. *NR 467.D1 DT R12 E—Part D—Service Notations*; BV Marine & Offshore Division: Neuilly-sur-Seine, France, 2020.
13. BV Marine & Offshore Division. *NR529 DT R03 E—Gas Fuelled Ships*; BV Marine & Offshore Division: Neuilly-sur-Seine, France, 2020.
14. Zender-Świercz, E. A Review of Heat Recovery in Ventilation. *Energies* **2021**, *14*, 1759. [CrossRef]
15. International Maritime Organization—IMO. *SOLAS—International Convention for the Safety of Life at Sea*; Lloyd's Register Rulefinder 2020—Version 9.33—Fix; International Maritime Organization: London, UK, 2018.
16. BV Marine & Offshore Division. *NR 467.B1 DT R12 E—Part B—Hull and Stability*; BV Marine & Offshore Division: Neuilly-sur-Seine, France, 2020.
17. RINA. *Part B Vol. 3—Hull and Stability. Rules for the Classification of Ships*; RINA: Chita, Japan, 2021.
18. RINA. *Part E Vol.2—Service Notation. Rules for the Classification of Ships*; RINA: Chita, Japan, 2021.
19. International Association of Classification—IACS. S27—Strength Requirements for Fore Deck Fittings and Equipment. 2013. Available online: https://www.iacs.org.uk/publications/unified-requirements/ur-s/ur-s27-rev6-cln/ (accessed on 10 July 2021).
20. Det Norske Veritas—Germanischer Lloyd. *SI-0364E—STATUTORY INTERPRETATIONS, SOLAS Interpretations*; 2018. Available online: https://rules.dnv.com/docs/pdf/DNV/SI/2018-02/DNVGL-SI-0364.pdf (accessed on 10 July 2021).
21. Romanian Standardization Association. *SR ISO 8861, Shipbuilding—Engine Room Ventilation in Diesel-Engined Ships—Design Requirements and Basis of Calculations*; Romanian Standardization Association: Bucharest, Romania, 2002.
22. International Association of Classification—IACS. *UR-M28 Requirements Concerning Machinery Installations—Ambient Reference Conditions*; International Association of Classification: Kolkata, India, 1978.
23. International Association of Classification—IACS. *UR-M40, Ambient Conditions—Temperatures*; International Association of Classification: Kolkata, India, 1986.
24. Mihai, V.; Rusu, L.; Presura, A. Ventilation of engine rooms in diesel engines ships. In *Analele University 'Dunărea Jos' of Galaţi*; Fascicle XI, Shipbuild; Romania, 2020; Volume 43, pp. 69–78. Available online: https://www.gup.ugal.ro/ugaljournals/index.php/fanship/article/view/4010 (accessed on 3 July 2021).
25. RINA. *Part C Vol. 1—Machinery, Systems and Fire Protection. Rules for the Classification of Ships*; RINA: Chita, Japan, 2021.
26. Romanian Standardization Association. *SR ISO 8862—Air-Conditioning and Ventilation of Machinery Control Rooms on Board Ships—Design Conditions and Basis of Calculation*; Romanian Standardization Association: Bucharest, Romania, 1995.
27. RINA. *Part C Vol. 2—Machinery, Systems and Fire Protection. Rules for the Classification of Ships*; RINA: Chita, Japan, 2021.
28. Romanian Standardization Association. *SR ISO 9943—Shipbuilding—Ventilation and Air-Treatment of Galleys and Pantries with Cooking Appliances*; Romanian Standardization Association: Bucharest, Romania, 1995.
29. Romanian Standardization Association. *SR ISO 8864—Air Conditioning and Ventilation of Weelhouse and Board Ships. Design Conditions and Basis of Calculations*; Romanian Standardization Association: Bucharest, Romania, 1995.
30. RINA. *Part E Vol. 3—Service Notations. Rules for the Classification of Ships*; RINA: Chita, Japan, 2021.
31. International Maritime Organization—IMO. *MSC/Circular.1034—Guidelines for Smoke Control and Ventilation Systems for Internal Assembly Stations and Atriums on New Passenger Ships*; Lloyd's Register Rulefinder 2020—Version 9.33—Fix; International Maritime Organization: London, UK, 2008.

32. European Committee for Drawing up Standards in the Field of Inland Navigation (CESNI). *European Standard laying down Technical Requirements for Inland Navigation Vessels (ES-TRIN)*; CESNI: Strasbourg, France, 2021; pp. 1–516.
33. Romanian Standardization Association. *SR ISO 9785—Shipbuilding, Ventilation of Cargo Spaces where Internal Combustion Engine Vehicles May be Driven—Calculation of Theoretical Total Airflow Required*; Romanian Standardization Association: Bucharest, Romania, 1995.
34. BV Marine & Offshore Division. *NR527 DT R04 E—Rules for the Classification of Ships Operating in Polar Waters and Icebreakers*; BV Marine & Offshore Division: Neuilly-sur-Seine, France, 2021.
35. BV Marine & Offshore Division. *NI 673 DT R00 E-Guidelines for Management of COVID-19 and Infectious Diseases*; BV Marine & Offshore Division: Neuilly-sur-Seine, France, 2020.
36. Wikipedia Rotary Air-to-Air Enthalpy Wheel. Available online: https://en.wikipedia.org/wiki/Thermal_wheel#/media/File:Rotary-heat-exchanger.svg (accessed on 3 July 2021).
37. Hoval. Energy Recovery Plate Heat Exchangers Design Handbook. Available online: www.hoval-energyrecovery.com (accessed on 3 July 2021).
38. Wikipedia Plate Heat Recovery. Available online: https://en.wikipedia.org/wiki/Recuperator#/media/File:Heat_exchanger.svg (accessed on 3 July 2021).
39. Carrier Airovision Air Handling Unit. Available online: http://www.alarko-carrier.com.tr/english (accessed on 3 July 2021).
40. Zheng, W.; Hu, J.; Wang, Z.; Li, J.; Fu, Z.; Li, H.; Jurasz, J.; Chou, S.K.; Yan, J. COVID-19 Impact on Operation and Energy Consumption of Heating, Ventilation and Air-Conditioning (HVAC) Systems. *Adv. Appl. Energy* **2021**, *31*, 100040. [CrossRef]
41. Nakazawa, E.; Ino, H.; Akabayashi, A. Chronology of COVID-19 Cases on the Diamond Princess Cruise Ship and Ethical Considerations: A Report from Japan. *Disaster Med. Public Health Prep.* **2020**, *14*, 506–513. [CrossRef] [PubMed]
42. Gupta, A.; Kunte, R.; Goyal, N.; Ray, S.; Singh, K. A comparative analysis of control measures on-board ship against COVID-19 and similar novel viral respiratory disease outbreak: Quarantine ship or disembark suspects? *Med. J. Armed Forces India* **2020**. [CrossRef] [PubMed]
43. Azimi, P.; Keshavarz, Z.; Cedeno Laurent, J.G.; Stephens, B.; Allen, J.G. Mechanistic transmission modeling of COVID-19 on the Diamond Princess cruise ship demonstrates the importance of aerosol transmission. *Proc. Natl. Acad. Sci. USA* **2021**, *118*, e2015482118. [CrossRef] [PubMed]
44. Kulkarni, H.; Smith, C.M.; Lee, D.D.H.; Hirst, R.A.; Easton, A.J.; O'Callaghan, C. Evidence of respiratory syncytial virus spread by aerosol time to revisit infection control strategies? *Am. J. Respir. Crit. Care Med.* **2016**, *194*, 308–316. [CrossRef] [PubMed]
45. Melikov, A.K. COVID-19: Reduction of airborne transmission needs paradigm shift in ventilation. *Build. Environ. J.* **2020**, *186*, 107336. [CrossRef] [PubMed]
46. Correia, G.; Rodrigues, L.; Gameiro da Silva, M.; Gonçalves, T. Airborne route and bad use of ventilation systems as non-negligible factors in SARS-CoV-2 transmission. *Med. Hypotheses* **2020**, *141*, 109781. [CrossRef]
47. Concept, F. HEPA Filter—Technical Specification. Available online: https://www.filter-concept.com/wp-content/uploads/2016/07/32.-Panel-Hepa-Filter.pdf (accessed on 5 July 2021).
48. ASHRAE. Core Recommendations for Reducing Airborne Infectious Aerosol Exposure. 2021. Available online: https://www.ashrae.org/filelibrary/technicalresources/covid-19/ashrae-filtration_disinfection-c19-guidance.pdf (accessed on 3 July 2021).
49. THENOW. HVAC Air Filters Catalogue. Available online: https://resourcewebsite.singoo.cc/14732352510385836/en/pdf/1621928318573/Updated%20HVAC%20air%20filters.pdf (accessed on 3 July 2021).
50. Kinglandclean-KLC HEPA Filter_DOP HEPA Filter With Hood. Available online: http://www.klcintl.com/product_type.aspx?code=010102&id=230 (accessed on 3 July 2021).
51. Heinen&Hopman. Centrifugal Fans Catalogue. Available online: https://heinenhopman.com/en/boat-engine-room-ventilation/centrifugal-fans-wpm-and-wpma/ (accessed on 3 July 2021).
52. U.S. Environmental Protection Agency. *Guide to Air Cleaners in the Home Portable Air Cleaners and Furnace or HVAC Filters in the Home*, 2nd ed.; U.S. Environmental Protection Agency: Washington, DC, USA, 2018; pp. 1–7. Available online: https://www.epa.gov/sites/production/files/2018-07/documents/guide_to_air_cleaners_in_the_home_2nd_edition.pdf (accessed on 3 July 2021).
53. Pinheiro, M.D.; Luís, N.C. COVID-19 could leverage a sustainable built environment. *Sustainability* **2020**, *12*, 5863. [CrossRef]
54. Jones, D. UV-C for HVAC Air and Surface Disinfection. *Inmotion* **2020**. Available online: https://www.researchgate.net/publication/344633657 (accessed on 5 July 2021).
55. Kowalski, W.J.; Walsh, T.J.; Petraitis, V. *COVID-19 Coronavirus Ultraviolet Susceptibility 2020 COVID-19 Coronavirus Ultraviolet Susceptibility*; PurpleSun Inc.: New York, NY, USA, 2020; pp. 1–4. [CrossRef]
56. ASHRAE. Ultraviolet Air and Surface Treatment. *ASHRAE Handbook-HVAC Appl.* **2019**, 62.1–62.17. Available online: https://www.ashrae.org/filelibrary/technicalresources/covid-19/i-p_a19_ch62_uvairandsurfacetreatment.pdf (accessed on 5 July 2021).
57. Clordisys. Ultraviolet Light Disinfection Data Sheet—Application Note 12. 2020. Available online: https://www.clordisys.com/pdfs/misc/UVDataSheet.pdf (accessed on 5 July 2021).
58. Morawska, L.; Tang, J.W.; Bahnfleth, W.; Bluyssen, P.M.; Boerstra, A.; Buonanno, G.; Cao, J.; Dancer, S.; Floto, A.; Franchimon, F.; et al. How can airborne transmission of COVID-19 indoors be minimised? *Environ. Int.* **2020**, *142*, 105832. [CrossRef] [PubMed]

59. Federation of European Heating Ventilation and Air Conditioning Associations. REHVA Covid19 HVAC Guidance—How to operate HVAC and Other Building Service Systems to Prevent the Spread of the Coronavirus (SARS-CoV-2) Disease (COVID-19) in Workplaces, 220AD. Available online: https://www.rehva.eu/fileadmin/user_upload/REHVA_COVID-19_guidance_document_V3_03082020.pdf (accessed on 5 July 2021).
60. Canadian Government. COVID-19: Guidance on Indoor Ventilation During the Pandemic. Available online: https://www.canada.ca/en/public-health/services/diseases/2019-novel-coronavirus-infection/guidance-documents/guide-indoor-ventilation-covid-19-pandemic.html#a6 (accessed on 3 June 2021).
61. American Ultraviolet. MRS45-12 Mobile UV Unit, UVC Germicidal Solutions. Available online: 5https://www.americanultraviolet.com/documents/MRS45-12-brochure.pdf (accessed on 5 July 2021).
62. American Ultraviolet. MRS33-8 Mobile UV Unit, UVC Germicidal Solutions. 1960. Available online: https://www.americanultraviolet.com/documents/MRS33-8-brochure.pdf (accessed on 6 July 2021).

MDPI
St. Alban-Anlage 66
4052 Basel
Switzerland
Tel. +41 61 683 77 34
Fax +41 61 302 89 18
www.mdpi.com

Inventions Editorial Office
E-mail: inventions@mdpi.com
www.mdpi.com/journal/inventions

www.ingramcontent.com/pod-product-compliance
Lightning Source LLC
Chambersburg PA
CBHW061830210926
51850CB00006B/848
* 9 7 8 3 0 3 6 5 5 3 9 5 5 *